Advanced Materials for Solar Energy Utilization

Advanced Materials for Solar Energy Utilization

Guest Editors

Xingwang Zhu
Tongming Su

MDPI Basel • Beijing • Wuhan • Barcelona • Belgrade • Novi Sad • Cluj • Manchester

Guest Editors

Xingwang Zhu
College of Environmental Science and Engineering
Yangzhou University
Yangzhou
China

Tongming Su
School of Chemistry and Chemical Engineering
Guangxi University
Nanning
China

Editorial Office
MDPI AG
Grosspeteranlage 5
4052 Basel, Switzerland

This is a reprint of the Special Issue, published open access by the journal *Materials* (ISSN 1996-1944), freely accessible at: https://www.mdpi.com/journal/materials/special_issues/276T0I76NB.

For citation purposes, cite each article independently as indicated on the article page online and as indicated below:

Lastname, A.A.; Lastname, B.B. Article Title. *Journal Name* **Year**, *Volume Number*, Page Range.

ISBN 978-3-7258-5321-2 (Hbk)
ISBN 978-3-7258-5322-9 (PDF)
https://doi.org/10.3390/books978-3-7258-5322-9

Contents

About the Editors

Xingwang Zhu

Xingwang Zhu is an Associate Professor at the College of Environmental Science and Engineering, Yangzhou University, China. His research focuses on photocatalysis, biomass valorization, and artificial photosynthesis, with an emphasis on the design of grain boundary–rich transition metal-oxide catalysts and their mechanisms in solar-to-chemical energy conversion. He has led several national and provincial research projects, including grants from the National Natural Science Foundation of China and the Jiangsu Natural Science Foundation. His representative works have been published in journals such as *Angewandte Chemie International Edition*, *ACS Catalysis*, and *Advanced Science*. He is actively involved in academic service as a youth Editorial Board Member of *Rare Metals*, a Guest Editor for *Polymers* and *Materials*, and Secretary-General of the *Photovoltaic Committee of the Jiangsu Energy Research Society*. He also serves as a peer reviewer for more than 300 manuscripts annually from leading journals in the fields of chemistry and materials science.

Tongming Su

Tongming Su is an Associate Professor at the School of Chemistry and Chemical Engineering, Guangxi University, China. He obtained his Ph.D. in industrial catalysis from the School of Chemistry and Chemical Engineering, Guangxi University, China, in 2018. From Sept. 2016 to Mar. 2018, he conducted scientific research on nanomaterials at Oak Ridge National Lab (ORNL) and the University of Tennessee, Knoxville, as a joint PhD student supported by the China Scholarship Council. His research interests include the synthesis of nanomaterials and their applications in catalysis, energy conversion, and environmental remediation. Over the years, he has undertaken multiple projects supported by the National Natural Science Foundation of China and the Guangxi Natural Science Foundation, and he have been included in the World's Top 2% Scientists by Stanford University for five consecutive years (2021–2025).

Preface

This Reprint, entitled Advanced Materials for Solar Energy Utilization, brings together a collection of high-quality research and review articles that reflect the current progress and emerging trends in solar-driven technologies. The subject of this Reprint spans a broad interdisciplinary scope, covering the synthesis, characterization, and application of advanced materials for photocatalysis, photovoltaics, solar fuel production, and environmental remediation.

The motivation behind compiling this Reprint stems from the global urgency to develop clean, sustainable energy solutions and environmentally friendly technologies. Solar energy, as one of the most promising renewable resources, offers immense potential for addressing critical challenges related to energy scarcity and pollution. However, its efficient conversion and utilization require the development of novel functional materials with tailored properties and enhanced performance.

This Reprint is intended for researchers, engineers, and graduate students working in materials science, chemistry, environmental science, and renewable energy. It provides readers with a comprehensive overview of recent advances in solar energy utilization, highlighting innovative approaches to material design and system integration. We hope this collection will inspire further exploration and collaboration in the field.

We thank all contributing authors for their valuable submissions and the editorial team for their support in making this Reprint possible. Financial support from the Publication Fund of Yangzhou University is gratefully acknowledged.

Xingwang Zhu and Tongming Su
Guest Editors

materials

Editorial

Advanced Materials for Solar Energy Utilization

Xingwang Zhu [1,*] and Tongming Su [2,*]

[1] School of Environmental Science and Engineering, Institute of Technology for Carbon Neutralization, Yangzhou University, Yangzhou 225009, China
[2] School of Chemistry and Chemical Engineering, Guangxi Colleges and Universities Key Laboratory of New Technology and Application in Resource Chemical Engineering, Guangxi University, Nanning 530004, China
* Correspondence: zxw@yzu.edu.cn (X.Z.); sutm@gxu.edu.cn (T.S.)

In the context of sustainable societal development, exploring novel, environmentally sustainable energy sources has emerged as a pivotal global concern [1–3]. As an inexhaustible, environmentally friendly, renewable and clean energy source, solar energy has been studied by a wide range of researchers [4–8]. Photocatalytic technology is predicated on advanced light-conversion materials, which efficiently drive the conversion of solar into chemical energy. They generate active species under light to degrade pollutants [9–12], convert energy [13–17], pursue environmental remediation [18–21], etc., so as to promote green industry chain development.

This Special Issue comprises a total of eleven original research articles authored by scientists from a variety of international backgrounds. The research presented here focuses on photocatalytic carbon dioxide (CO_2) reduction, photocatalytic hydrogen production from water decomposition, the development of new and advanced photocatalytic materials, photocatalytic pollutant degradation, the preparation of photovoltaic materials, and photoelectron catalytic oxidation. This Special Issue provides a platform for scientists to present their original research on "Advanced Materials for Solar Energy Utilization". The following brief synopses outline the papers that we have been honored to include, with the aim of highlighting advanced materials that have recently enabled solar energy conversion for use.

The substantial discharge of industrial CO_2 has been identified as pivotal in the genesis of numerous ecological issues, with the mitigation of CO_2 through photocatalysis emerging as a compelling technological solution in addressing the prevailing challenge of carbon emissions. Xu et al. [22] synthesized oxygen vacancy-modified MIL-125(Ti) with a high specific surface area via calcination in a hydrogen–argon mixture and applied it in photocatalytic CO_2 conversion. The introduction of oxygen vacancies (OVs) resulted in an increase in the surface area and internal pores of the material, as well as in the Ti^{3+}/Ti^{4+} ratio. The presence of OVs facilitated electron transfer from the Ti sites on MIL-125(Ti) to CO_2, while lowering its activation energy barrier. The results demonstrate that the OV-modified MIL-125(Ti) performs exceptionally, at almost 100% selectivity in photocatalytic CO_2 reduction.

The efficiency of visible light-driven photocatalytic CO_2 reduction is relatively low when photogenerated carriers in photocatalysts are recombined. Li et al. [23] prepared WO_3/BiOBr composite catalysts with fast charge separation performance to increase the efficiency of photogenerated carrier separation. This was attributed to the formation of a compact S-scheme heterojunction between WO_3 and BiOBr, which has been demonstrated to accelerate photogenerated charge separation. The yield of CO_2 photoreduced to CO by WO_3/BiOBr was found to be 1.56 times higher than that of BiOBr in the visible light range and without the assistance of a sacrificial agent.

Crystalline phase engineering represents a highly efficacious method for modifying electron transfer pathways and regulating the internal electronic structure and separation efficiency of electron photocatalysts. Sun et al. [24] prepared Ru cocatalysts with face-centered cubic (fcc) and hexagonal close-packed (hcp) structures, modifying C_3N_4 by precisely adjusting the ratio of the Ru precursor and reducing solvent. Compared with fcc-Ru, hcp-Ru/C_3N_4 exhibited superior charge separation and transfer efficiency. Moreover, Gibbs free energy calculations demonstrated that the hydrogen adsorption energy ($\Delta G_{H^*} = -0.14$ eV) of hcp-Ru was more closely aligned with the optimal value than that of fcc-Ru (-0.32 eV). The process of hydrogen production by hydrolysis achieved through the utilization of hcp-Ru has been demonstrated to be both efficient and highly stable.

Chalcogenide quantum dots have been shown to exhibit fluorescence properties due to their ultrasmall size, with surface and quantum confinement effects being the main factors in this regard. Li et al. [25] successfully prepared uniformly dispersed and size-tunable $CsPbBr_3$ QDs by changing the halogens in metal halide chalcogenides on the basis of the theory of LaMer nucleation and growth, and creating a Br^--rich reaction environment using the thermal injection method. This study demonstrated that $CsPbBr_3$ QDs exhibit specific optical properties when their size reaches the critical Bohr exciton radius value. A further decrease in size results in a photoluminescence wavelength shift to the deep blue band, as well as the manifestation of distinct fluorescence properties. This creates potential for the construction of novel composite catalysts for applications in photocatalysis.

The extensive utilization of antibiotics has the potential to compromise ecological balance and human health. Consequently, researchers have explored the photocatalytic degradation of antibiotics, a process that offers numerous environmental advantages, among others. Su et al. [26] synthesized a $Cd_xMn_{1-x}S$ solid solution with precise control over the ratio of Mn^{2+} and Cd^{2+} ions and prepared C@$Cd_xMn_{1-x}S$ composite catalysts by compositing $Cd_xMn_{1-x}S$ with biomass-gasified carbon slag using a hydrothermal method. The active species of $\bullet O_2^-$ and H^+ produced in C@$Cd_xMn_{1-x}S$ were directly involved in the photocatalytic degradation reactions of tetracycline (TC), with the degradation efficiency reaching 90.35% within 60 min.

Wang et al. [27] prepared an environmentally friendly polyacrylonitrile (PAN)-based Janus-structured composite membrane by adjusting the asymmetric wettability through electrostatic spinning; they improved the thickness of both by employing TiO_2-modified PAN as a hydrophilic base layer and PCL as a hydrophobic layer. The prepared PAN/TiO_2-PCL20 composite membranes demonstrated optimal oil–water separation performance for oil emulsions stabilized with various surfactants. In addition, these membranes exhibited both excellent and stable Rhodamine B (RhB) adsorption performance and removal capacity.

Rychtowski et al. [28] prepared TNR@Ni-foam structures via the alkaline hydrothermal method, which uses TiO_2 with two different structures. In situ FTIR spectroscopy was used to assess the photodegradation properties of acetaldehyde in the TNR@Ni-foam structures. The results demonstrated that P25 facilitated the binding of Na species, thereby contributing to the formation of a layered $Na_2Ti_3O_7$ structure and $Ni(OH)_2$ species. This process resulted in a reduction in the electron transfer barrier and an increase in the separation of charge carriers, consequently achieving efficient photocatalytic acetaldehyde degradation.

Wang et al. [29] prepared n^+-poly-Si layers with thicknesses ranging from 30 to 100 nm via low-pressure chemical vapor deposition (LPCVD). The thickness of the n^+-poly-Si layer substantially suppressed metallization-induced recombination under metal interface contact, as well as on the contact resistivity of the cells. To circumvent the detrimental effects of Ag particle corrosion, low passivation metal values and resistivity under metal contact were ensured by optimizing the thickness of the n^+-poly-Si layer to 70 nm with a surface atomic concentration of $5 \times 10^{20}/cm^3$. This optimization reduces the cost of n^+-poly-Si layers for commercial applications in the photovoltaic industry.

Perovskite solar cell (PSC) stability can be improved through reducing the number of interfacial defects present at the interface between the perovskite and the electron transport layer. Du et al. [30] mixed two self-assembled molecules with different functional groups to form SnO_2/perovskite interlayers, where -H_2PO_3 and -COOH improved the anchoring and carrier transport at the buried interface, respectively, as well as enhancing the photovoltaic performance. Furthermore, the amine group (-NH_2) of the two small molecules has been shown to interact effectively with the uncoordinated Pb_2^+ in the perovskite layer, thereby enhancing the quality of the perovskite film and potentially reducing interfacial defects. This complementary synergistic passivation strategy has been demonstrated to enhance the stability of PSCs while concomitantly increasing their power conversion efficiency.

Selective emitter (SE) technology substantially influences the passivation and contact performance of n-TOPCon solar cells. To achieve the objective of clean production without the introduction of nitrogen, SiO_x is the optimal choice for large-scale production. n-TOPCon solar cells with different back-side phosphorus (P)-SE structures were prepared by Liu et al. [31] via three- and four-step methods, respectively. A comparison revealed that the four-step method exhibited higher performance and stability than that of the three-step method. However, with respect to full-scale electrical properties, both methods could yield results comparable results. This is highly important for improving the backside technology of n-TOPCon solar cells in mass production, improving the efficiency of solar cells and lowering energy usage.

Sn_3O_4 has emerged as a promising semiconductor material due to its excellent visible light absorption properties. Gribov et al. [32] prepared Sn_3O_4, SnO_2 and Sn_3O_4/SnO_2 using a hydrothermal method and investigated the photo-electrooxidation of a sequence of organic substrates with RHE under UV and visible light irradiation over a range of potentials from 0.6 to 1.4 V. The Sn_3O_4 exhibited a high level of activity in photo-electrooxidation reactions involving acetone and formic acid under visible-light conditions. The photo-electrooxidation of these substances exhibited high activity, whereas Sn_3O_4/SnO_2 only demonstrated significant activity in formic acid oxidation. The occurrence of SnO_2 particles within the Sn_3O_4/SnO_2 composite enhances the photocurrent when subjected to UV irradiation. However, this phenomenon is accompanied by a substantial decline in oxidation efficiency under visible-light conditions.

Author Contributions: X.Z.: Writing—original draft, Supervision, Resources, Funding acquisition. T.S.: Writing—review & editing, Resources, Funding acquisition. All authors have read and agreed to the published version of the manuscript.

Funding: This Special Issue was funded by the National Natural Science Foundation of China (22308300, 22208065), the Natural Science Foundation of Jiangsu Province (BK20220598), and the Key Laboratory of Electrochemical Energy Storage and Energy Conversion of Hainan Province (KFKT2022001).

Conflicts of Interest: The authors declare no conflicts of interest.

References

1. Dong, H.; Zhao, J.; Lu, Y.; Tian, Z.; Wang, S.; Bai, X.; Gong, G.; Wang, J.; Wang, L.; Chen, S. Construction of supramolecular metal-halogen bonded organic frameworks for efficient solar energy conversion. *J. Energy Chem.* **2025**, *108*, 527–535. [CrossRef]
2. Chung, H.Y.; Wong, R.J.; Wu, H.; Gunawan, D.; Amal, R.; Ng, Y.H. Scalable and Integrated Photocatalytic Reactor Systems for Solar-to-Fuel Production: Photoredox and Photoreforming Processes. *Adv. Energy Mater.* **2025**, *15*, 2404956. [CrossRef]
3. Lyu, W.; Liu, Y.; Chen, D.; Wang, F.; Li, Y. Engineering the electron localization of metal sites on nanosheets assembled periodic macropores for CO_2 photoreduction. *Nat. Commun.* **2024**, *15*, 10589. [CrossRef]

4. Wang, J.; Liao, W.; Tan, Y.; Henrotte, O.; Kang, Y.; Liu, K.; Fu, J.; Lin, Z.; Chai, L.; Cortes, E.; et al. Transfer dynamics of photo-generated carriers in catalysis. *Chem. Soc. Rev.* **2025**, *54*, 6553–6596. [CrossRef]
5. Jia, P.; Yu, Y.; Chen, T.; Huang, H. "Electricity"-Assisted Catalytic Solar-to-Fuel Processes. *Angew. Chem. Int. Ed.* **2025**, *64*, e202508809. [CrossRef]
6. Huang, Z.-J.; Liu, Y.-L.; Yang, J.-H.; Sun, D.-X.; Qi, X.-D.; Wang, Y. Pine needle-like hierarchical copper foam-based high-performance phase change composites for all-weather solar-thermal-electric conversion. *Chem. Eng. J.* **2025**, *514*, 163268. [CrossRef]
7. Chen, K.; Li, Y.; Feng, Y.; Li, X.; Li, Z.; Liu, S.; Ge, C.; Chen, X. In situ vertical alignment of MoS_2 on Co/C dodecahedron boosting phase change materials for solar-thermoelectric generation. *J. Energy Chem.* **2025**, *107*, 548–557. [CrossRef]
8. Hernandez, R.R.; Armstrong, A.; Burney, J.; Ryan, G.; Moore-O'Leary, K.; Diédhiou, I.; Grodsky, S.M.; Saul-Gershenz, L.; Davis, R.; Macknick, J.; et al. Techno–ecological synergies of solar energy for global sustainability. *Nat. Sustain.* **2019**, *2*, 560–568. [CrossRef]
9. Suo, J.; Dai, Y.; Yin, L.; Hua, Q. Stable cobalt-doped MOF/PAN electrospun membrane: Enhanced tetracycline degradation via adsorption and dual free radical pathways. *Chem. Eng. J.* **2025**, *519*, 164956. [CrossRef]
10. Sambyal, S.; Sharma, R.; Priye, A.; Kumar, M.; Chauhan, V.; Shandilya, P. Nanocellulose supported $ZnWO_4/SrTiO_3/MoO_3$ heterojunction: Highly efficient visible light photocatalyst for ciprofloxacin degradation. *Chem. Eng. J.* **2025**, *516*, 164167. [CrossRef]
11. Pan, M.; Gao, R.; Liang, S.; Hu, X.; Xie, M.; Chen, W.; Wang, S. Oxygen vacancies-rich bimetallic Zr/Co-UiO-66 materials for bifunctional adsorption-photodegradation of tetracycline antibiotics in water. *Chem. Eng. J.* **2025**, *517*, 164397. [CrossRef]
12. Li, S.; Li, X.; Liu, Y.; Zhang, P.; Zhang, J.; Zhang, B. Interfacial engineering of a plasmonic $Ag/Ag_2CO_3/C_3N_5$ S-scheme heterojunction for high-performance photocatalytic degradation of antibiotics. *Chin. J. Catal.* **2025**, *72*, 130–142. [CrossRef]
13. Wang, X.; Li, Y.; Chu, R.; Zhang, Z.; Hou, W.; Wang, L.; Li, L.; Zhang, P. Engineered an organic Z-scheme heterojunction of cu porphyrin-COF/g-C_3N_4 with planar covalent interaction for sustainable solar energy conversion. *Chem. Eng. J.* **2025**, *519*, 164922. [CrossRef]
14. Wang, C.; Xie, Z.; Xie, P.; Liang, H.; Ding, Y.; Wang, Y.; Leung, M.K.H.; Zeng, G.; Ho, J.C.; Bandaru, S.; et al. Phase Transition in Halide Double Perovskites for Solar-To-Chemical Energy Conversion. *Adv. Energy Mater.* **2025**, *15*, 2500921. [CrossRef]
15. Rath, B.B.; Fuchs, L.; Stemmler, F.; Rodríguez-Camargo, A.; Wang, Y.; Dorfner, M.F.X.; Olbrich, J.; van Slageren, J.; Ortmann, F.; Lotsch, B.V. Insights into Decoupled Solar Energy Conversion and Charge Storage in a 2D Covalent Organic Framework for Solar Battery Function. *J. Am. Chem. Soc.* **2025**, *147*, 18492–18503. [CrossRef]
16. Jiao, F.; Yang, S.; Weng, Y.; Mei, L.; Shi, X.; Yu, H.; Liu, Q. Solar methane carbon cyclic reforming with over 40% solar-to-fuel efficiency and nearly 100% selectivity in kWth scale solar fuel production prototype. *Chem. Eng. J.* **2025**, *512*, 162435. [CrossRef]
17. Feng, C.; Raziq, F.; Huang, H.; Wu, Z.P.; Alqahtani, H.S.; Alqahtani, R.; Rahman, M.Z.; Chang, B.; Gascon, J.; Zhang, H. Shining Light on Hydrogen: Solar-Powered Catalysis with Transition Metals. *Adv. Mater.* **2025**, *37*, 2410387. [CrossRef]
18. Luo, Y.; Chen, G.; Wang, Z.; Zhang, S.; Zheng, X.; Meng, S.; Chen, S. Cu–S Covalent Bonds Enable the Anchoring of Single-atom Cu on Layered MoS_2 for Highly Selective and Active Photothermal Catalytic Conversion of CO_2-H_2O to Ethanol. *Adv. Sci.* **2025**, e04167. [CrossRef]
19. Ding, C.; Yang, L.; Lu, X.; Chi, H.; Yang, Y.; Zhang, Y.; Yuan, J.; Wang, X.; Zhou, Y.; Zou, Z. Synergistic effect of asymmetric Ru-Pd dual single-atom-sites and Pd nanoparticles on carbon nitride for efficient CO_2 photoreduction. *Appl. Catal. B Environ.* **2025**, *365*, 124917. [CrossRef]
20. Silerio-Vázquez, F.D.J.; Kakavandi, B.; Tayyab, M.; Baigenzhenov, O.; Hosseini-Bandegharaei, A.; Proal-Nájera, J.B. Sustainability insights into photocatalytic CO_2-to-CH_4 conversion utilizing g-C_3N_4-based photocatalysts: Flaws, progress, and prospectives. *Coord. Chem. Rev.* **2025**, *542*, 216901. [CrossRef]
21. Guan, Q.; Ran, W.; Zhang, D.; Li, W.; Li, N.; Huang, B.; Yan, T. Non-Metal Sulfur Doping of Indium Hydroxide Nanocube for Selectively Photocatalytic Reduction of CO_2 to CH_4: A "One Stone Three Birds" Strategy. *Adv. Sci.* **2024**, *11*, 2401990. [CrossRef]
22. Xu, H.; Song, H.; Wang, X.; Zhu, X. Oxygen Vacancy Modification MIL-125(Ti) Promotes CO_2 Photoreduction to CO with Near 100% Selectivity. *Materials* **2025**, *18*, 1343. [CrossRef]
23. Li, C.; Lu, X.; Chen, L.; Xie, X.; Qin, Z.; Ji, H.; Su, T. WO_3/BiOBr S-Scheme Heterojunction Photocatalyst for Enhanced Photocatalytic CO_2 Reduction. *Materials* **2024**, *17*, 3199. [CrossRef]
24. Sun, X.; Cao, X.; Zhou, G.; Lv, T.; Xu, J.; Zhou, Y.; Wang, Z.; Yi, J. Phase-Controlled Synthesis of Ru Supported on Carbon Nitride and the Application in Photocatalytic H_2 Evolution. *Materials* **2025**, *18*, 1259. [CrossRef]
25. Li, H.; He, J.; Wang, X.; Liu, Q.; Luo, X.; Wang, M.; Liu, J.; Liu, C.; Liu, Y. Synthesis of Size-Adjustable $CsPbBr_3$ Perovskite Quantum Dots for Potential Photoelectric Catalysis Applications. *Materials* **2024**, *17*, 1607. [CrossRef]
26. Su, Y.; Zeng, Z.; Chen, H.; Lv, Z.; Tan, C.; Chen, C. Preparation and Photocatalytic Degradation Performance of C@$Cd_xMn_{1-x}S$ to Tetracycline Hydrochloride. *Materials* **2025**, *18*, 1062. [CrossRef]
27. Wang, Y.; Huang, J.; Zhang, Y.; Zhang, S.; Li, L.; Pang, X. The Design of PAN-Based Janus Membrane with Adjustable Asymmetric Wettability in Wastewater Purification. *Materials* **2024**, *17*, 417. [CrossRef]

28. Rychtowski, P.; Prowans, B.; Miadlicki, P.; Trzeciak, M.; Tryba, B. Preparation of TiO_2 Nanorods@Ni-Foam for Photocatalytic Decomposition of Acetaldehyde-In Situ FTIR Surface Investigation. *Materials* **2025**, *18*, 986. [CrossRef]
29. Wang, Q.; Gao, B.; Wu, W.; Guo, K.; Huang, W.; Ding, J. Effects on Metallization of n^+-Poly-Si Layer for N-Type Tunnel Oxide Passivated Contact Solar Cells. *Materials* **2024**, *17*, 2747. [CrossRef]
30. Du, K.; Wang, A.; Li, Y.; Xu, Y.; Li, L.; Yuan, N.; Ding, J. The Synergistic Effect of Phosphonic and Carboxyl Acid Groups for Efficient and Stable Perovskite Solar Cells. *Materials* **2023**, *16*, 7306. [CrossRef]
31. Liu, Z.; Guo, C.; Liu, Y.; Wang, J.; Su, X.; Wang, Q. Performance of Large Area n-TOPCon Solar Cells with Selective Poly-Si Based Passivating Contacts Prepared by PECVD Method. *Materials* **2024**, *17*, 849. [CrossRef]
32. Gribov, E.; Koshevoy, E.; Kuznetsov, A.; Mikhnenko, M.; Losev, E.; Lyulyukin, M. Flat-Band Potential Determination and Catalytical Properties of Sn_3O_4/SnO_2 Heterostructures in the Photo-Electrooxidation of Small Organic Molecules under Ultraviolet (370 nm) and Blue (450 nm) Light. *Materials* **2023**, *16*, 7300. [CrossRef]

materials

MDPI

Article

All-Day Freshwater Harvesting Using Solar Auto-Tracking Assisted Selective Solar Absorption and Radiative Cooling

Jing Luo, Haining Ji *, Runteng Luo, Xiangkai Zheng and Tianjian Xiao

School of Physics and Optoelectronics, Xiangtan University, Xiangtan 411105, China; 202205710216@smail.xtu.edu.cn (R.L.)
* Correspondence: sdytjhn@xtu.edu.cn

Abstract: The shortage of freshwater resources has become the core bottleneck of global sustainable development. Traditional freshwater harvesting technologies are restricted by geographical conditions and environmental limitations, making them increasingly difficult to satisfy the growing water demand. In this study, based on the synergistic coupling mechanism of photothermal conversion and radiative cooling, a solar auto-tracking assisted selective solar absorber and radiative cooling all-weather freshwater harvesting device was innovatively developed. The prepared selective solar absorber achieved a high absorptivity of 0.91 in the solar spectrum (0.3–2.5 μm) and maintained a low emissivity of 0.12 in the mid-infrared range (2.5–20 μm), significantly enhancing the photothermal conversion efficiency. The radiative cooling film demonstrated an average cooling effect of 7.62 °C during typical daytime hours (12:00–13:00) and 7.03 °C at night (22:00–23:00), providing a stable low-temperature environment for water vapor condensation. The experimental results showed that the experimental group equipped with the solar auto-tracking system collected 0.79 kg m^{-2} of freshwater in 24 h, representing a 23.4% increase compared to the control group without the solar auto-tracking system. By combining theoretical analysis with experimental validation, this study presents technical and economic advantages for emergency water and island freshwater supply, offering an innovative solution to mitigate the global freshwater crisis.

Keywords: freshwater harvesting; selective solar absorber; radiative cooling; solar tracking

1. Introduction

The scarcity of freshwater resources [1,2], crucial for supporting the sustainable development of human societies and maintaining ecosystem stability, has become a major global issue [3]. Although approximately 70% of the Earth's surface is covered by water, only 2.5% of the world's total water resources are freshwater [4]. The World Health Organization predicts that by 2030, half of the global population will experience water shortages, as the demand for freshwater is expected to increase by 40%, exacerbating the imbalance between supply and demand [5].

Self-powered freshwater harvesting devices are a key technological solution to water scarcity [6,7]. Traditional methods, such as rainwater harvesting and groundwater pumping, face significant limitations due to geographic and climatic factors, making it challenging to provide a large-scale and sustainable freshwater supply [8,9]. Seawater, the most abundant water resource on Earth, makes up about 96.53% of the planet's total water [10,11]. In recent years, seawater distillation technologies powered by electricity or fossil fuels have become widely used [12,13]. However, these technologies face challenges due to their high costs and environmental pollution [14,15]. To overcome the bottlenecks in freshwater harvesting technology, the research team has conducted systematic and innovative research [16]. In solar-driven technology, researchers have improved collector photothermal conversion efficiency using material innovations and structural optimizations [17]. In the research and development of collector materials [18], a breakthrough has been achieved in the preparation technology of selective solar absorbers (SSAs) [19,20]. Prof. Tian and Prof. Xiao et al. [21] successfully prepared a Cu-rich plasma nanostructured SSA with broadband absorption and low mid-infrared emission properties through selective leaching reaction using solution treatment. The SSA achieves over 95% absorptivity in the solar spectrum (0.3–2.5 μm) and less than 5% emissivity in the mid-infrared range (2.5–20 μm), thereby enhancing the efficiency of solar thermal collectors through effective light capture and reduced thermal radiation [22–24]. Additionally, researchers have successfully prepared a floating magnetic hydrogel composed of multi-layer graphite nanosheets, polyvinyl alcohol, SiO_2 aerogel, and Fe_3O_4 nanoparticles, providing an innovative solution for solar desalination and water purification with high photothermal conversion efficiency, easy recovery, and sustainability [25]. At the collector structural level, researchers enhanced photothermal conversion efficiency and hydrodynamic performance by designing photonic crystal optical structures, microchannel heat dissipation systems, and optimized flow channel layouts. These results show that the multi-path serpentine photovoltaic-thermal system achieves optimal power output under identical wetted area conditions, with an average total power of 423.84 W m^{-2} [26]. Additionally, sky radiation cooling technology, which relies on the thermal radiation transmission through the atmospheric window, has gained attention due to its passive heat dissipation and zero energy consumption [27–29]. Sky radiation cooling relies on heat radiation through the atmospheric window, using radiative cooling coatings with high solar reflectivity and infrared emissivity to achieve efficient thermal dissipation to the cosmic cooling source [30–33]. During the day, radiative cooling coatings leverage high solar reflectivity to reduce surface temperature, establishing a thermal gradient that drives water vapor condensation. At night, the coatings radiate energy to space via mid-infrared high emissivity, dropping the surface temperature below the dew point to capture atmospheric water vapor for freshwater collection [34,35]. The above technological breakthroughs provide innovative solutions for realizing high-efficiency solar-powered seawater distillation and all-weather freshwater collection.

However, solar collectors are limited to daytime operation due to the periodic changes in the azimuth and altitude of solar radiation, preventing real-time tracking of the sun's trajectory. For instance, in the northern hemisphere mid-latitudes, the collector incidence angle exceeds 30° outside of noon, causing a photothermal conversion efficiency reduction of over 30% [23]. In addition, salt crystallization from seawater evaporation reduces its evaporation efficiency [36,37]. Sky radiation cooling technology depends on ambient humidity and natural convection. Without active evaporation driving, water collection relies solely on dew condensation, leading to low average daily yields that fail to meet practical needs. Therefore, developing an integrated system combining high-efficiency solar thermal conversion with radiative cooling is crucial to overcoming existing technological bottlenecks [38–40].

This paper presents a novel all-weather freshwater harvesting device by integrating interfacial photothermal evaporation, radiative cooling, and solar auto-tracking technologies, with its day-night performance systematically investigated. During the day, the solar tracking device drives the SSA to track the sun in real time, optimizing the incident angle between the photothermal interface and solar radiation to maximize absorption efficiency while minimizing salt deposition. Water vapor from seawater evaporation condenses on the surface formed by the radiative cooling film. At night, the film radiates heat to space, lowering its temperature below the ambient dew point to capture atmospheric moisture. With dual innovations of photothermal-radiative collaboration and dynamic tracking, this device shows great potential in desalination and freshwater collection.

2. System Summarization

The all-weather freshwater collection system designed in this study integrates radiative cooling, solar photothermal conversion, and solar auto-tracking technologies to build an efficient freshwater collection system. Among them, the solar auto-tracking device adjusts the main optical axis of the solar collector in real time, aligning it parallel to the sun's rays based on feedback from the light sensing mechanism, ensuring maximum light and heat absorption efficiency. In the daytime operation mode, the solar collector device induces seawater evaporation at the interface. The resulting water vapor rises to the surface of the radiative cooling film, where it condenses into liquid water and flows into the collection chamber. At night, the radiative cooling film utilizes its high infrared emission properties to dissipate heat into outer space via long-wave radiation. This process reduces the surface temperature of the film below the ambient dew point, causing atmospheric water vapor to condense and enabling continuous freshwater collection day and night. The workflow of the all-weather freshwater collection device is shown in Figure 1 [10,41].

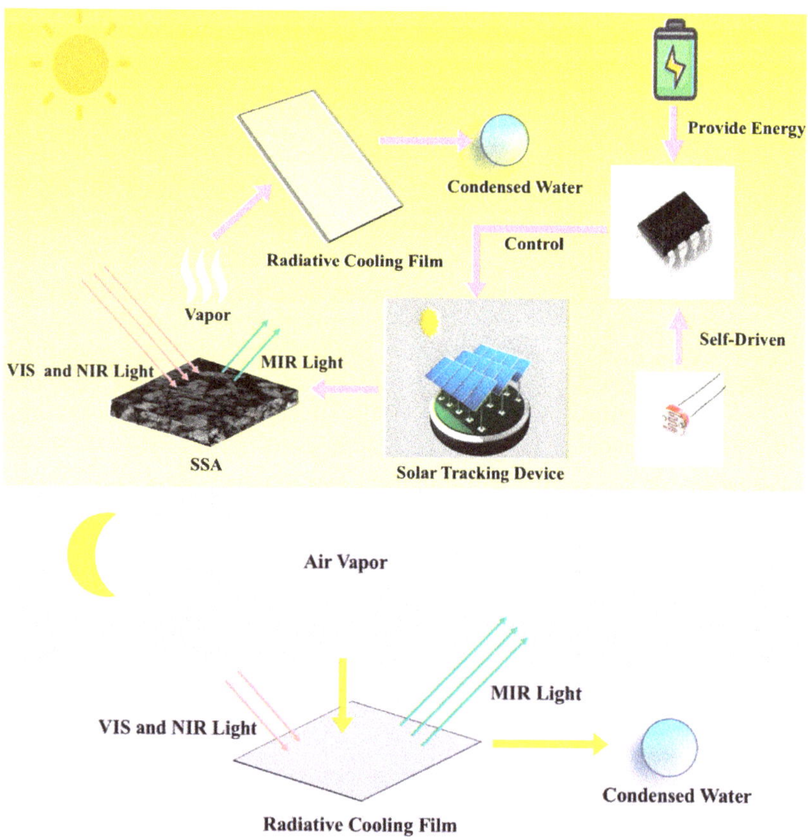

Figure 1. Flow chart of the all-weather freshwater collection device.

In the integrated all-weather freshwater collection unit, a radiative cooling film is attached to the inclined surface at the top of the device, forming a highly efficient condensation interface. Below it, a suspended SSA module—featuring a hollowed-out design—facilitates upward vapor transport from evaporating seawater. Due to gravity, the condensed water flows along the inclined film and collects in a side tank, enabling continuous freshwater harvesting and storage. The outer tanks are primarily designed to collect fresh water generated on the condensation side, while the inner tanks are dedicated to gathering fresh water produced on the evaporation side. The structure of the freshwater collection device is shown in Figure 2.

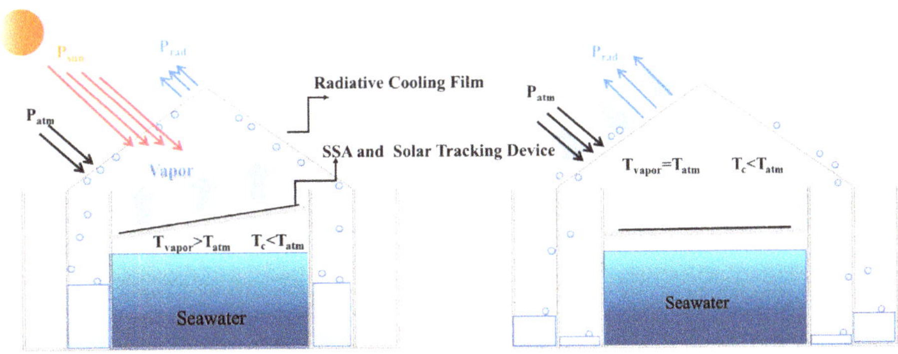

Figure 2. Schematic diagram of the structure of the freshwater collection device and energy flow. P_{sun} is the incident solar power, P_{atm} is the atmospheric radiant power, P_{rad} is the radiant power of the radiatively cooled film, T_{atm} is the atmospheric and ambient temperatures, T_{vapor} is the temperature of the vapors, and T_c is the temperature of the polymer coating.

2.1. Preparation of Solar Collector Module

The solar collector module is fabricated using zinc sheets as the substrate. The substrate pretreatment involves three steps: (1) etching with dilute hydrochloric acid to remove surface oxides and contaminants, (2) sequential ultrasonic cleaning with isopropanol and deionized water to eliminate residual etchant and impurities, and (3) surface drying using compressed air.

Copper nanoparticle deposition was performed using a dip-dry method with a 0.01 mol L^{-1} $CuSO_4$ solution prepared by dissolving $CuSO_4 \cdot 5H_2O$ in deionized water under constant magnetic stirring. The solution was then heated to 75 °C in a vacuum drying oven. The pretreated zinc substrates were immersed in the heated copper sulfate solution for 2 min. During this process, a controlled temperature and reaction time facilitated the reduction of copper ions on the zinc surface, resulting in the formation of a uniform copper nanoparticle layer. After deposition, the zinc sheet was rinsed with isopropyl alcohol and deionized water several times to terminate the surface chemical reaction, followed by air drying to complete the SSA sample preparation [42].

2.2. Preparation of the Solar Auto-Tracking System

The solar auto-tracking system employs a photoelectric sensing mechanism coupled with a closed-loop control architecture. The core components include the TDA2822 dual-channel power amplifier chip designed by STMicroelectronics (Geneva, Switzerland), geared motors, photoresistors, 10 K resistors, and a power module. Utilizing photoresistors as light sensors, the system continuously monitors variations in solar azimuth and elevation angles. The detected light intensity is converted into voltage signals using a voltage divider circuit. The signals are then amplified and compared differentially using the TDA2822. When the light intensity difference exceeds a predetermined threshold, the control circuit activates the geared motors. The gear transmission mechanism adjusts both the pitch and azimuth angles of the solar collector, maintaining optimal alignment between the SSA's primary optical axis and incident sunlight. This active tracking mechanism ensures optimal light capture and thermal collection efficiency throughout operational periods. A detailed schematic diagram of the solar auto-tracking system is presented in Figure S1.

2.3. Preparation of the Radiative Cooling Module

The radiative cooling module incorporates a composite functional film developed by Ningbo Radi-Cool Advanced Energy Technologies Co., Ltd. (Ningbo, China) [43]. The composite functional film is made of a polymer matrix embedded with inorganic nano-functional fillers, engineered to achieve selective spectral modulation. The film demonstrates high infrared emissivity within the atmospheric window (8–13 μm), enabling efficient heat dissipation to outer space through long-wave radiation. Simultaneously, it exhibits high solar reflectivity across the solar spectrum (0.3–2.5 μm), effectively minimizing solar heat gain. This dual-function spectral selectivity allows the film to achieve remarkable passive cooling performance without external energy input, maintaining a sub-ambient temperature even under direct sunlight. By establishing a stable, low-temperature condensation interface, the film significantly enhances the efficiency of freshwater collection systems.

3. Experimental Section

3.1. Performance Testing of the Evaporation and Condensation Side

A scanning electron microscope (SEM ZEISS Sigma 300, Oberkochen, Germany) was used to observe the microscopic morphology of the samples. We characterized the SSA's copper nanostructures (pore size ~50–200 nm), obtaining microstructural parameters including nanoparticle size, spatial arrangement, and surface roughness.

The emissivity spectra of SSA and radiative cooling film in the 2.5–20 μm band were measured using a Fourier transform infrared spectrometer (Nicolet iS50, Thermo Fisher Scientific Inc., Cleveland, OH, USA) at 8 cm^{-1} resolution (0.005 μm wavelength precision), with 128 scans per measurement to enhance the signal-to-noise ratio. Meanwhile, a PerkinElmer Lambda 1050 UV-VIS-NIR spectrophotometer (PerkinElmer, Inc., Shelton, CT, USA) equipped with a 60 mm integrating sphere was used to determine the absorptivity and reflectivity in the 0.3–2.5 μm band. The instrument scanned from 200 to 2500 nm at 200 nm/min with 1 nm resolution and was calibrated against a $BaSO_4$ white plate before each test to ensure accuracy. All samples were pretreated in a vacuum drying oven before testing to effectively remove adsorbed water and other volatile impurities from the surface, ensuring that environmental factors did not influence the test results. To reflect the overall optical properties of materials across broad spectral ranges, we employed average optical parameters for analysis. The average absorptivity and average reflectivity were calculated as follows:

$$\bar{\varepsilon}_{\lambda_1-\lambda_2} = \frac{\int_{\lambda_1}^{\lambda_2} \varepsilon(\lambda)B(\lambda,T)\mathrm{d}\lambda}{\int_{\lambda_1}^{\lambda_2} B(\lambda,T)\mathrm{d}\lambda} \quad (1)$$

$$\bar{\gamma}_{\lambda_1-\lambda_2} = \frac{\int_{\lambda_1}^{\lambda_2} \gamma(\lambda)B(\lambda,T)\mathrm{d}\lambda}{\int_{\lambda_1}^{\lambda_2} B(\lambda,T)\mathrm{d}\lambda} \quad (2)$$

$$B(\lambda,T) = \frac{2\pi \mathrm{h} c_0^2 \lambda^{-5}}{e^{\frac{hc_0}{k\lambda T}} - 1} \quad (3)$$

where $\overline{\varepsilon}_{\lambda_1-\lambda_2}$ is the average emissivity, $\overline{\gamma}_{\lambda_1-\lambda_2}$ is the average reflectivity, and $B(\lambda, T)$ is the irradiance of the blackbody at a wavelength and temperature of T.

A systematic investigation of the radiative cooling film's condensation performance was conducted using a custom-designed experimental apparatus. The test platform was installed on a school building rooftop to ensure an unobstructed sky view and minimize environmental interference. Temperature measurements were simultaneously recorded for both the film surface and ambient air using high-precision type K thermocouples (±0.1 °C accuracy). Data collection focused on two critical periods: midday (12:00–13:00) during peak solar irradiance and nighttime (22:00–23:00) when ambient temperatures reached their daily minimum. By calculating the temperature difference between the film and the ambient temperature, the condensation efficiency of the radiative cooling film was evaluated under different operating conditions.

3.2. All-Weather Freshwater Collection Experiments

The all-weather freshwater harvesting performance experiment was carried out on 1 May 2025, in Xiangtan City, Hunan Province (27.87° N, 112.91° E) under clear sky conditions. During the experimental period, ambient temperatures recorded by the meteorological station ranged from 18 °C to 31 °C. A controlled experimental design was implemented to assess the impact of the solar auto-tracking device on the efficiency of solar thermal collection. The experimental group was equipped with an auto-tracking system based on the TDA2822 chip, ensuring real-time alignment of the SSA's main optical axis with the sun. The control group, on the other hand, used a traditional fixed collector. All device materials, structural parameters, and environmental conditions were kept consistent between the two groups, with the only variation being the tracking system. During the experimental process, a load cell was utilized to record the amount of freshwater collected.

4. Results and Discussion

4.1. Evaporation Side Performance Evaluation

The three-dimensional porous copper nanostructures, formed by the solution leaching process of SSA, are crucial to its excellent optical properties. The SSA exhibits a homogeneous black appearance, with SEM characterization (Figure 3a) revealing uniformly distributed copper nanoparticles across the surface. Quantitative analysis demonstrates excellent particle dispersion with minimal agglomeration, as evidenced by the narrow size distribution. The micro-nanocomposite morphology observed in the SEM image significantly enhances the material's specific surface area, facilitating multiple reflections and absorptions of light. The expanded absorption spectral range imparts broadband absorption properties to the SSA. This innovative structural design overcomes the optical limitations of traditional planar materials, allowing the material to capture solar radiation energy more efficiently.

Under ideal conditions, the average absorptivity of SSA peaks at 1 in the solar wavelength range (0.3–2.5 μm). In contrast, it drops to 0 in the mid-infrared region (2.5–20 μm).

The spectral analysis results of SSA (Figure 3b) demonstrate exceptional optical performance across different wavelength ranges. In the solar spectrum (0.3–2.5 μm), the material exhibits an average absorptivity of 0.91, significantly surpassing the absorption capabilities of conventional materials such as wood and ceramics [41]. In the long-wave infrared band, SSA shows a remarkably low emissivity of 0.12, indicating effective suppression of thermal radiation losses. The spectrally selective properties of the material enable the SSA to efficiently absorb solar radiation while minimizing heat loss to the environment through infrared radiation. This efficient light-to-heat conversion mechanism provides a solid energy foundation for seawater evaporation, greatly enhancing the evaporation rate and strongly supporting the efficient operation of the freshwater collection system.

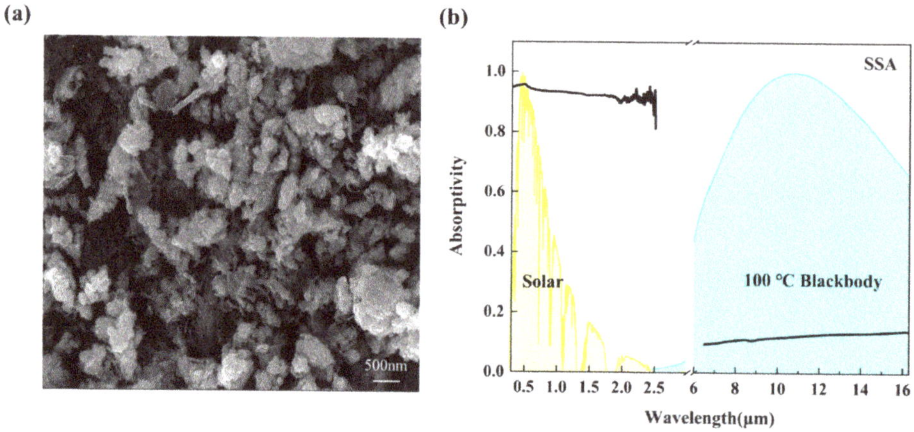

Figure 3. (**a**) SEM image of the selective solar absorber. (**b**) Actual spectral absorptivity of the selective solar absorber.

4.2. Condensation Side Performance Evaluation

For optimal cooling performance, the radiative cooling film should have an average reflectivity of 1 in the solar wavelength range and an average emissivity of 1 in the long-wave infrared region (Figure 4). Spectral analysis results indicate an average reflectivity of up to 0.98 in the near-infrared (NIR) solar spectral band (0.3–2.5 μm) and an average emissivity of 0.75 in the long-wave infrared (LWIR) band. These results confirm the material's high reflectivity in the NIR band and its efficient thermal radiation in the LWIR band (Figure 5a), both of which contribute to the formation of a stable condensation surface.

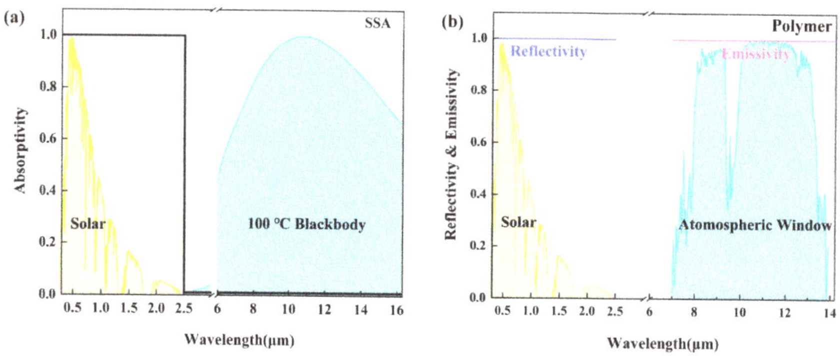

Figure 4. (**a**) Ideal spectral absorptivity for selective solar absorptivity. (**b**) Ideal reflectivity and emissivity of polymer coating.

Experimental measurements from the temperature monitoring system (Figure 5b) highlight the exceptional performance of the radiative cooling film under clear-sky conditions, achieving efficient sub-ambient cooling throughout the day. Notably, during peak solar irradiance at noon, when ambient temperatures reached their highest, the film achieved an average temperature reduction of 7.62 °C, with a maximum cooling of 10.4 °C (Figure 6a). This breakthrough stems from the film's spectrally selective optical design, which integrates high solar reflectance (0.3–2.5 μm) with strong thermal emissivity (8–13 μm atmospheric window). The design facilitates net radiative heat loss to outer space, enabling sustained surface cooling even under intense solar irradiation. This daytime sub-ambient cooling capability ensures the film surface temperature stays below the ambient dew point, creating stable thermodynamic conditions for water vapor condensation. It effectively mitigates the negative impact of high daytime temperatures on the condensation process.

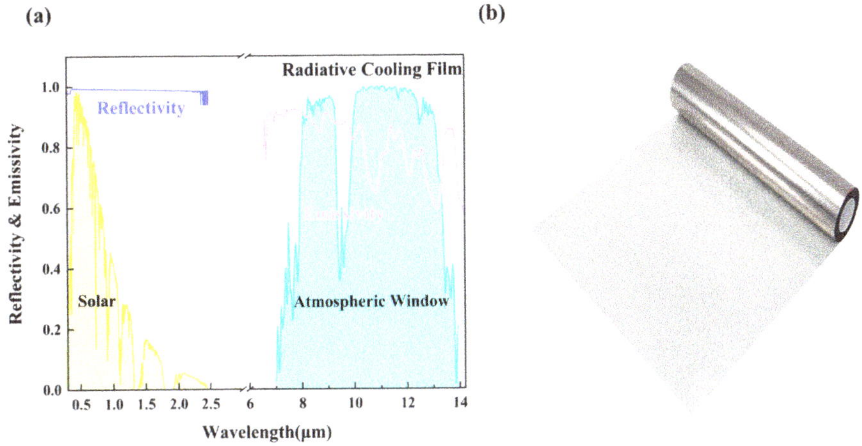

Figure 5. (**a**) Physical view of the radiative cooling film. (**b**) Actual solar spectral reflectivity and thermal LWIR spectral emissivity of radiative cooling film.

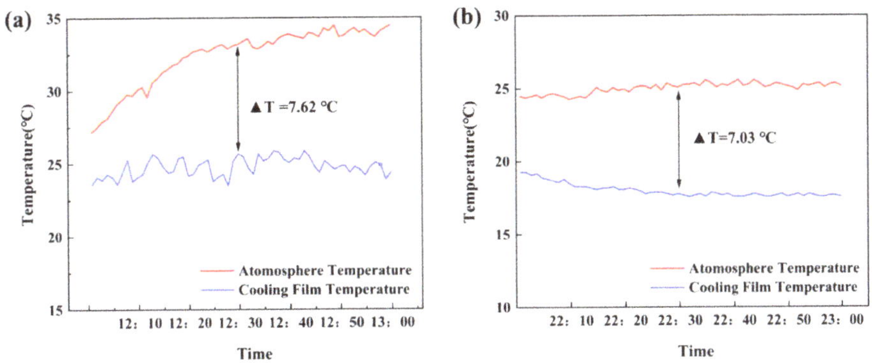

Figure 6. (**a**) Radiative cooling film and atmospheric temperature in the 12:00–13:00 outdoor experiment. (**b**) Radiative cooling film and atmospheric temperature in the 22:00–23:00 outdoor experiment.

During the nighttime, the radiative cooling film achieved an average temperature reduction of 7.03 °C, with a maximum cooling of 7.8 °C (Figure 6b). This further reinforces its ability to operate continuously and efficiently throughout the day. As the ambient temperature decreases at night, the film continues to radiate energy into outer space, further lowering its surface temperature and promoting the formation of the condensation interface. The low-humidity conditions at night pose new challenges to the condensation process compared to daytime. However, the film continues to exhibit a significant cooling effect, demonstrating its high adaptability to varying environmental conditions. This feature enables the radiative cooling module to operate stably at day and night, ensuring continuous condensation efficiency in the freshwater harvesting system throughout the entire day, which is essential for increasing overall freshwater yield.

By integrating both daytime and nighttime experimental data, the radiative cooling film exhibits stable all-weather cooling performance, establishing a solid thermodynamic foundation for water vapor condensation. This further confirms the feasibility and high efficiency of the film as a core component in freshwater collection systems. Moreover, these data lay the groundwork for applying radiative cooling technology in water resource acquisition and provides an essential experimental basis for optimizing radiative cooling materials and expanding their potential applications.

4.3. Freshwater Collection Efficiency and Discussion

The results of the outdoor tests are shown in Figure 7a. The experimental data demonstrate that the freshwater collection of the control group, using the conventional fixed collector, was 0.64 kg m^{-2} over 24 h. In contrast, the experimental group equipped with the solar auto-tracking device achieved a freshwater collection of 0.79 kg m^{-2}, representing a 23.4% increase compared to the control group. Notably, compared with solely relying on solar thermal collection (0.42 kg m^{-2}) [44], the freshwater collection capacity of this device has increased by 88%. It also demonstrates a significant growth compared to pure film condensation (0.1 kg m^{-2}) [45]. Moreover, during the three-month testing period, the fluctuation range of freshwater collected by the device was maintained within ±5%, demonstrating its excellent stability.

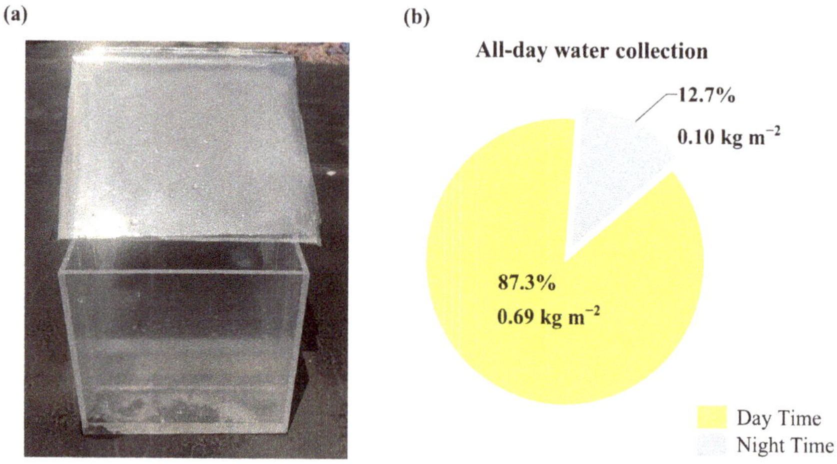

Figure 7. (**a**) Snapshot of the experimental setup. (**b**) Detailed data on daytime and nighttime freshwater collection on the outdoor test day.

This notable enhancement demonstrates the solar auto-tracking device's superior performance in boosting solar thermal efficiency. By dynamically adjusting the collector angle to maintain optimal solar incidence, the device maximizes energy capture and significantly improves daytime solar-thermal conversion efficiency. This finding is consistent with existing theories on solar tracking-enhanced energy efficiency, verifying that dynamic light capture strategies play a pivotal role in freshwater harvesting systems.

Time-division data analysis showed the experimental group harvested 0.69 kg m^{-2} of freshwater during daytime (06:00–18:00), comprising 87.3% of the daily total, and 0.10 kg m^{-2} at night (18:00–06:00), accounting for 12.7% (Figure 7b). This indicates that daytime photothermal-driven evapotranspiration dominates freshwater production, while highlighting the critical role of radiative cooling technology in extending freshwater harvesting duration. At night, with decreasing ambient temperature, the radiative cooling module emits long-wave infrared radiation to outer space, reducing its surface temperature below the ambient dew point and facilitating water vapor condensation. This continuous condensation mechanism, which does not rely on traditional energy sources, ensures the sustained collection of freshwater and enables the potential for an all-weather freshwater supply.

5. Conclusions

In this study, an all-weather freshwater collection device was successfully designed and fabricated, integrating selective solar absorption, photoelectric sensing, and sky radiative cooling mechanisms. The SSA, fabricated using room-temperature "dip-dry" technology based on a chemical substitution reaction, shows spectral selectivity with 0.95 solar absorptivity and 0.12 infrared emissivity, enabling efficient photothermal conversion. When solar tracking is integrated into the system, freshwater production is increased by 23.4%. The radiative cooling film maintains substantial sub-ambient cooling of 7.62 °C (daytime) and 7.03 °C (nighttime), ensuring continuous vapor condensation. The experimental data showed that the average daily freshwater collection of the device was 0.79 kg m^{-2}, with an energy consumption of 0.033 kg m^{-2} h^{-1} for the produced water. The freshwater collection efficiency of the device fluctuated within ±5% over three months of continuous testing, demonstrating excellent stability. The device operates solely on solar energy, effectively overcoming the high energy consumption and carbon emission limitations of traditional desalination technologies. This provides a groundbreaking solution to the global freshwater crisis.

Supplementary Materials: The following supporting information can be downloaded at: https://www.mdpi.com/article/10.3390/ma18132967/s1, Figure S1: Schematic diagram related to the solar auto-tracking system.

Author Contributions: Software, R.L.; Validation, J.L.; Investigation, J.L., H.J., X.Z. and T.X.; Data curation, R.L.; Writing—original draft, J.L. and H.J.; Writing—review & editing, J.L., H.J., R.L., X.Z. and T.X.; Supervision, H.J. All authors have read and agreed to the published version of the manuscript.

Funding: This work was financially supported by the Hunan Provincial College Student Innovation and Entrepreneurship Training Program (No. S202310530170), the National Natural Science Foundation of China (No. 51902276), the Natural Science Foundation of Hunan Province (Nos. 2019JJ50583 and 2023JJ30585), and the Scientific Research Fund of Hunan Provincial Education Department (No. 21B0111).

Institutional Review Board Statement: Not applicable.

Informed Consent Statement: Not applicable.

Data Availability Statement: The original contributions presented in this study are included in the article and Supplementary Materials. Further inquiries can be directed to the corresponding author.

Conflicts of Interest: The authors declare no conflict of interest.

References

1. Grison, C.; Koop, S.; Eisenreich, S.; Hofman, J.; Chang, I.S.; Wu, J.; Savic, D.; van Leeuwen, K. Integrated Water Resources Management in Cities in the World: Global Challenges. *Water Resour. Manag.* **2023**, *37*, 2787–2803. [CrossRef]
2. Mekonnen, M.M.; Hoekstra, A.Y. Four billion people facing severe water scarcity. *Sci. Adv.* **2016**, *2*, e1500323. [CrossRef] [PubMed]
3. Zhong, R.; Chen, A.; Zhao, D.; Mao, G.; Zhao, X.; Huang, H.; Liu, J. Impact of international trade on water scarcity: An assessment by improving the Falkenmark indicator. *J. Clean. Prod.* **2023**, *385*, 135740. [CrossRef]
4. Kalogirou, S. Seawater desalination using renewable energy sources. *Prog. Energy Combust. Sci.* **2005**, *31*, 242–281. [CrossRef]

5. Green, P.A.; Voeroesmarty, C.J.; Harrison, I.; Farrell, T.; Saenz, L.; Fekete, B.M. Freshwater ecosystem services supporting humans: Pivoting from water crisis to water solutions. *Glob. Environ. Change-Hum. Policy Dimens.* **2015**, *34*, 108–118. [CrossRef]
6. Nishad, S.; Elmoughni, H.M.; Krupa, I. Advancements in radiative cooling structures for atmospheric water harvesting: A comprehensive review. *Appl. Energy* **2025**, *377*, 124576. [CrossRef]
7. Gude, V.G. Desalination and water reuse to address global water scarcity. *Rev. Environ. Sci. Bio-Technol.* **2017**, *16*, 591–609. [CrossRef]
8. Lasage, R.; Verburg, P.H. Evaluation of small scale water harvesting techniques for semi-arid environments. *J. Arid Environ.* **2015**, *118*, 48–57. [CrossRef]
9. An, K.J.; Lam, Y.F.; Hao, S.; Morakinyo, T.E.; Furumai, H. Multi-purpose rainwater harvesting for water resource recovery and the cooling effect. *Water Res.* **2015**, *86*, 116–121. [CrossRef]
10. Xi, Z.; Li, S.; Yu, L.; Yan, H.; Chen, M. All-Day Freshwater Harvesting by Selective Solar Absorption and Radiative Cooling. *ACS Appl. Mater. Interfaces* **2022**, *14*, 26255–26263. [CrossRef]
11. Chen, M.; Wang, J.; Li, S.; Chen, W.; Yan, H.; Sheldon, B.W.; Li, Q.; Shi, C. An All-Passive and Macropatterned Architecture Design for Water Harvesting. *Nano Lett.* **2024**, *24*, 16143–16150. [CrossRef] [PubMed]
12. Xiang, T.; Xie, S.; Chen, G.; Zhang, C.; Guo, Z. Recent advances in atmospheric water harvesting technology and its development. *Mater. Horiz.* **2025**, *12*, 1084–1105. [CrossRef] [PubMed]
13. Jarimi, H.; Powell, R.; Riffat, S. Review of sustainable methods for atmospheric water harvesting. *Int. J. Low-Carbon Technol.* **2020**, *15*, 253–276. [CrossRef]
14. Thavalengal, M.S.; Jamil, M.A.; Mehroz, M.; Xu, B.B.; Yaqoob, H.; Sultan, M.; Imtiaz, N.; Shahzad, M.W. Progress and Prospects of Air Water Harvesting System for Remote Areas: A Comprehensive Review. *Energies* **2023**, *16*, 2686. [CrossRef]
15. Wang, M.; Liu, E.; Jin, T.; Zafar, S.-u.; Mei, X.; Fauconnier, M.-L.; De Clerck, C. Towards a better understanding of atmospheric water harvesting (AWH) technology. *Water Res.* **2024**, *250*, 121052. [CrossRef]
16. Sathyamurthy, R.; El-Agouz, S.A.; Nagarajan, P.K.; Subramani, J.; Arunkumar, T.; Mageshbabu, D.; Madhu, B.; Bharathwaaj, R.; Prakash, N. A Review of integrating solar collectors to solar still. *Renew. Sustain. Energy Rev.* **2017**, *77*, 1069–1097. [CrossRef]
17. Cervantes-Rendon, E.; Ibarra-Bahena, J.; Cervera-Gomez, L.E.; Romero, R.J.; Cerezo, J.; Rodriguez-Martinez, A.; Dehesa-Carrasco, U. Rural Application of a Low-Pressure Reverse Osmosis Desalination System Powered by Solar-Photovoltaic Energy for Mexican Arid Zones. *Sustainability* **2022**, *14*, 10958. [CrossRef]
18. Lyu, M.; Lin, J.; Shi, D. Solar Desalination via Multilayers of Transparent Photothermal Fe_3O_4@ $Cu_{2-x}S$ Thin Films. *Energy Technol.* **2021**, *9*, 2100590. [CrossRef]
19. Zakaria, M.; Sharaky, A.M.; Al-Sherbini, A.-S.; Bassyouni, M.; Rezakazemi, M.; Elhenawy, Y. Water Desalination Using Solar Thermal Collectors Enhanced by Nanofluids. *Chem. Eng. Technol.* **2022**, *45*, 15–25. [CrossRef]
20. Xu, Y.-X.; Han, J.-C.; Tang, Q.; Gu, L.; Wu, Q.-Y. A leaf-like photothermal evaporator with enhanced thermal conduction in reversed solar-driven desalination for efficient freshwater harvest. *Sep. Purif. Technol.* **2025**, *354*, 129453. [CrossRef]
21. Tian, Y.; Liu, X.; Caratenuto, A.; Li, J.; Zhou, S.; Ran, R.; Chen, F.; Wang, Z.; Wan, K.-t.; Xiao, G.; et al. A new strategy towards spectral selectivity: Selective leaching alloy to achieve selective plasmonic solar absorption and infrared suppression. *Nano Energy* **2022**, *92*, 106717. [CrossRef]
22. Chen, M.; Li, S.; Pang, D.; Yan, H. Selective absorber and emitter boost water evaporation and condensation toward water collection. *Mater. Today Energy* **2022**, *28*, 101072. [CrossRef]
23. Alqaed, S.; Mustafa, J.; Almehmadi, F.A.; Jamil, B. Estimation of ideal tilt angle for solar-PV panel surfaces facing south: A case study for Najran City, Saudi Arabia. *J. Therm. Anal. Calorim.* **2023**, *148*, 8641–8654. [CrossRef]
24. Ou, Y.; Ji, H.; Wang, Y.; Liu, B.; Chen, Y.; Tao, J.; Huang, Y.; Wang, J. Photothermal synergistic modulation of patterned VO2-Based composite films for smart windows. *Sol. Energy* **2024**, *279*, 112808. [CrossRef]
25. Liu, Y.; Tian, J.; Xu, L.; Wang, Y.; Fei, X.; Li, Y. Multilayer graphite nano-sheet composite hydrogel for solar desalination systems with floatability and recyclability. *New J. Chem.* **2020**, *44*, 20181–20191. [CrossRef]
26. Kazemian, A.; Ma, T.; Hongxing, Y. Evaluation of various collector configurations for a photovoltaic thermal system to achieve high performance, low cost, and lightweight. *Appl. Energy* **2024**, *357*, 122422. [CrossRef]
27. Wang, Y.; Ji, H.; Chen, Y.; Liu, B.; Huang, J.; Lu, M.; Ou, Y.; Zhao, Y.; Tao, J.; Huang, Y.; et al. Artificially adjustable radiative cooling device with environmental adaptability. *Ceram. Int.* **2023**, *49*, 40297–40304. [CrossRef]
28. Zhao, Y.; Ji, H.; Ou, Y.; Wang, Y.; Chen, Y.; Tao, J.; Liu, B.; Lu, M.; Huang, Y.; Wang, J. Novel sunlight-driven Cu_7S_4/VO_2 composite films for smart windows. *J. Mater. Chem. C* **2024**, *12*, 2534–2543. [CrossRef]
29. Wang, Y.; Ji, H.; Liu, B.; Tang, P.; Chen, Y.; Huang, J.; Ou, Y.; Tao, J. Radiative cooling: Structure design and application. *J. Mater. Chem. A* **2024**, *12*, 9962–9978. [CrossRef]
30. Yang, X.; Wang, S.; Zhang, D.; Yu, J. Theoretical investigation of two ideal radiative cooling materials and radiative sky water-cooling module with ideal selective radiative materials. *Energy Build.* **2024**, *323*, 114768. [CrossRef]

31. Qi, G.; Tan, X.; Jiao, Y.; Dai, Q.; Qiao, Y.; Yang, X.; Wang, Y.; Chen, S.; Shi, C.; Yan, K.; et al. A scalable and aging-resistant film for radiative cooling. *Appl. Therm. Eng.* **2024**, *257*, 124310. [CrossRef]
32. Wang, C.; Chen, H.; Wang, F. Passive daytime radiative cooling materials toward real-world applications. *Prog. Mater. Sci.* **2024**, *144*, 101276. [CrossRef]
33. Shao, H.; Niu, J.; Zhang, Y.; Wang, H.; Lu, C.; Ma, Y.; Chen, H.; Li, S.; Qian, H. Mass-Producible Transparent Flexible Passive-Cooling Film. *ACS Appl. Mater. Interfaces* **2024**, *16*, 57246–57252. [CrossRef]
34. Lv, S.; Guo, Y.; Lv, W.; Wang, Z.; Pan, Y.; Ren, J. Modeling and performance evaluation of radiative cooling-assisted interfacial evaporation for all-day freshwater harvesting. *Desalination* **2025**, *593*, 118241. [CrossRef]
35. Huang, Y.; Li, Q.; Chen, Z.; Chen, M. Sorbent-coupled radiative cooling and solar heating to improve atmospheric water harvesting. *J. Colloid Interface Sci.* **2024**, *655*, 527–534. [CrossRef]
36. Ramful, R.; Sowaruth, N. Low-cost solar tracker to maximize the capture of solar energy in tropical countries. *Energy Rep.* **2022**, *8*, 295–302. [CrossRef]
37. Wu, C.-H.; Wang, H.-C.; Chang, H.-Y. Dual-axis solar tracker with satellite compass and inclinometer for automatic positioning and tracking. *Energy Sustain. Dev.* **2022**, *66*, 308–318. [CrossRef]
38. Liu, X.; Beysens, D.; Bourouina, T. Water Harvesting from Air: Current Passive Approaches and Outlook. *ACS Mater. Lett.* **2022**, *4*, 1003–1024. [CrossRef]
39. Xu, J.; Huo, X.; Yan, T.; Wang, P.; Bai, Z.; Chao, J.; Yang, R.; Wang, R.; Li, T. All-in-one hybrid atmospheric water harvesting for all-day water production by natural sunlight and radiative cooling. *Energy Environ. Sci.* **2024**, *17*, 4988–5001. [CrossRef]
40. Lyu, T.; Han, Y.; Chen, Z.; Fan, X.; Tian, Y. Hydrogels and hydrogel derivatives for atmospheric water harvesting. *Mater. Today Sustain.* **2024**, *25*, 100693. [CrossRef]
41. Arunkumar, T.; Parbat, D.; Lee, S.J. Comprehensive review of advanced desalination technologies for solar-powered all-day, all-weather freshwater harvesting systems. *Renew. Sustain. Energy Rev.* **2024**, *199*, 114505. [CrossRef]
42. Mandal, J.; Wang, D.; Overvig, A.C.; Shi, N.N.; Paley, D.; Zangiabadi, A.; Cheng, Q.; Barmak, K.; Yu, N.; Yang, Y. Scalable, "Dip-and-Dry" Fabrication of a Wide-Angle Plasmonic Selective Absorber for High-Efficiency Solar-Thermal Energy Conversion. *Adv. Mater.* **2017**, *29*, 1702156. [CrossRef]
43. Yin, X.; Yang, R.; Tan, G.; Fan, S. Terrestrial radiative cooling: Using the cold universe as a renewable and sustainable energy source. *Science* **2020**, *370*, 786–791. [CrossRef]
44. Tian, Y.; Jiang, Y.; Zhu, R.; Yang, X.; Wu, D.; Wang, X.; Yu, J.; Li, Y.; Gao, T.; Li, F. Solar-Driven Multistage Device Integrating Dropwise Condensation and Guided Water Transport for Efficient Freshwater and Salt Collection. *Environ. Sci. Technol.* **2024**, *58*, 7335–7345. [CrossRef]
45. Xu, J.; Zhang, J.; Fu, B.; Song, C.; Shang, W.; Tao, P.; Deng, T. All-Day Freshwater Harvesting through Combined Solar-Driven Interfacial Desalination and Passive Radiative Cooling. *ACS Appl. Mater. Interfaces* **2020**, *12*, 47612–47622. [CrossRef]

materials

MDPI

Article

Oxygen Vacancy Modification MIL-125(Ti) Promotes CO_2 Photoreduction to CO with Near 100% Selectivity

Hangmin Xu [1], Hao Song [1], Xiaozhi Wang [1,2,*] and Xingwang Zhu [1,*]

[1] College of Environmental Science and Engineering, Institute of Technology for Carbon Neutralization, Yangzhou University, Yangzhou 225009, China; xhm@zxw.ac.cn (H.X.); sh@zxw.ac.cn (H.S.)
[2] Jiangsu Collaborative Innovation Center for Solid Organic Waste Resource Utilization, Nanjing 210095, China
* Correspondence: xzwang@yzu.edu.cn (X.W.); zxw@yzu.edu.cn (X.Z.)

Abstract: The substantial release of industrial carbon dioxide has been identified as a key factor in the development of various environmental issues. In addressing these concerns, the utilization of photocatalytic technology for carbon reduction has garnered significant attention. The disadvantage of CO_2 photoreduction is the problem of product yield and selectivity. It is known that MIL-125(Ti) with a high specific surface area (S_{BET}) possesses more active sites using Ti as a node. The calcination of MIL-125(Ti) in a reducing atmosphere has been shown to introduce oxygen vacancies (O_V), thereby enhancing the material's surface and internal pores. This process has been demonstrated to result in a significant increase in the S_{BET} and an enhancement of the Ti^{3+}/Ti^{4+} ratio. The increased Ti^{3+} centers have been found to improve the material's reducing properties. The results demonstrate that the O_V-rich MIL-125-2H material exhibits the high-performance and highly selective photoreduction in CO_2.

Keywords: MIL-125(Ti); oxygen vacancy; Ti^{3+} centers; CO_2 photoreduction

1. Introduction

The issue of industrial carbon dioxide (CO_2) emissions represents a significant challenge to global environmental improvement [1–3]. The utilization of advanced technology for the mitigation of carbon emissions has emerged as a primary focus for scholars in the field. The development of artificial photocatalytic technology in this field has been a long-standing endeavor. Nevertheless, the implementation of photocatalytic technology remains encumbered by several challenges, including light absorption, CO_2 adsorption and desorption, and catalyst stability [4,5].

Metal–organic frameworks (MOFs) are structures that possess internal porosity supported by a metallic backbone, formed by the coordination of metal ions/clusters and organic ligands [6]. The attention that is now being paid to MOFs in a variety of fields, such as gas adsorption, is due to their multifunctional structure and high specific surface area [7]. In particular, nano-sized MOFs have been shown to expose a greater number of active sites, thereby enhancing their performance [8,9]. It has been established that the size of MIL-125(Ti) is in the nanoscale, which has been demonstrated to promote an increase in the specific surface area and active sites of the MOFs, thereby facilitating gas photocatalytic activation. Concurrently, the elevated specific surface area of MIL-125(Ti) facilitates the promotion of Ti-O-C connections, thereby enhancing the capacity for CO_2 adsorption on the MIL-125(Ti) surface and facilitating CO_2 activation [10]. The adsorption fixation of CO_2 by nanoscale titanium-based MOF has been reported in the literature, suggesting that MIL-125(Ti) has the potential to act as a photocatalytic activator of CO_2 [11,12]. However, the further modification of MIL-125(Ti) to enhance its intrinsic properties and effectively improve its gas adsorption capacity at the interface, increase the active sites, and shorten the diffusion path of the guest molecules is challenging.

In recent years, there has been a growing body of research focusing on the impact of oxygen vacancies in MOFs on electron transfer in semiconductors [13,14]. These O_V are known to act as sites for immobilizing metal atoms, facilitating acid–base interactions, and enabling uniform loading of donors/acceptors [15]. In light of these observations, we propose a hypothesis that the introduction of O_V around the Ti metal of MIL-125(Ti) allows the excited state electrons to first accumulate within O_V and, later, further transfer to the Ti sites to enhance charge separation and transfer [16]. Sun et al. demonstrated that, by introducing oxygen vacancies, the concentration of Ti^{3+} in the active site could be regulated, thereby affecting the electron transfer pathway during its photocatalytic nitrogen fixation and ammonia production [17]. However, the presence of excess O_V has been shown to inhibit the separation of photogenerated carriers, which has a detrimental effect on the catalytic performance of the catalysts [18]. Therefore, the introduction of an appropriate amount of O_V in MIL-125(Ti) for the application of photocatalytic reduction in CO_2 is particularly important.

In this study, MOFs comprising Ti as the metal node were synthesized and calcined under a reducing gas atmosphere. This process resulted in an augmentation of the specific surface area (S_{BET}) of MIL-125(Ti) without compromising the structural integrity. Concurrently, it facilitated a reduction in the proportion of Ti^{4+} to Ti^{3+}, thereby enhancing the reduction capacity of the Ti sites. Concurrently, the oxygen vacancies that were introduced promoted the aggregation of nearby electrons to the Ti sites and further transferred them to the adsorbed CO_2. Incorporating theoretical calculations reveals that the O_V introduced is instrumental in reducing the energy barrier associated with the activated CO_2 step of the MIL-125-2H, thereby enhancing its efficiency. In conclusion, the modified MIL-125-2H reduced CO_2 to CO in yields of up to 771.22 $\mu mol\ g^{-1}\ h^{-1}$, as well as the selectivity of up to near 100%. The investigation revealed that oxygen-rich vacancies (O_V-rich) and a high specific surface area may play a significant role in the photocatalytic CO_2 reduction process.

2. Experimental Section

2.1. Chemical Reagent

Terephthalic acid ($C_8H_6O_4$, ≥99%) and tetra-butyl ortho-titanate ($Ti(OC_4H_9)_4$, ≥99%) were purchased from Aladdin Biochemical Technology Co., Ltd. (Shanghai, China). N,N-dimethylformamide (DMF, ≥99.5%) and methanol (CH_3OH, ≥99.5%) were purchased from Sinopharm Chemical Reagent Co., Ltd. (Shanghai, China) These reagents were purchased directly for use and have not been handled in any way. Ultrapure water (18.25 MΩ/cm) was prepared by a laboratory ultrapure water machine (YL-400BU, Shenzhen YL Electronics Co., Ltd., Shenzhen, China).

2.2. Preparation of MIL-125(Ti)

The preparation of MIL-125(Ti) was conducted in accordance with methods previously documented in the extant literature [17]. The $C_8H_6O_4$ (1.5 g, 9 mmol) was dissolved in 27 mL of DMF and ultrasonicated for a period of 5 min until complete dissolution. Concurrently, $Ti(OC_4H_9)_4$ (0.78 mL, 2.25 mmol) was gradually added to 3 mL of methanol under vigorous stirring to yield a clear solution. The methanol solution containing titanate was subsequently transferred to the organic linker that has been previously described. Following a 30 min stirring period, a gradual transition of the color of the mixture from yellow to milky white was observed. Subsequently, the milky mixture was transferred to a flask and heated in an oil bath at 130 °C for 24 h. After cooling to room temperature, the suspension was subjected to centrifugation at 6000 rpm for 3 min. The precipitate was collected and subsequently washed several times with DMF in order to remove unreacted organic ligands. Thereafter, the unreacted Ti ions were removed by washing with methanol, after which the precipitate was collected by centrifugation. The solid was then dried under vacuum at 80 °C for 5 h to yield a white solid with a yield of approximately 0.5 g.

2.3. Preparation of MIL-125-xH

The MIL-125-xH was prepared by subjecting MIL-125(Ti) to heat treatment at 250 °C for various durations under the protection of a mixture of Ar and H_2 (Ar 95%, H_2 5%). The heating rate was set to 5 °C/min, the gas flow rate was set to 0.1 L/min, and the samples were allowed to cool naturally to room temperature before the aeration process was halted. The duration of the heat treatment was 0.5, 2, and 5 h and corresponded to MIL-125-0.5H, MIL-125-2H, and MIL-125-5H, respectively.

2.4. Photocatalytic Measurements

A quantity of 10 mg of photocatalyst was added to a photoreactor, together with 4 mL of ultrapure water, 2 mL of lactic acid, and 6 mL of acetonitrile. The mixture was then sonicated in order to ensure uniform dispersion. The temperature of the photoreactor was controlled at 10 °C by circulating cooling water, with the aim of preventing the decomposition of the solution. The reaction was continuously stirred with a magnetic stirrer at 500 rpm. The photoreactor was then connected to the photocatalytic carbon dioxide reduction system, after which the entire system was evacuated to vacuum using a vacuum pump. The system was then charged with 20 kPa of high-purity CO_2, and the lines were rinsed and vacuumed to eliminate other residual gases. Finally, high-purity CO_2 was reintroduced until a pressure of 80 kPa was reached. A 300 W xenon lamp (PLS-SEX300, Beijing Perfectlight, Beijing, China) was utilized as a light source during the reaction, and at the hourly interval, 1 mL of gas was withdrawn from the reaction system into a gas chromatograph (GC9790II) to detect the products (5 h as a reaction cycle).

3. Results and Discussion

The original MIL-125(Ti) was a strong skeleton synthesized using Ti as the metal junction and terephthalic acid as the organic ligand [19]. In this study, a low-temperature calcination approach was employed to introduce oxygen vacancies in the vicinity of the metal nodes without compromising the integrity of the MIL-125(Ti) backbone. Prior to calcination, the morphology of MIL-125(Ti) as observed under the scanning electron microscope (SEM) exhibited a round cake-like shape and a relatively smooth surface (Figure 1a). Following calcination at a low temperature of 250 °C for different durations, the skeletons of MIL-125-0.5H (Figure S1a), MIL-125-2H (Figure 1b), and MIL-125-5H (Figure S1b) remained unchanged, as evidenced by SEM analysis. However, an increase in the number of pores on the catalyst surface was observed, indicating that calcination led to an enhancement in the S_{BET} of MIL-125-xH. In the transmission electron microscope (TEM), the size of MIL-125-2H was found to be approximately 300–500 nm (Figure 1c). As demonstrated in Figure 1d (high resolution- TEM, HR-TEM), no additional impurities were identified on MIL-125-2H, and the interior was found to be sustained by a skeleton rather than a solid structure. In the elemental distribution map (Figure 1e–h), Ti, O, and C demonstrate a uniform distribution.

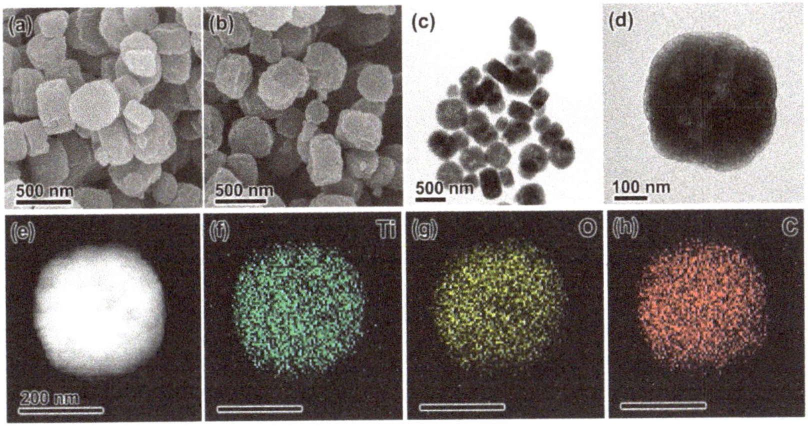

Figure 1. The SEM images of (**a**) MIL-125(Ti) and (**b**) MIL-125-2H. (**c**) TEM image of MIL-125-2H. (**d**) The HR-TEM image of MIL-125-2H. (**e**) STEM image and (**f–h**) elemental distribution map of MIL-125-2H.

To explore the crystal structure of MIL-125-xH (x = 0.5, 2, 5), X-ray diffraction (XRD) was employed. As demonstrated in Figure 2a, the diffraction peaks of MIL-125-xH remain largely unchanged in comparison to those of MIL-125(Ti), suggesting that the phase structure of MIL-125-xH remains unchanged following calcination [20,21]. The crystalline size was calculated based on XRD, and as demonstrated in Table S1, the crystalline size increased in the calcined samples but decreased after an extended period of calcination. However, it is noteworthy that the crystalline size of all the samples was below 100 nm, which is in the nanoscale. This facilitates an increase in the specific surface area of the sample and improves the capture of CO_2 [22]. In order to comprehend the concentration of O_V in MIL-125-xH, electron paramagnetic resonance (EPR) analysis (Figure 2b) was conducted, revealing that oxygen vacancies were not detected in MIL-125(Ti) prior to calcination. Subsequent to calcination, the concentration of oxygen vacancies exhibited a gradual enhancement with increasing calcination time [23]. In the N_2 adsorption–desorption test, the S_{BET} of MIL-125(Ti) was 910.05 m^2/g. In comparison, the S_{BET} of MIL-125-xH increased gradually with the increase in calcination time, which was mainly due to the increase in calcination time, which increased the pore space of the sample. However, when the calcination duration reached 5 h, the S_{BET} of MIL-125-5H was only 882.54 m^2/g. This was due to the long calcination time, which led to the partial decomposition of organic ligands and blocked the pores in the sample. The MIL-125-2H prepared after 2 h of calcination had the largest specific surface area (1481.29 m^2/g), providing more active sites for CO_2 reduction. Therefore, the functional groups in the samples were analyzed using Fourier transform infrared (FTIR) spectroscopy. As demonstrated in Figure S2, MIL-125(Ti) and MIL-125-xH exhibit analogous characteristic peaks. The range of 400–800 cm^{-1} is attributed to the stretching vibration of (O-Ti-O). The 1380–1654 cm^{-1} range is attributed to the stretching vibration of the backbone O-C-O and C=O, which is characteristic of the carboxylate band, thus confirming the presence of carboxylate linkers in the material [17,21]. Meanwhile, these characteristic peaks did not change with the increase in calcination duration. However, in the context of Raman spectroscopy (Figure S3), the characteristic peaks of MIL-125-xH appeared to diminish with the prolongation of the calcination duration, and in particular, almost no signal was detected in MIL-125-5H. Combined FTIR and Raman spectroscopy has been demonstrated to show that a small amount of ligand decomposition occurs in MIL-125-5H.

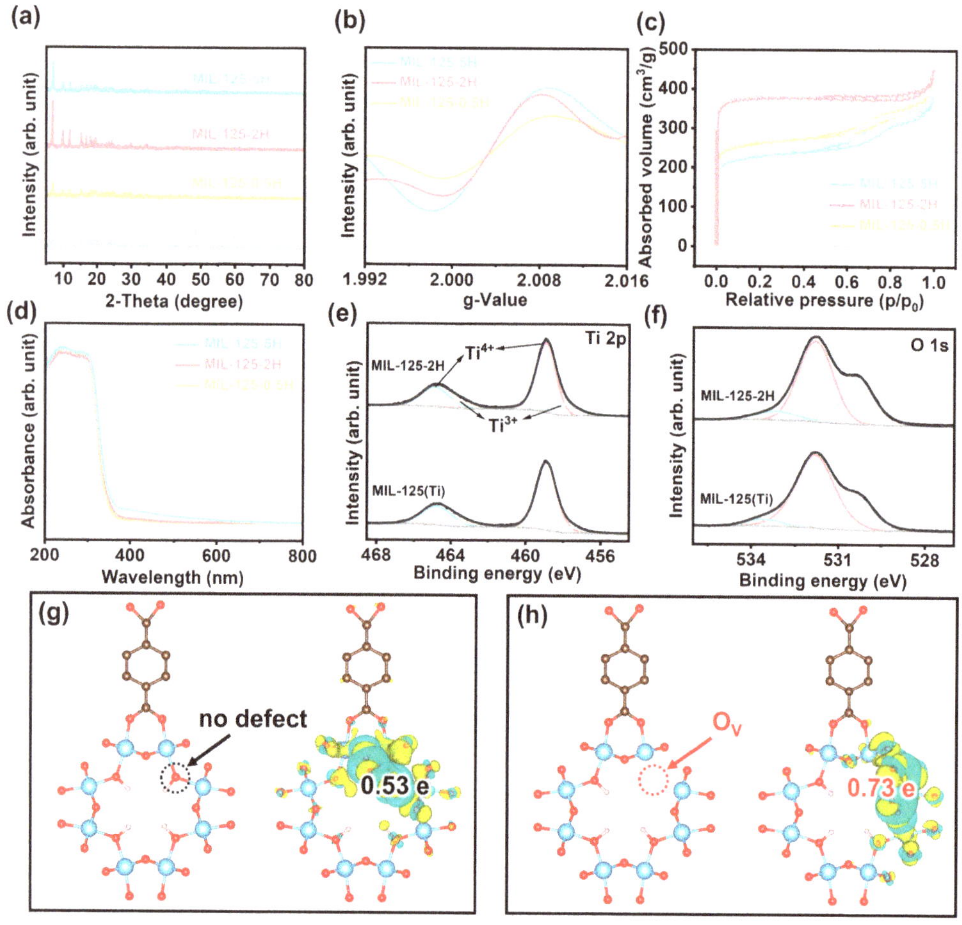

Figure 2. (**a**) XRD profiles, (**b**) EPR spectra, (**c**) N_2 adsorption and desorption curves, and (**d**) UV-vis DRS spectra of samples. The high-resolution XPS spectrum of (**e**) Ti 2p and (**f**) O 1s. The Bader charge of (**g**) MIL-125(Ti) and (**h**) MIL-125-2H. The green, red, yellow, and blue curves in panels (**e**,**f**) are the XPS splitting curves in MIL-125(Ti) and MIL-125-2H. (**e**,**f**) show the upper part of the curves as MIL-125-2H and the lower part of the curves as MIL-125(Ti). Therefore, to avoid confusion, we have changed the nomenclature of MIL-125(Ti) and MIL-125-2H in panels (**e**,**f**) to black.

In the UV-visible diffuse reflectance spectra (UV-vis DRS), the MIL-125-xH light absorption edge demonstrates a slight redshift with increasing calcination time, which is primarily attributable to the deepening of the color of MIL-125-xH as a consequence of calcination, thereby augmenting its light absorption range [17]. Subsequently, the light absorption capacity of the samples was calculated by constructing models for MIL-125(Ti) and MIL-125-2H. As demonstrated in Figure S4a,b, within the range of ~370–400 nm, MIL-125-2H exhibits a more pronounced light absorption capacity in comparison to MIL-125(Ti), thereby substantiating the assertion that the incorporation of O_V extends the light absorption and establishes the basis for efficient photoexcitation of electrons [24]. Consequently, the density of states for the samples was subsequently calculated and compared to that of MIL-125(Ti). MIL-125-2H was found to exhibit a higher density of states at the Fermi level (Figure S5a,b). The introduction of O_V was found to increase the semiconductor intrinsic semi-conductivity and facilitate electron transfer [25]. X-ray photoelectron spectroscopy (XPS) was applied to reveal the local electronic structure and valence states of the prepared samples. Three major peaks belonging to Ti, O, and C were detected in the full XPS spectrum (Figure S6). In the XPS spectrum of Ti 2p (Figure 2e), the peaks with binding energies at positions 464.90 eV and 458.95 eV are attributed to Ti^{4+}, while the peaks with binding energies at positions 463.65 eV and 457.50 eV are ascribed to Ti^{3+} [26]. However, the peak area of the Ti^{3+} in MIL-125-2H increased significantly after 2 h of calcination. This is due to the increase in oxygen vacancies, which partially convert Ti^{4+} to Ti^{3+}. As shown in Table S2, the ratio of Ti^{3+}/Ti^{4+} increased to 0.30 for MIL-125-2H compared to 0.23 for MIL-125(Ti). Subsequently, the change in O_V content was further determined from the XPS spectrum of O 1s. As demonstrated in Figure 2f, the binding energies situated at 530.20, 531.85, and 533.65 eV are attributed to adsorbed oxygen, oxygen vacancies, and the lattice oxygen of MIL-125(Ti), respectively [27]. The peak intensity of the oxygen vacancies belonging to MIL-125-2H was slightly enhanced in comparison with MIL-125(Ti), indicating that calcination did enhance the concentration of O_V, which is consistent with the Ti^{4+}/Ti^{3+} conversion in MIL-125-2H and the results of the EPR tests.

Based on the structural analysis, theoretical computational models of MIL-125(Ti) without defects and MIL-125-2H containing O_V near the Ti nodes were constructed, as shown in Figure 2g,h. The results of the calculation of the Bader charge of CO_2 adsorbed at the sample interface (Figure 2g,h) demonstrate that, in the absence of O_V, electrons are more likely to accumulate at the Ti node. Concurrently, a greater number of electrons are transferred to CO_2 by MIL-125-2H (0.73 e) in comparison to MIL-125(Ti) (0.53 e), thereby promoting CO_2 photoreduction [25,28].

In order to analyze whether the band structure of the samples meets the thermodynamic requirements for photocatalytic reduction in CO_2, ultraviolet photoelectron spectroscopy (UPS) and valence band-XPS (VB-XPS) were employed. As demonstrated in Figure 3a,d, the UPS cut-off edges (E_C) of MIL-125(Ti) and MIL-125-2H are 17.15 eV and 17.44 eV, respectively. The work functions of these two materials are 4.07 eV and 3.78 eV, as calculated based on $\varphi = hv - E_C$ (hv = 21.22 eV) [29]. The results of the VB-XPS tests (Figure 3b,e) demonstrate that the VB of MIL-125(Ti) and MIL-125-2H are 2.82 eV and 2.96 eV, respectively. Consequently, the valence band maximum (VBM, $E_{VB,\,Vac}$) of the two relative to the vacuum level is calculated to be −6.89 eV and −6.74 eV, respectively. The potentials of the samples were converted, according to the standard hydrogen electrode ($E_{VB,NHE} = -E_{VB,Vac} - 4.44$) [30]. The band gaps of MIL-125(Ti) and MIL-125-2H are 3.75 eV and 3.70 eV, respectively, as determined by the Tauc plots of the UV-vis DRS transformations. The conduction band minimum (CBM) was determined through calculation of the band gap, and the band structure of the sample was found to satisfy the thermodynamic requirements for CO_2 photoreduction (Figure 3f). Concurrently, the CBM of MIL-125-2H is higher than that of MIL-125(Ti), which renders MIL-125-2H more capable of reduction than MIL-125(Ti).

The CO_2 photoreduction properties of the samples under investigation were evaluated by performing the reaction in a system of 6 mL acetonitrile, 4 mL H_2O to provide the proton source, and 2 mL triethanolamine as a sacrificial agent. The reaction was completed in the μGAS1000 system (Figure S7). The CO_2 reduction rate test results (Figure 4a) demonstrated that the calcined MIL-125-xH facilitated the reduction in CO_2 to CO at a higher rate than MIL-125(Ti) (545.54 μmol $g^{-1}\cdot h^{-1}$). Among the samples, MIL-125-2H (771.22 μmol $g^{-1}\cdot h^{-1}$) exhibited the highest rate of reduction to CO, which is hypothesized to be attributable to its augmented S_{BET}, which provides an increased number of active sites and facilitates greater electron transfer from Ti^{3+} active sites to CO_2. It is also noteworthy that, despite the presence of CH_4 and H_2 in the products, all samples exhibit the selective conversion of up to near 100% for CO. The calculation of the number of moles of consumed effective photogenerated electrons throughout the reaction indicates that almost all of the consumed photogenerated electrons are involved in the photoreduction in CO_2 to CO, which is consistent with the gas product selectivity results (Figure 4b). A series of comparative experiments were conducted in order to explore the factors that influence the reaction process. As demonstrated in Figure 4c, in the absence of a catalyst addition to the reaction system, no gas was produced. Similarly, in the absence of light, no gas was produced. No production of CO and CH_4 was observed under dark conditions for the initial 2 h; however, a linear increase in CO production was immediately evident upon the activation of the lights (Figure 2d). These findings indicate that the complete reaction must occur in the presence of a catalyst and must be catalyzed by light, both of which are essential components. Following the substitution of Ar for CO_2, it was determined that a negligible amount of H_2 (1.75 μmol g^{-1} h^{-1}) was produced. This finding indicates that the C source of CO and CH_4 is derived from the reduction in CO_2, rather than from the decomposition of C in the catalyst or from the decomposition of organic reagents in solution. This outcome serves to reinforce the conclusion that only a minimal number of molar photogenerated electrons are implicated in the hydrogen evolution reaction (HER). In order to establish that CO and CH_4 are derived from CO_2, $^{12}CO_2$ gas was replaced by $^{13}CO_2$ gas (Figure 4e). A high-intensity signal with a m/z = 29 was detected in the GC-MS. This signal is believed to be ^{13}CO. However, no signal was detected for $^{13}CH_4$, which was attributable to the low yield of $^{13}CH_4$ [31,32]. Consequently, it can be concluded that CO is indeed produced by CO_2 activation. Finally, an investigation was conducted into the stability of MIL-125-2H. As demonstrated in Figure 4f, it was determined that there was no substantial decline in its activity after five cycles. Post-reaction SEM (Figure S8) revealed that the backbone of MIL-125-2H remained unaltered and did not undergo collapse. Subsequently, an investigation was conducted into the ratio of Ti^{3+}/Ti^{4+} in MIL-125-2H following the stability test. The results, as illustrated in Figure S9 and Table S3, demonstrated that there was no alteration in the ratio of Ti^{3+}/Ti^{4+} peak areas. This finding suggests that MIL-125-2H exhibits excellent stability. The performance of CO_2 photoreduction in the present study was subsequently compared with a similar study. It emerged that the introduction of O_V in MIL-125(Ti) resulted in superior performance (Table S4).

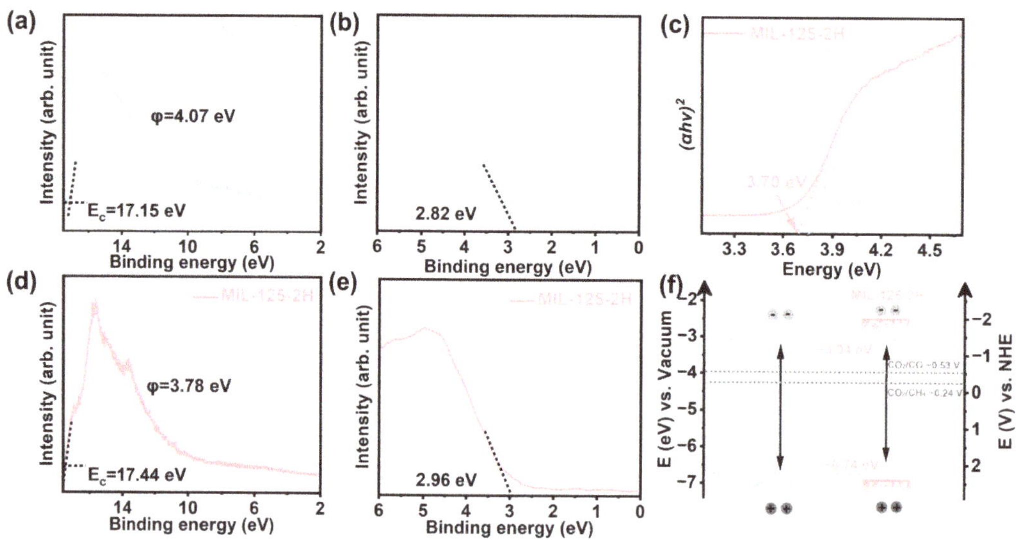

Figure 3. The UPS spectrum and VB-XPS of (**a**,**b**) MIL-125(Ti) and (**d**,**e**) MIL-125-2H. (**c**) The band gap diagram and (**f**) the positions of band structure of samples.

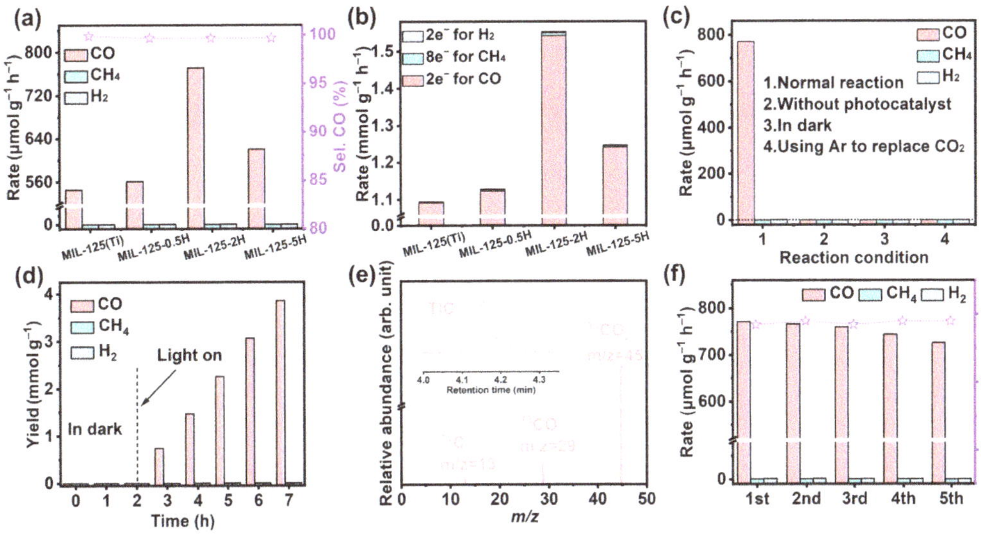

Figure 4. (**a**) Comparison of the rate of photocatalytic reduction in CO_2 products from samples and the selectivity of CO. (**b**) Number of molar electrons consumed in reaction. (**c**) Comparison of CO_2 photoreduction performance under different reaction conditions. (**d**) Gas products of CO_2 photoreduction as a function of time. (**e**) GC-MS spectra of $^{13}CO_2$ photoreduction products by MIL-125-2H. (**f**) Stability testing of catalyst.

Photogenerated electron transport properties were investigated by means of a series of sample photoelectric property tests. As shown in Figure 5a, MIL-125-2H exhibits electrochemical impedance spectroscopy (EIS) radius in comparison to MIL-125(Ti), indicating its enhanced efficacy in facilitating photogenerated electron transfer [33]. Subsequently, the photocurrent test revealed that MIL-125-2H exhibited a photocurrent of up to ~1.80 μA under light excitation, while MIL-125(Ti) demonstrated a maximum of ~1.0 μA. This finding suggests that MIL-125-2H is able to promote more photogenerated electron transfer than MIL-125(Ti) [34]. The steady-state fluorescence lifetime (PL) measurements demonstrate that the PL peak intensity of MIL-125-2H is considerably lower than that of MIL-125(Ti) at an excitation wavelength of 350 nm (Figure S10). This finding indicates that the introduction of oxygen vacancies has resulted in a substantial suppression of carrier complexation in the sample [35]. Next, the transient photoluminescence (TR-PL) lifetime of the photogenerated electrons transferred was examined. As demonstrated in Figure 5c, the TR-PL lifetime of MIL-125-2H (5.41 ns) is greater than that of MIL-125(Ti) (4.51 ns), which suggests a higher utilization of photogenerated electrons in MIL-125-2H [36,37]. Finally, the surface photovoltage of the samples was tested, and as demonstrated in Figure S11, compared to MIL-125(Ti), MIL-125-2H has a significant potential difference at 310–370 nm because of the presence of O_V, which promotes the enrichment of electrons on O_V and their migration towards the Ti sites. The combined photoelectronic characterization demonstrates that the incorporation of O_V substantially hinders the photogenerated carriers from undergoing bulk and surface phase complexation on MIL-125-2H. Furthermore, the introduction of O_V enhances the lifetime of photogenerated carriers and facilitates the attainment of a higher photogenerated carrier separation efficiency in MIL-125-2H.

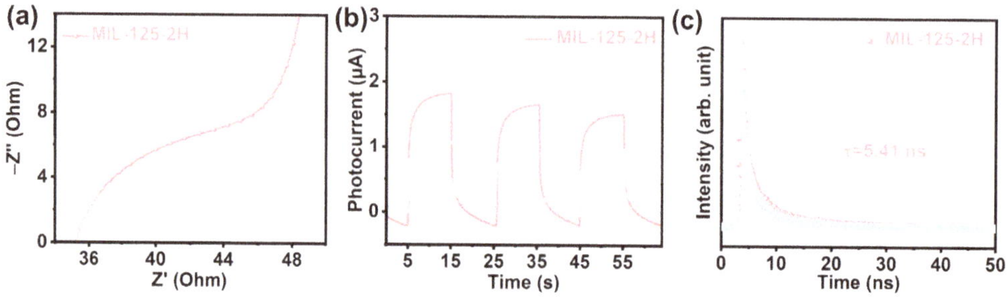

Figure 5. (**a**) The EIS spectra, (**b**) the transient photocurrent response, and (**c**) the TR-PL spectra of MIL-125(Ti) and MIL-125-2H.

The present study employed in situ characterization techniques in conjunction with DFT calculations to explore the mechanism of the dynamic photoreduction process of CO_2. The present study investigates the simulation of MIL-125(Ti) and MIL-125-2H photocatalytic activation of CO_2 for in situ Fourier transform infrared spectroscopy (FT-IR) testing. As shown in Figure 6a,b, the wavenumbers at ~1298, ~1355, ~1462, ~1621, and ~1648 cm^{-1} correspond to m-CO_3^{2-}, b-CO_3^{2-}, CO_2^-, *COOH, and H_2O, respectively. The wavenumbers at ~1215 cm^{-1} and ~1385 cm^{-1} correspond to HCO_3^- [38,39]. These belong to the intermediates of CO_2 reduction to CO, with no intermediates belonging to the CH_4 intermediates. Meanwhile, the peaks of these intermediates were significantly enhanced with increasing light duration. The peaks belonging to H_2O are also significantly enhanced, which is caused by the continuous decomposition of H_2O to provide more proton sources. Compared with MIL-125(Ti), the intensity of the in situ FT-IR peaks of MIL-125-2H was stronger, indicating that MIL-125-2H has a greater ability to reduce CO_2 to CO [40].

In consideration of the results obtained from in situ infrared spectroscopy, the following pathways for the photoreduction in CO_2 can be identified:

$$CO_2 + * \rightarrow * CO_2 \tag{1}$$

$$* CO_2 + H^+ + e^- \rightarrow * COOH \tag{2}$$

$$* COOH + H^+ + e^- \rightarrow * CO + H_2O \tag{3}$$

$$* CO \rightarrow CO \uparrow + * \tag{4}$$

As depicted in Figure 6c (* represents photocatalyst), the key rate steps on MIL-125(Ti) and MIL-125-2H are identified as CO_2 adsorption and CO desorption at the interface with ΔG values of 1.22 eV and 0.53 eV, respectively. The presence of O_V in MIL-125-2H leads to a substantial reduction in the overall activation energy barriers and the formation energies of key intermediates [24]. Consequently, the strategic addition of O_V can greatly increase activity by lowering the energy barrier required for the overall reaction.

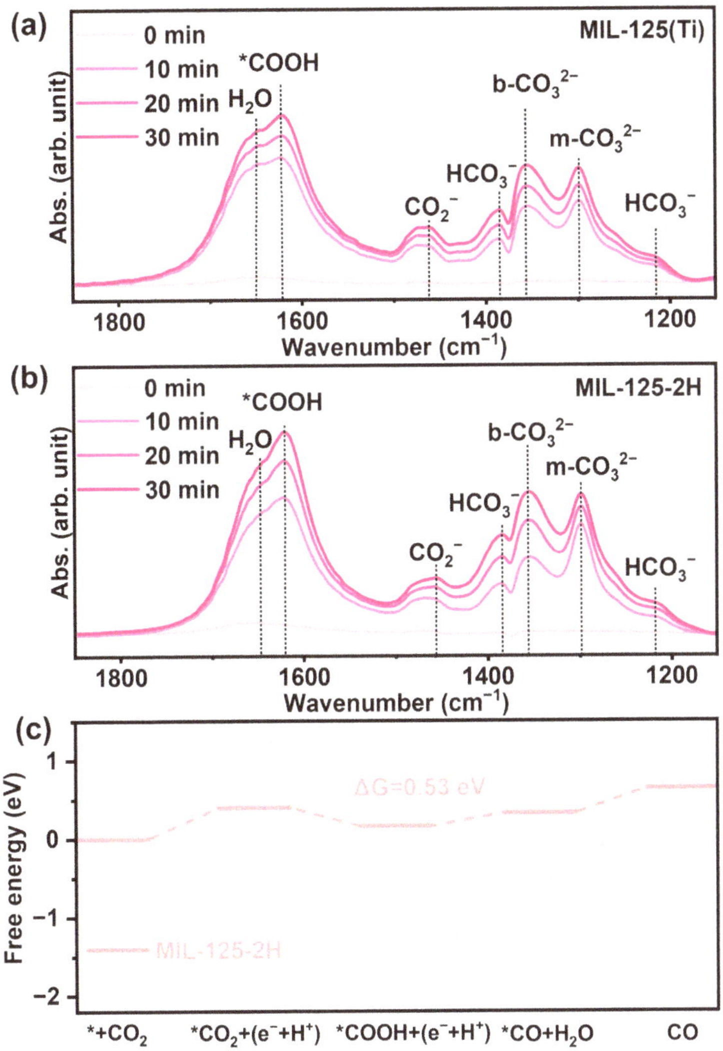

Figure 6. The in situ FTIR spectra under CO_2 reaction conditions with increasing light time of (**a**) MIL-125(Ti) and (**b**) MIL-125-2H. (**c**) Calculation of the Gibbs free energy for the activation of CO_2 to CO by MIL-125(Ti) and MIL-125-2H.

4. Conclusions

In the present study, an O_V-rich strategy was developed, designed, and introduced on MIL-125(Ti). The introduction of O_V resulted in an increase in the specific gravity of Ti^{3+}/Ti^{4+} on MIL-125-2H and the reduction capacity of Ti^{3+}, thereby providing a greater number of active sites for CO_2 photoreduction. The presence of O_V has been shown to facilitate the transfer of electrons from the Ti site on MIL-125-2H to CO_2, in addition to lowering the activation energy barrier of CO_2. In consideration of the aforementioned advantages, MIL-125-2H demonstrated a reduction in CO_2 to CO with a high yield and high selectivity of 771.22 μmol g^{-1} h^{-1} and near 100%, respectively. The findings of the study culminated in the synthesis of an efficient and stable catalyst, thus providing a novel strategy for the application of vacancy engineering to artificial carbon photoreduction.

Supplementary Materials: The following supporting information can be downloaded at: https://www.mdpi.com/article/10.3390/ma18061343/s1, Figure S1: The SEM images of (a) MIL-125-0.5H and (b) MIL-125-5H. Figure S2: FTIR spectroscopy of the sample. Figure S3: Raman spectroscopy of the sample. Figure S4: Theoretical calculation of light absorption spectrum of (a) MIL-125(Ti) and (b) MIL-125-2H. Figure S5: The density of state of (a) MIL-125(Ti) and (b) MIL-125-2H. Figure S6: The full XPS spectrum of MIL-125-0.5H and MIL-125-5H. Figure S7: CO_2 photocatalytic reaction system. Figure S8: SEM image of MIL-125-2H after stability test. Figure S9: The XPS spectrum of MIL-125-2H after stability test. Figure S10: The PL spectrum of MIL-125(Ti) and MIL-125-2H. Figure S11: The SPV image of MIL-125(Ti) and MIL-125-2H. Table S1: The crystallite size of the sample. Table S2: Ratio of Ti^{3+} and T^{4+} content. Table S3: The ratio of Ti^{3+}/Ti^{4+} before and after the MIL-125-2H reaction Table S4: Comparison of the performance of this work with similar literature on photocatalytic reduction of CO_2 to CO. References [41–47] have been cited in Supplementary Materials file.

Author Contributions: H.X.: methodology, investigation, conceptualization, data curation, writing—original draft. H.S.: investigation, writing—original draft. X.W.: resources, funding acquisition. X.Z.: supervision, methodology, resources, funding acquisition. All authors have read and agreed to the published version of the manuscript.

Funding: This research was funded by National Natural Science Foundation of China, grant number 22308300; Natural Science Foundation of Jiangsu Province, grant number BK20220598; Postgraduate Research & Practice Innovation Program of Jiangsu Province, grant number SJCX23_1925.

Institutional Review Board Statement: Not applicable.

Informed Consent Statement: Not applicable.

Data Availability Statement: The original contributions presented in this study are included in the article/Supplementary Materials. Further inquiries can be directed to the corresponding authors.

Acknowledgments: The authors would like to express their gratitude to Jian Xu for his linguistic and grammatical guidance in this work.

Conflicts of Interest: The authors declare that they have no known competing financial interests or personal relationships that could have appeared to influence the work reported in this paper.

References

1. Chang, X.; Wang, T.; Gong, J. CO_2 photo-reduction: Insights into CO_2 activation and reaction on surfaces of photocatalysts. *Energy Environ. Sci.* **2016**, *9*, 2177–2196. [CrossRef]
2. Müller, L.J.; Kätelhön, A.; Bringezu, S.; McCoy, S.; Suh, S.; Edwards, R.; Sick, V.; Kaiser, S.; Cuéllar-Franca, R.; El Khamlichi, A.; et al. The carbon footprint of the carbon feedstock CO_2. *Energy Environ. Sci.* **2020**, *13*, 2979–2992. [CrossRef]
3. Sullivan, I.; Goryachev, A.; Digdaya, I.A.; Li, X.; Atwater, H.A.; Vermaas, D.A.; Xiang, C. Coupling electrochemical CO_2 conversion with CO_2 capture. *Nat. Catal.* **2021**, *4*, 952–958. [CrossRef]
4. Fang, S.; Rahaman, M.; Bharti, J.; Reisner, E.; Robert, M.; Ozin, G.A.; Hu, Y.H. Photocatalytic CO_2 reduction. *Nat. Rev. Methods Primers* **2023**, *3*, 61. [CrossRef]
5. Ran, J.; Jaroniec, M.; Qiao, S.Z. Cocatalysts in Semiconductor-Based Photocatalytic CO_2 Reduction: Achievements, Challenges, and Opportunities. *Adv. Mater.* **2018**, *30*, 1704649. [CrossRef]
6. Zhou, H.-C.J.; Kitagawa, S. Metal–Organic Frameworks (MOFs). *Chem. Soc. Rev.* **2014**, *43*, 5415–5418. [CrossRef]
7. Liang, J.; Gvilava, V.; Jansen, C.; Öztürk, S.; Spieß, A.; Lin, J.; Xing, S.; Sun, Y.; Wang, H.; Janiak, C. Cucurbituril-Encapsulating Metal–Organic Framework via Mechanochemistry: Adsorbents with Enhanced Performance. *Angew. Chem. Int. Ed.* **2021**, *60*, 15365–15370. [CrossRef]
8. Lu, Y.; Zhang, G.; Zhou, H.; Cao, S.; Zhang, Y.; Wang, S.; Pang, H. Enhanced Active Sites and Stability in Nano-MOFs for Electrochemical Energy Storage Through Dual Regulation by Tannic Acid. *Angew. Chem. Int. Ed.* **2023**, *62*, e202311075. [CrossRef]
9. Wen, L.; Sun, K.; Liu, X.; Yang, W.; Li, L.; Jiang, H.L. Electronic State and Microenvironment Modulation of Metal Nanoparticles Stabilized by MOFs for Boosting Electrocatalytic Nitrogen Reduction. *Adv. Mater.* **2023**, *35*, 2210669. [CrossRef]
10. Cano, F.J.; Coste, S.; Reyes-Vallejo, O.; Makowska-Janusik, M.; Velumani, S.; Olvera, M.d.l.L.; Kassiba, A. Influence of GO oxidation degrees on the organization and physical features of TiO_2–GO-based nanocomposites for water dye removal. *Surf. Interfaces* **2024**, *46*, 104004. [CrossRef]

11. Wu, S.; Xing, X.; Wang, D.; Zhang, J.; Chu, J.; Yu, C.; Wei, Z.; Hu, M.; Zhang, X.; Li, Z. Highly Ordered Hierarchically Macroporous MIL-125 with High Specific Surface Area for Photocatalytic CO_2 Fixation. *ACS Sustain. Chem. Eng.* **2019**, *8*, 148–153. [CrossRef]
12. Liu, L.; Chen, Y.; Zhang, J.; Zeng, F.; Bao, J.; Li, W.; Tian, G. Engineering defect-rich H-$W_{18}O_{49}$ nanowire/MIL-125(Ti) nanosheet S-scheme heterostructure for remarkable CO_2 photoconversion. *Chem. Eng. J.* **2024**, *501*, 157310. [CrossRef]
13. Hussain, M.B.; Kang, B.; Cheng, X.; Ma, C.; Wang, X.; Mehmood, R.; Iqbal, S. Oxygen vacancy induced Pt-decorated MOF photocatalyst for hydrogen production. *Int. J. Hydrog. Energy* **2023**, *48*, 13780–13790. [CrossRef]
14. Liu, Y.; Ye, C.; Chen, L.; Fan, J.; Liu, C.; Xue, L.; Sun, J.; Zhang, W.; Wang, X.; Xiong, P.; et al. High Entropy-Driven Role of Oxygen Vacancies for Water Oxidation. *Adv. Funct. Mater.* **2024**, *34*, 2314820. [CrossRef]
15. Chen, Y.; Liu, Y.; Zhai, W.; Liu, H.; Sakthivel, T.; Guo, S.; Dai, Z. Metastabilizing the Ruthenium Clusters by Interfacial Oxygen Vacancies for Boosted Water Splitting Electrocatalysis. *Adv. Energy Mater.* **2024**, *14*, 2400059. [CrossRef]
16. Xin, Y.; Tian, J.; Xiong, X.; Wu, C.; Carabineiro, S.A.C.; Yang, X.; Chen, Z.; Xia, Y.; Jin, Y. Enhanced Photocatalytic Efficiency Through Oxygen Vacancy-Driven Molecular Epitaxial Growth of Metal–Organic Frameworks on $BiVO_4$. *Adv. Mater.* **2025**, *37*, 2417589. [CrossRef]
17. Sun, Y.; Ji, H.; Sun, Y.; Zhang, G.; Zhou, H.; Cao, S.; Liu, S.; Zhang, L.; Li, W.; Zhu, X.; et al. Synergistic Effect of Oxygen Vacancy and High Porosity of Nano MIL-125(Ti) for Enhanced Photocatalytic Nitrogen Fixation. *Angew. Chem. Int. Ed.* **2023**, *63*, e202316973. [CrossRef]
18. Song, Q.; Sun, C.; Wang, Z.; Bai, X.; Wu, K.; Li, Q.; Zhang, H.; Zhou, L.; Pang, H.; Liang, Y.; et al. Directed charge transfer in all solid state heterojunction of Fe doped MoS_2 and C–TiO_2 nanosheet for enhanced nitrogen photofixation. *Mater. Today Phys.* **2021**, *21*, 100563. [CrossRef]
19. Han, Y.; Huang, W.; He, M.; An, B.; Chen, Y.; Han, X.; An, L.; Kippax-Jones, M.; Li, J.; Yang, Y.; et al. Trace benzene capture by decoration of structural defects in metal–organic framework materials. *Nat. Mater.* **2024**, *23*, 1531–1538. [CrossRef]
20. Fu, Y.; Sun, D.; Chen, Y.; Huang, R.; Ding, Z.; Fu, X.; Li, Z. An amine-functionalized titanium metal-organic framework photocatalyst with visible-light-induced activity for CO_2 reduction. *Angew. Chem. Int. Ed.* **2012**, *51*, 3364–3367. [CrossRef]
21. Dai, Q.; Gao, G.; Tang, J.; Jiang, R.; Sun, S.; Ye, Y.; Li, S.; Xie, R.; Zhang, J. The MIL-125(Ti)/Co_3O_4 towards efficiently removing tetracycline by synergistic adsorption-photocatalysis roles. *Mater. Des.* **2025**, *250*, 113608. [CrossRef]
22. Wang, Z.; Teramura, K.; Hosokawa, S.; Tanaka, T. Photocatalytic conversion of CO_2 in water over Ag-modified $La_2Ti_2O_7$. *Appl. Catal. B Environ.* **2015**, *163*, 241–247. [CrossRef]
23. Jiang, J.; Wang, X.; Xu, Q.; Mei, Z.; Duan, L.; Guo, H. Understanding dual-vacancy heterojunction for boosting photocatalytic CO_2 reduction with highly selective conversion to CH_4. *Appl. Catal. B Environ.* **2022**, *316*, 121679. [CrossRef]
24. Zhu, X.; Xu, H.; Bi, C.; Song, H.; Zhou, G.; Zhong, K.; Yang, J.; Yi, J.; Xu, H.; Wang, X. Piezo-photocatalysis for efficient charge separation to promote CO_2 photoreduction in nanoclusters. *Ultrason. Sonochem.* **2023**, *101*, 106653. [CrossRef]
25. Xu, H.; Song, H.; Bi, C.; Zhou, G.; Liu, X.; Zhong, K.; Jiang, W.; Yang, J.; Shen, W.; Hao, N.; et al. Breaking the intrinsic activity barriers of bilayer metal oxides for catalytic CO_2 reduction. *J. Colloid Interface Sci.* **2024**, *675*, 419–428. [CrossRef]
26. Wu, T.; Zhao, H.; Zhu, X.; Xing, Z.; Liu, Q.; Liu, T.; Gao, S.; Lu, S.; Chen, G.; Asiri, A.M.; et al. Identifying the Origin of Ti^{3+} Activity Toward Enhanced Electrocatalytic N_2 Reduction over TiO_2 Nanoparticles Modulated by Mixed-Valent Copper. *Adv. Mater.* **2020**, *32*, 2000299. [CrossRef]
27. Lu, L.; Zhang, H.; Sun, Z.; Wang, J.; Wang, H.; Xue, J.; Shen, Q.; Li, Q. Creation of robust oxygen vacancies in 2D ultrathin BiOBr nanosheets by irradiation through photocatalytic memory effect for enhanced CO_2 reduction. *Chem. Eng. J.* **2023**, *477*, 146892. [CrossRef]
28. Zhu, X.; Xu, H.; Liu, J.; Bi, C.; Tian, J.; Zhong, K.; Wang, B.; Ding, P.; Wang, X.; Chu, P.K.; et al. Stacking Engineering of Heterojunctions in Half-Metallic Carbon Nitride for Efficient CO_2 Photoreduction. *Adv. Sci.* **2023**, *10*, 2307192. [CrossRef]
29. Zhao, J.; Ji, M.; Chen, H.; Weng, Y.-X.; Zhong, J.; Li, Y.; Wang, S.; Chen, Z.; Xia, J.; Li, H. Interfacial chemical bond modulated $Bi_{19}S_{27}Br_3$/g-C_3N_4 Z-scheme heterojunction for enhanced photocatalytic CO_2 conversion. *Appl. Catal. B Environ.* **2022**, *307*, 121162. [CrossRef]
30. Liu, Q.; Cheng, H.; Chen, T.; Lo, T.W.B.; Ma, J.; Ling, A.; Wang, F. Boosted CO desorption behaviors induced by spatial dyadic heterostructure in polymeric carbon nitride for efficient photocatalytic CO_2 conversion. *Appl. Catal. B Environ.* **2021**, *295*, 120289. [CrossRef]
31. Ning, C.; Yang, J.; Bai, S.; Chen, G.; Liu, G.; Shen, T.; Zheng, L.; Xu, S.M.; Kong, X.; Liu, B.; et al. An Efficient Intercalation Supramolecular Structure for Photocatalytic CO_2 Reduction to Ethylene Under Visible Light. *Adv. Funct. Mater.* **2023**, *33*, 2300365. [CrossRef]
32. Wang, J.; Yang, C.; Mao, L.; Cai, X.; Geng, Z.; Zhang, H.; Zhang, J.; Tan, X.; Ye, J.; Yu, T. Regulating the Metallic Cu–Ga Bond by S Vacancy for Improved Photocatalytic CO_2 Reduction to C_2H_4. *Adv. Funct. Mater.* **2023**, *33*, 2213901. [CrossRef]
33. Zhou, G.; Chen, Y.; Chen, G.; Xu, H.; Yin, W.; Wang, B.; Zhu, X.; Ning, X.; Chu, P.K.; Wang, X.; et al. Enhanced CO_2 photoreduction in pure water systems by surface-reconstructed asymmetric Mn–Cu sites. *Appl. Catal. B Environ. Energy* **2025**, *361*, 124617. [CrossRef]

34. Ma, M.; Huang, Z.; Doronkin, D.E.; Fa, W.; Rao, Z.; Zou, Y.; Wang, R.; Zhong, Y.; Cao, Y.; Zhang, R.; et al. Ultrahigh surface density of Co-N_2C single-atom-sites for boosting photocatalytic CO_2 reduction to methanol. *Appl. Catal. B Environ.* **2022**, *300*, 120695. [CrossRef]
35. Zhang, Y.; Song, T.; Zhou, X.; Yang, Y. Oxygen-vacancy-boosted visible light driven photocatalytic oxidative dehydrogenation of saturated N-heterocycles over Nb_2O_5 nanorods. *Appl. Catal. B Environ.* **2022**, *316*, 121622. [CrossRef]
36. Miao, Z.; Wang, Q.; Zhang, Y.; Meng, L.; Wang, X. In situ construction of S-scheme AgBr/BiOBr heterojunction with surface oxygen vacancy for boosting photocatalytic CO_2 reduction with H_2O. *Appl. Catal. B Environ.* **2022**, *301*, 120802. [CrossRef]
37. Wu, J.; Xiong, L.; Hu, Y.; Yang, Y.; Zhang, X.; Wang, T.; Tang, Z.; Sun, A.; Zhou, Y.; Shen, J.; et al. Organic half-metal derived erythroid-like $BiVO_4$/hm-C_4N_3 Z-Scheme photocatalyst: Reduction sites upgrading and rate-determining step modulation for overall CO_2 and H_2O conversion. *Appl. Catal. B Environ.* **2021**, *295*, 120277. [CrossRef]
38. Yang, H.; Hou, H.; Yang, M.; Zhu, Z.; Fu, H.; Zhang, D.; Luo, Y.; Yang, W. Engineering the S-scheme heterojunction between NiO microrods and MgAl-LDH nanoplates for efficient and selective photoreduction of CO_2 to CH_4. *Chem. Eng. J.* **2023**, *474*, 145813. [CrossRef]
39. Yang, J.; Zhu, X.; Yu, Q.; He, M.; Zhang, W.; Mo, Z.; Yuan, J.; She, Y.; Xu, H.; Li, H. Multidimensional In_2O_3/In_2S_3 heterojunction with lattice distortion for CO_2 photoconversion. *Chin. J. Catal.* **2022**, *43*, 1286. [CrossRef]
40. Zhu, X.; Zhou, G.; Wang, Z.; Zhong, K.; Ding, P.; Song, Y.; Yuan, J.; She, Y.; Li, H.; Xu, H. Nanostructure and functional group engineering of black phosphorus via plasma treatment for CO_2 photoreduction. *J. CO_2 Util.* **2021**, *54*, 101745. [CrossRef]
41. Liang, J.; Yu, H.; Shi, J.; Li, B.; Wu, L.; Wang, M. Dislocated Bilayer MOF Enables High-Selectivity Photocatalytic Reduction of CO_2 to CO. *Adv. Mater.* **2023**, *35*, e2209814. [CrossRef] [PubMed]
42. Zhang, L.; Zhou, G.; Chen, G.; Wang, H.; Zhao, Q.; Yin, W.; Yi, J.; Zhu, X.; Wang, X.; Ning, X. Bimetallic NiCu catalyst derived from spent MOF adsorbent for efficient photocatalytic CO_2 reduction. *Chem. Eng. J.* **2024**, *497*, 154701. [CrossRef]
43. Gao, X.; Guo, B.; Guo, C.; Meng, Q.; Liang, J.; Liu, J. Zirconium-Based Metal-Organic Framework for Efficient Photocatalytic Reduction of CO_2 to CO: The Influence of Doped Metal Ions. *ACS Appl. Mater. Interfaces* **2020**, *12*, 24059–24065. [CrossRef]
44. Zhao, X.; Sun, L.; Jin, X.; Xu, M.; Yin, S.; Li, J.; Li, X.; Shen, D.; Yan, Y.; Huo, P. Cu media constructed Z-scheme heterojunction of UiO-66-NH2/Cu_2O/Cu for enhanced photocatalytic induction of CO_2. *Appl. Surf. Sci.* **2021**, *545*, 148967. [CrossRef]
45. Jiang, H.; Wang, L.; Yu, X.; Sun, L.; Li, J.; Yang, J.; Liu, Q. Precise regulation of built-in electric field over NH_2-MIL-125-Ti/WO_{3-x} S-scheme heterojunction for achieving simultaneous formation of CO and H_2O_2 from CO_2 and H_2O. *Chem. Eng. J.* **2023**, *466*, 143129. [CrossRef]
46. Li, Z.; Bai, W.; Liu, D.; Han, B.; Liang, Y.; Qi, J. Preloaded oxygen vacancy conditioning Ni/TiO_2 to enhance photocatalytic CO_2 reduction. *Sep. Purif. Technol.* **2024**, *330*, 125250. [CrossRef]
47. Zhang, Y.; Wang, Y.; Hu, Z.; Huang, J.; Yang, S.; Li, H. High-efficiency photocatalytic CO_2 reduction enabled by interfacial Ov and isolated Ti^{3+} of g-C_3N_4/TiO_2 Z-scheme heterojunction. *J. Colloid Interface Sci.* **2024**, *663*, 891–901. [CrossRef]

materials

MDPI

Article

Phase-Controlled Synthesis of Ru Supported on Carbon Nitride and the Application in Photocatalytic H_2 Evolution

Xiaohu Sun [1], Xiangyang Cao [1], Ganghua Zhou [1], Tiaolong Lv [1], Jian Xu [1], Yubo Zhou [2], Zhigang Wang [1,*] and Jianjian Yi [1,2,3,*]

1 College of Environmental Science and Engineering, Yangzhou University, Yangzhou 225127, China; mz120211186@stu.yzu.edu.cn (X.S.)
2 Ningbo Solartron Technology Co., Ltd., Ningbo 315034, China
3 School of Material Science and Engineering, Jiangsu University, Zhenjiang 212013, China
* Correspondence: wangzg@yzu.edu.cn (Z.W.); jjyi@yzu.edu.cn (J.Y.)

Abstract: This work aims to explore the influence of crystal phase engineering on the photocatalytic hydrogen evolution activity of Ru/C_3N_4 systems. By precisely tuning the combination of Ru precursors and reducing solvents, we successfully synthesized Ru co-catalysts with distinct crystal phases (hcp and fcc) and integrated them with C_3N_4. The photocatalytic hydrogen evolution experiments demonstrated that hcp-Ru/C_3N_4 achieved a significantly higher hydrogen evolution rate (24.23 μmol h^{-1}) compared to fcc-Ru/C_3N_4 (7.44 μmol h^{-1}), with activity reaching approximately 42% of Pt/C_3N_4 under the same conditions. Photocurrent and electrochemical impedance spectroscopy analyses revealed that hcp-Ru/C_3N_4 exhibited superior charge separation and transfer efficiency. Moreover, Gibbs free energy calculations indicated that the hydrogen adsorption energy of hcp-Ru ($\Delta G_{H^*} = -0.14$ eV) was closer to optimal compared to fcc-Ru (−0.32 eV), enhancing the hydrogen generation process. These findings highlight that crystal-phase engineering plays a critical role in tuning the electronic structure and catalytic properties of Ru-based systems, offering insights for the design of highly efficient noble metal catalysts for photocatalysis.

Keywords: phase engineering; ruthenium; photocatalysis; hydrogen evolution; co-catalyst

1. Introduction

The development of more efficient nanomaterial catalysts is crucial in fields such as environmental protection, food processing, petrochemicals and energy utilization. Catalyst performance can be enhanced by controlling the nanostructures through strategies involving size [1], shape [2], alloys [3], crystal phase [4], and defects [5]. Phase engineering of nanomaterials (PEN), as an effective strategy for the rational design and precise synthesis of nanomaterials with controllable phases, has gained significant attention in recent years [6–9]. Phase engineering modulates performance, functionality, and applications by altering the catalyst's surface electronic and geometric structures through adjustments in reaction kinetics and surface energy [10–12]. For instance, bulk Nickel (Ni) typically shows the face-centered cubic (fcc) phase, but non-conventional hexagonal close-packed (hcp) Ni nanoparticles can be synthesized via chemical methods or one-pot procedures [13,14]. Experimental results indicate that hcp Ni exhibits superior catalytic performance compared to fcc Ni across various reactions. Additionally, studies have explored the impact of crystallinity on catalytic performance, revealing that the amorphous 1T phase of $MoSe_2$ rich in unsaturated coordination sites, facilitates proton coupling to form hydrogen [15]. This provides new insights into phase engineering for fine-tuning growth modes. Nevertheless, the practical application of phase engineering is still in its early stages, and there is an urgent need for the development of environmentally friendly and versatile synthetic methods for phase regulation. Furthermore, the influence mechanisms of template, reduction kinetics, and capping agents on atomic stacking require further investigation.

Photocatalytic hydrogen production is considered a green, clean, safe, and low-cost renewable energy technology. Due to their high electron capture efficiency, excellent chemical stability, and abundant surface active sites, noble metals play an essential role in the preparation of photocatalysts for hydrogen production [16–19]. Among them, platinum (Pt), with its optimal Fermi level and zero-approaching hydrogen adsorption Gibbs free energy, is one of the most promising co-catalysts for photocatalytic hydrogen production [20]. However, due to its high cost, large-scale use remains impractical. To reduce the dependence on expensive Pt catalysts for photocatalytic water splitting, researchers have explored various types of catalysts and optimized their nanostructures to develop high-activity, low-cost co-catalysts. Ruthenium (Ru) has attracted significant attention in the field of photocatalysis due to its lower cost and the ability to achieve activity comparable to Pt under certain optimized conditions [21–23]. Zhu et al. optimized the asymmetric electronic properties of Ru to lower the energy barrier for Ru^{n+} in water splitting, and the resulting Ru/NC catalyst exhibited high alkaline hydrogen evolution reaction (HER) activity (21.9 mV @ 10 mA cm^{-2}, 29.03 mV dec^{-1}) [24]. Dong et al. discovered that the incorporation of copper (Cu) in the alloy altered the reaction pathway for Ru in photocatalysis, achieving a 96% selectivity for CO_2 to CH_4 [25]. Despite numerous successful cases of Ru in various catalytic reactions, studies on the synthesis and optimal crystal phase of Ru co-catalysts for hydrogen production are scarce, and the mechanisms underlying the impact of Ru crystal phase on photocatalytic performance remain largely unexplored. Therefore, it is essential to investigate the feasibility of phase control in Ru for enhancing hydrogen production.

In this study, we employ a rational and controllable chemical reduction method to prepare hcp and fcc Ru nanoparticles and successfully load them onto C_3N_4 ultrathin nanosheets (hcp-Ru/C_3N_4, fcc-Ru/C_3N_4) to demonstrate the feasibility of improving the photocatalytic hydrogen production performance of Ru-based co-catalysts using phase engineering. Experimental results show that the Ru-based co-catalysts exhibit distinct phase-dependent behavior. hcp-Ru/C_3N_4 displays more efficient photocatalytic hydrogen production activity (24.23 μmol h^{-1}) and stability, significantly outperforming fcc-Ru/C_3N_4 (7.44 μmol h^{-1}). This superior performance is attributed to its higher charge separation efficiency and lower Gibbs free energy, as supported by detailed physicochemical characterization and theoretical calculations. This study highlights the vast potential and versatility of Ru phase-controlled synthesis for improving photocatalytic hydrogen production.

2. Materials and Methods

2.1. Materials

Ruthenium acetylacetonate (Ru(acac)$_3$, $C_{15}H_{21}O_6Ru$, ≥97%), triethylene glycol ($C_6H_{14}O_4$, >99%), triethanolamine (TEOA, $C_6H_{15}NO_3$, >99%), polyvinylpyrrolidone (($C_6H_9NO)_n$, K30, MW ≈ 40,000) were purchased from Shanghai Aladdin Biochemical Technology (Shanghai, China). Melamine ($C_3H_6N_6$, ≥99.0%), ruthenium (III) chloride hydrate ($RuCl_3 \cdot xH_2O$), ethanol absolute (C_2H_6O, ≥99.7%), ethylene glycol ($C_2H_6O_2$, ≥99.5%), potassium bromide (KBr, ≥99.0%) and sodium sulfate (Na_2SO_4, ≥99.0%) were purchased from Sinopharm Chemical Reagent Co., Ltd. (Shanghai, China).

2.2. Methods

2.2.1. Synthesis of C_3N_4 Nanosheets

An amount of 2 g of melamine was weighed and added to a covered ceramic crucible. The crucible was then placed in a muffle furnace and heated to 550 °C at a rate of 2 °C/min, where it was held for 4 h. After the reaction, the resulting yellow block was obtained as the bulk material. The bulk material was ground into a powder, and 800 mg of the powder was spread evenly in a partially open ceramic boat. The boat was then placed in the muffle furnace and heated to 550 °C at a rate of 10 °C/min, where it was held for 40 min. The resulting white powder was the desired C_3N_4 for the experiment.

2.2.2. Synthesis of C_3N_4-Ru Samples

In the synthesis of hcp-Ru/C_3N_4, 10 mL of an ethylene glycol (EG) solution containing 50 mg of C_3N_4 nanosheets and 0.5 mmol of polyvinylpyrrolidone (PVP) was sonicated in a 25 mL round-bottom flask for 30 min. Then, 0.025 mmol of $RuCl_3 \cdot xH_2O$ was added to the flask and sonicated until well dispersed. The flask containing the reaction mixture was then transferred to an oil bath and heated to 200 °C for 3 h. After the reaction was complete, the sample was washed with ethanol and deionized water, and the product was separated by centrifugation, followed by freeze-drying to obtain hcp-Ru/C_3N_4.

In the synthesis of fcc-Ru/C_3N_4, 10 mL of a triethylene glycol (TEG) solution containing 50 mg of C_3N_4 nanosheets and 0.5 mmol of PVP was sonicated in a 25 mL round-bottom flask for 30 min. Then, 0.025 mmol of Ru(acac)$_3$ was added to the flask and sonicated until well dispersed. The flask containing the reaction mixture was then transferred to an oil bath and heated to 200 °C for 3 h. After the reaction was complete, the sample was washed with ethanol and deionized water, and the product was separated by centrifugation, followed by freeze-drying to obtain fcc-Ru/C_3N_4.

2.3. Photocatalytic Hydrogen Activity Measurement

The photocatalytic hydrogen evolution reaction was carried out in an online reaction detection system (CEL-SPH2N, China Education Au-Light, Beijing, China). The reaction solution consisted of 50 mL of TEOA/H_2O (10% vol TEOA). The photocatalyst used in the measurement was 10 mg. The reaction was irradiated with a 300 W xenon lamp (CEL-HX F300/CEL-HX UV300, China Education Au-Light, Beijing, China) without any filters. Prior to irradiation, the system was evacuated to a vacuum state, and the reaction temperature was maintained at 15 °C using a circulating cooling water system. The hydrogen production rate was analyzed by gas chromatography (GC 7920) equipped with a TCD detector and a 5Å molecular sieve column (all components purchased from China Education Au-light, Beijing, China), using argon as the carrier gas.

The external quantum efficiency (*EQE*) values were determined using the following equation:

$$EQE = \frac{the\ number\ of\ reacted\ electrons}{the\ number\ of\ incident\ photons} \times 100\%$$

A more detailed calculation process can be found in the Supplementary Materials, including the information of the area of light spot, light intensities per unit area, and wavelength-dependent hydrogen evolution rates per unit area.

3. Results

The premise for studying the impact of crystal phase structure on photocatalytic reactions is the development of methods for controlling the crystal phase synthesis of Ru. We first prepared ultrathin C_3N_4 nanosheets as a support using a top-down thermal exfoliation method. Their coiled two-dimensional structure and large surface area are expected to provide an ideal environment for Ru loading. Different precursors and reducing agents can synergistically regulate the reduction kinetics rate and play an important role in selectively forming hcp and fcc phases of Ru [26,27]. The strong reducing nature of EG and the more easily dissociable chloride ligands in $RuCl_3$ molecules favor the formation of hcp Ru seeds. In contrast, TEG, with its weaker reducing ability, and the more stable acetylacetonate ligands in $Ru(acac)_3$ act to uniformly release Ru atoms, promoting the formation of the fcc structure. Therefore, in the synthesis of hcp-Ru/C_3N_4, $RuCl_3{\cdot}xH_2O$ was used as the Ru source, EG as the reducing agent and solvent, and PVP as the stabilizer, to chemically reduce and load hcp phase Ru nanoparticles onto the C_3N_4 nanosheets. The synthesis of fcc-Ru/C_3N_4 followed a similar method, with the only difference being the use of $Ru(acac)_3$ as the Ru source and TEG as the reducing agent and solvent (Figure 1a).

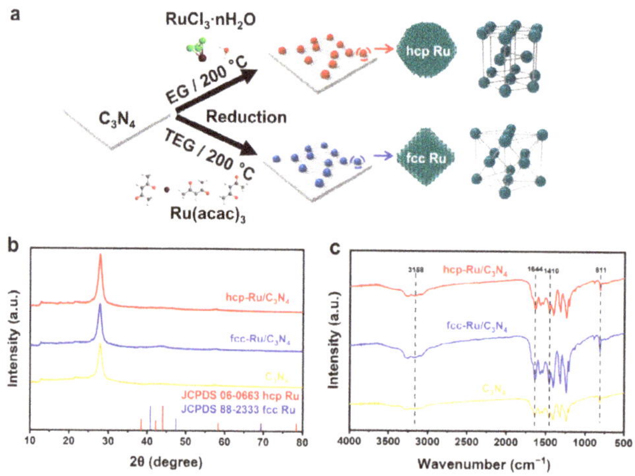

Figure 1. (**a**) Schematic illustrating of the synthesis process of hcp-Ru/C_3N_4 and fcc-Ru/C_3N_4. (**b**) XRD patterns and (**c**) FT-IR spectra.

To understand the basic chemical structure of the prepared catalysts, we studied their crystal structure using X-ray diffraction (XRD). The XRD patterns of the prepared hcp-Ru/C_3N_4 and fcc-Ru/C_3N_4 exhibited features similar to those of C_3N_4 (Figure 1b). Two characteristic diffraction peaks of C_3N_4 were detected at 2θ angles of 12.8° and 27.8°, corresponding to the stacking units of the in-planar repeating tri-*s*-triazine unit and the conjugated aromatic segments, respectively. These peaks correspond to the (100) and (002) crystal planes [28]. No characteristic peaks of Ru were detected in the XRD patterns of hcp-Ru/C_3N_4 and fcc-Ru/C_3N_4 due to the low content and small particle size, which were undetectable [29]. This was later confirmed through transmission electron microscope (TEM, Tecnai G2 F30 S-TWIN, purchased from FEI COMPANY, Hillsboro, OR, USA), which calculated the average particle size of Ru. Fourier transform infrared (FT-IR, Nicolet iS5, purchased from Thermo Fisher Scientific, Waltham, MA, USA) spectra were used to identify the surface functional groups of hcp-Ru/C_3N_4, fcc-Ru/C_3N_4, and C_3N_4 (Figure 1c). The characteristic peaks of C_3N_4 were clearly observed in all the catalysts. The absorption peak at 811 cm^{-1} corresponds to the characteristic vibration of the triazine units. The absorption peaks at 1410 and 1644 cm^{-1} are attributed to the stretching vibrations of C-N heterocyclic bonds, and the absorption peak at 3158 cm^{-1} corresponds to the stretching vibration of the O-H bond [30]. These results demonstrate that the introduction of Ru does not alter the original fundamental functional group structure of C_3N_4.

After determining the basic chemical structure of the prepared catalysts, we further investigated their microstructure and chemical composition through a series of electron microscopy characterizations. Scanning electron microscope (SEM, S-4800II, purchased from HITACHI, Tokyo, Japan) results show that the prepared C_3N_4 exhibits irregular porous coiled folds (Figure S1a,b). Atomic force microscope (AFM, SPM-9700HT, purchased from Shimadzu, Tokyo, Japan) observed that its thickness is approximately 2.8 nm (Figure 2d,e), providing further evidence of its typical ultrathin two-dimensional nanosheet structure. The larger lateral dimensions and Brunauer–Emmett–Teller (BET, ASAP2460, purchased from Micromeritics Instruments Corporation, Atlanta, GA, USA) surface area (Figure S7b–d) of C_3N_4 facilitate Ru loading, while the ultrathin thickness effectively shortens the charge migration path from the bulk to the surface, thereby reducing electron–hole recombination [31,32]. After loading with metallic Ru nanoparticles, the morphology of C_3N_4 did not show significant changes, and both hcp-Ru/C_3N_4 and fcc-Ru/C_3N_4 composite photocatalysts still maintained an irregular nanosheet morphology (Figure S1c–f). TEM results of hcp-Ru/C_3N_4 and fcc-Ru/C_3N_4 show that the Ru nanoparticles are uniformly dispersed on the flat surface of C_3N_4, with no apparent aggregation (Figure 2a–c), which can be attributed to the appropriate amount of reducing agent and PVP that control the uniform growth of metal Ru nanoparticles, preventing aggregation on C_3N_4 [33]. In addition, we have supplemented TEM-based size distribution analyses for both hcp-Ru (1.03 nm) and fcc-Ru (1.54 nm) samples (Figure S2a,b). While a minor size difference exists between the two phases (~0.5 nm), we propose that this subtle variation has negligible impacts on catalytic performance trends [34,35].

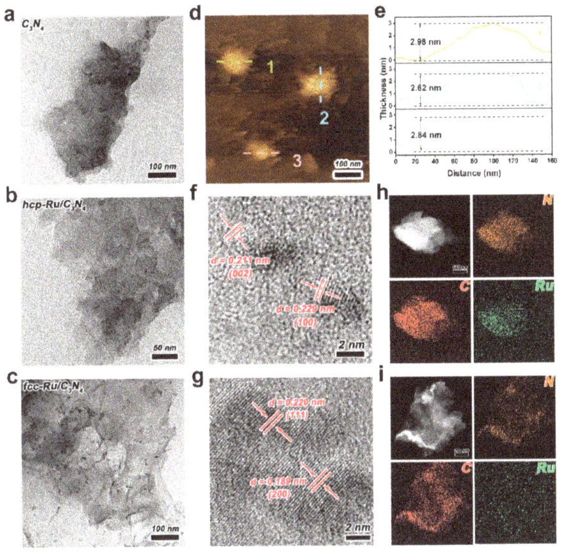

Figure 2. (**a**–**c**) TEM images of C_3N_4, hcp-Ru/C_3N_4 and fcc-Ru/C_3N_4. (**f**,**g**) HRTEM images and (**h**,**i**) STEM and elemental mapping images of hcp-Ru/C_3N_4 and fcc-Ru/C_3N_4. (**d**) AFM image of C_3N_4. (**e**) Thickness profile in (**d**).

To reveal the fine-phase structure of the Ru loaded on the catalyst, we performed high-resolution transmission electron microscope (HRTEM, Tecnai G2 F30 S-TWIN, purchased from FEI COMPANY, Hillsboro, OR, USA) characterization. The lattice spacings of 0.211 and 0.229 nm in Figure 2f correspond to the (002) and (100) planes of hcp Ru, while the lattice spacings of 0.189 and 0.220 nm in Figure 2g correspond to the (200) and (111) planes of fcc Ru. Additional HRTEM images are presented in Figure S3 to further clarify the phase purity of Ru nanoparticles in the catalysts. This indicates that the method used in this work successfully controls the crystal phase of Ru. Elemental mapping images further confirmed the uniform distribution of C, N, and Ru elements (Figure 2h,i), providing additional evidence that Ru nanoparticles of both crystal phases are uniformly loaded on C_3N_4. Furthermore, the actual Ru loading amount on hcp-Ru/C_3N_4 and fcc-Ru/C_3N_4 was determined to be 3.53 wt% and 3.64 wt% by ICP-MS (Elan DRC-e, purchased from PerkinElmer, Waltham, MA, USA) (Table S1).

X-ray photoelectron spectroscopy (XPS, ESCALAB 250Xi, purchased from Thermo Fisher Scientific, Waltham, MA, USA) was used to characterize the element valence states and surface composition of the catalysts. Full XPS spectra of hcp-Ru/C_3N_4 and fcc-Ru/C_3N_4 clearly showed signals corresponding to C, N, and Ru, further confirming the successful synthesis of the catalysts (Figure 3a). The C 1s spectrum of hcp-Ru/C_3N_4 (Figure 3b) can be deconvoluted into two peaks at 284.8 eV and 287.9 eV, corresponding to graphitic carbon (C-C) and sp^2-bonded aromatic structure (N-C=N), respectively. In the N 1s spectrum (Figure 3c), binding energy of 398.5 eV and 400.3 eV correspond to pyridinic N (N1) and pyrrolic N (N2) [36]. The Ru 3p spectrum (Figure 3d) reveals the Ru $3p_{3/2}$ and Ru $3p_{1/2}$ spin–orbit peaks, with fitting peaks at 461.2 eV and 483.5 eV assigned to metallic Ru (Ru^0) [37]. In contrast, the C 1s and N 1s spectra of fcc-Ru/C_3N_4 show a more negative binding energy shift (~0.3 eV), and the signal peaks in the Ru 3p spectrum are shifted positively by 0.3 eV. This suggests that C_3N_4 interacts more strongly with hcp Ru, with more electrons transferred to hcp Ru [38]. Despite the differences in the crystal phase structure of Ru loaded on the surfaces of hcp-Ru/C_3N_4 and fcc-Ru/C_3N_4, their similar XPS spectra indicate that both catalysts possess similar chemical valence states.

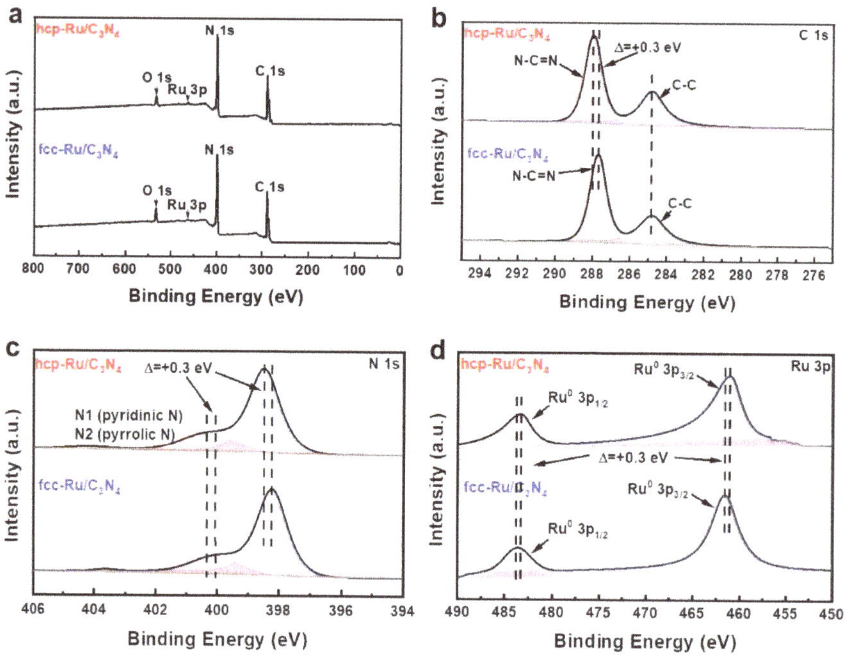

Figure 3. (**a**) XPS survey spectra and XPS spectra of (**b**) C 1s, (**c**) N 1s and (**d**) Ru 3p of hcp-Ru/C_3N_4 and fcc-Ru/C_3N_4.

After confirming the fine chemical structure of the prepared catalysts, we further discussed the rationality of constructing the Ru/C_3N_4 photocatalytic system using density functional theory (DFT). From a thermodynamic perspective, the direction of charge transfer is from the component with a higher Fermi level (i.e., lower work function) to the component with a lower Fermi level (i.e., higher work function) [39]. As shown in Figure 4a–c, we first calculated the theoretical work functions of C_3N_4, hcp-Ru, and fcc-Ru using DFT, which were found to be 4.69, 4.99, and 5.38 eV, respectively. The data indicate that after loading Ru onto C_3N_4, the photogenerated electrons produced upon light excitation can thermodynamically transfer to Ru, thereby enhancing charge separation efficiency and the electron concentration on the catalyst surface, providing theoretical support for Ru's role as a co-catalyst.

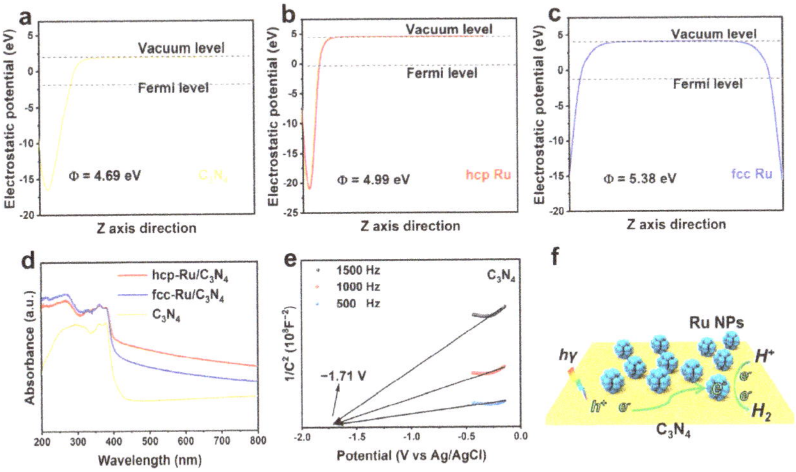

Figure 4. (**a**–**c**) Average potential profiles along the z-axis direction and calculated work function values of C_3N_4, hcp-Ru/C_3N_4 and fcc-Ru/C_3N_4. (**d**) UV-Vis diffuse reflection spectra of hcp-Ru/C_3N_4, fcc-Ru/C_3N_4 and C_3N_4. (**e**) Mott–Schottky plots of C_3N_4. (**f**) Schematic diagram of photocatalytic H_2 evolution process on the surface of hcp-Ru/C_3N_4 and fcc-Ru/C_3N_4.

Next, we assessed the light absorption ability of the catalysts using UV-Vis diffuse reflection spectra (DRS) (Figure 4d). It was found that C_3N_4 has a typical absorption edge of around 450 nm. After loading Ru, the optical absorption properties of the material were improved, and the visible light absorption range increased. This is mainly due to the synthesized catalyst being black in color. However, it should be noted that this enhancement in absorption is not significant, as the increased light absorption comes from Ru, which cannot be directly excited by light. The light absorption and charge generation capacity are solely determined by C_3N_4. Therefore, it can be concluded that both hcp-Ru/C_3N_4 and fcc-Ru/C_3N_4 exhibit similar charge generation behaviors.

Based on the theoretical analysis, we further determined the electric potential of C_3N_4 and speculated on its charge transfer pathway through Tauc curves and Mott–Schottky (M–S) plots. From the Tauc curve data, the bandgap of C_3N_4 was calculated to be 3.03 eV (Figure S4a). The flat band (FB) potential of C_3N_4 was determined to be −1.71 V vs. Ag/AgCl, pH = 7, using the M–S curves (Figure 4e) at 500, 1000, and 1500 Hz. Assuming that for n-type semiconductors, the gap between the flat band potential and conduction band (CB) potential can be neglected [40]. The conduction band potential of C_3N_4 was converted to −1.1 V vs. NHE, pH = 0 [41]. Thus, its valence band (VB) potential was determined to be 1.93 V vs. NHE, pH = 0, and the band structure was obtained (Figure S4b). Therefore, we can speculate on the charge transfer pathway in the Ru-C_3N_4 catalyst during photocatalytic hydrogen evolution (Figure 4f). Upon light irradiation, C_3N_4 generates electron–hole pairs. The photogenerated electrons migrate through the interface to the Ru surface. Due to the rich hydrogen evolution active sites at the edge and basal plane of Ru, the electrons gathered on its surface react with adsorbed protons to produce H_2. Meanwhile, the holes generated in C_3N_4 are consumed by the sacrificial agent, preventing the recombination of electron–hole pairs before charge separation and consumption.

In actual photocatalytic experiments, the hydrogen evolution activity of both Ru/C_3N_4 catalysts (with a theoretical loading of 5 wt%) was significantly improved compared to the pure C_3N_4. The average hydrogen evolution rates are shown in Figure 5a. Among them, hcp-Ru/C_3N_4 (24.23 μmol h^{-1}) exhibited much higher catalytic activity than fcc-Ru/C_3N_4 (7.44 μmol h^{-1}), indicating that the crystal structure of Ru significantly affects the photocatalytic hydrogen production performance of Ru/C_3N_4. To better understand the hydrogen production performance of hcp-Ru/C_3N_4, it is essential to evaluate its external quantum efficiency (*EQE*). As shown in Table S2, the *EQE* value of hcp-Ru/C_3N_4 is measured to be 5.28% at 420 nm. Clearly, compared to the catalysts listed in Table S3, hcp-Ru/C_3N_4 demonstrates superior performance in both hydrogen evolution rate and quantum efficiency, further highlighting its exceptional photocatalytic activity. We also tested the hydrogen evolution activity of Pt/C_3N_4 under the same conditions, which reached 57.46 μmol h^{-1}. This result indicates that the performance of the synthesized Ru co-catalyst is approximately 42% of that of Pt, showing a distinct advantage over most non-Pt co-catalysts. Additionally, we investigated the stability of hcp-Ru/C_3N_4 and fcc-Ru/C_3N_4 by conducting five photocatalytic reactions (25 h), as shown in Figure 5b,c. The results revealed that hcp-Ru/C_3N_4 demonstrated superior stability in hydrogen production compared to fcc-Ru/C_3N_4. Moreover, after the photocatalytic reaction, we performed characterizations on hcp-Ru/C_3N_4, which showed good chemical and structural stability (Figure S5). The slight decline in activity of hcp-Ru/C_3N_4 during the reaction could be attributed to the consumption of the sacrificial agent over prolonged reaction times.

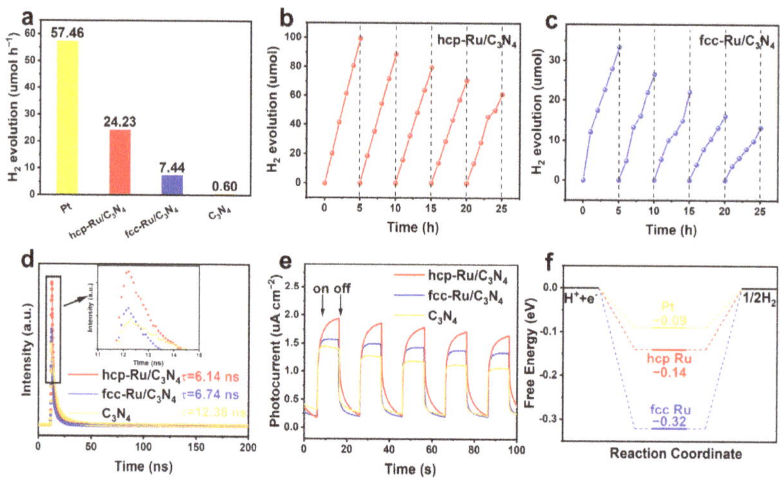

Figure 5. (**a**) Photocatalytic hydrogen evolution rates for C_3N_4-based hybrids loaded with phase-engineered Ru. (**b**) Cycle experiment of hcp-Ru/C_3N_4. (**c**) Cycle experiment of fcc-Ru/C_3N_4. (**d**) transient-state spectra of the catalysts. (**e**) Photocurrent vs. time (I-t) curves of the catalysts. (**f**) Calculated Gibbs free energy diagrams for hydrogen evolution at hcp Ru and fcc Ru surfaces.

To further explore the reasons for the significant difference in activity between the two Ru crystal phase catalysts, we continued to investigate their photoelectrochemical properties. We combined Linear sweep voltammetry (LSV) curves, photoelectrochemical measurements (photocurrent and electrochemical impedance spectroscopy (EIS)), steady-state fluorescence (PL), and transient-state fluorescence (FL) to further evaluate the charge dynamics of the two catalysts. As observed from the LSV curves (Figure S6b), the photocurrent density of hcp-Ru/C_3N_4 is higher than that of fcc-Ru/C_3N_4, and both are greater than that of C_3N_4. This indicates that Ru can significantly reduce the surface energy barrier of C_3N_4, enhances the consumption rate of photogenerated holes, and increases the density of free photogenerated electrons required for hydrogen evolution at lower potentials. The electrochemical hydrogen evolution activity is consistent with the photocatalytic performance, further demonstrating that hcp-Ru exhibits superior water reduction kinetics. As shown in Figure 5e, compared to C_3N_4, the introduction of Ru effectively enhanced the photocurrent response under light illumination, indicating that Ru loading promotes charge transfer in C_3N_4. hcp-Ru exhibited stronger photocurrent intensity than fcc-Ru, with the photocurrent response intensity order being: hcp-Ru/C_3N_4 > fcc-Ru/C_3N_4 > C_3N_4. The EIS results are consistent with the photocurrent results (Figure S6a), indicating that electrons in hcp-Ru/C_3N_4 exhibit more efficient transfer. Additionally, both materials showed a significant decrease in charge lifetime (Figure 5d) and a marked reduction in FL intensity (Figure S7a), further evidencing the enhanced charge separation efficiency [42,43]. Notably, although theoretical calculations of the Fermi level (Figure 4a–c) and charge difference distribution (Figure S6c) indicate that photogenerated electrons can effectively migrate from the conduction band of C_3N_4 to Ru nanoparticles, the electron density at the fcc-Ru/C_3N_4 interface is higher than that of hcp-Ru, which contradicts the actual photoelectric response and EIS results. We speculate that fcc-Ru can capture more electrons, but the water reduction kinetics on the surface of fcc-Ru is relatively poor, resulting in fewer electrons participating in surface hydrogen evolution, leading to a higher internal electron–hole recombination rate; thus, a reduced number of effective electrons transferred to the electrode. The comparison of Gibbs free energy results further supports our hypothesis.

The high photogenerated charge separation efficiency could be one of the reasons for the superior photocatalytic hydrogen evolution activity of hcp-Ru/C_3N_4. However, the surface molecular conversion barrier should also be considered an important factor [44]. It is evident that hcp-Ru possesses more active surface catalytic sites, corresponding to lower reaction energy barriers. Specifically, the rate-determining step for hydrogen evolution on the Ru surface is the formation of H* species. From the Gibbs free energy of this step, hcp-Ru ($\Delta G_{H^*} = -0.14$ eV) is significantly lower than fcc-Ru ($\Delta G_{H^*} = -0.32$ eV), and even approaches the efficient catalytic behavior of Pt (Figure 5f), making it easier to undergo reduction reactions at the CB edge to generate H_2. This indicates that, compared to fcc-Ru/C_3N_4, hcp-Ru/C_3N_4 not only exhibits higher charge separation efficiency but also demonstrates better surface molecular conversion efficiency. Based on these favorable physicochemical properties induced by phase engineering, hcp-Ru/C_3N_4 shows higher activity and better stability in photocatalytic hydrogen evolution reactions.

4. Conclusions

This study highlights the significant impact of crystal phase engineering on the photocatalytic performance of Ru/C_3N_4 systems for hydrogen evolution. By manipulating the choice of Ru precursors and reducing solvents, we achieved precise control over the synthesis of hcp- and fcc-phase Ru catalysts. Experimental results demonstrated that the hcp-Ru/C_3N_4 composite displayed markedly superior hydrogen evolution activity and stability compared to its fcc counterpart, benefiting from enhanced charge separation and transfer capabilities. Thermodynamic analysis further revealed that the surface catalytic properties of hcp-Ru, characterized by an optimized hydrogen adsorption energy, play a pivotal role in improving its photocatalytic efficiency. This work provides valuable insights into the rational design of photocatalysts via crystal phase engineering, emphasizing the synergistic effects of electronic structure and surface reactivity for effective hydrogen production.

Supplementary Materials: The following supporting information can be downloaded at: https://www.mdpi.com/article/10.3390/ma18061259/s1, Figure S1: images of C_3N_4, hcp-Ru/C_3N_4 and fcc-Ru/C_3N_4; Figure S2: (a,b) Size distribution histograms of Ru nanocrystals on the C_3N_4 nanosheets; Figure S3: (a–c) HRTEM images of hcp-Ru/C_3N_4. (d–f) HRTEM images of fcc-Ru/C_3N_4; Figure S4: (a) Tauc plots and (b) band strctures of C_3N_4; Figure S5: (a) TEM image and (d) HRTEM image of hcp-Ru/C_3N_4 after reaction. (b) XRD patterns of hcp-Ru/C_3N_4 after reaction. (c) Size distribution histograms of Ru nanocrystals on the hcp-C_3N_4 nanosheets after reaction. (d) STEM and elemental mapping images of hcp-Ru/C_3N_4; Figure S6: (a) EIS Nyquist plots and (b) LSV curves of the catralysts. (c) Theoretical simulated charge difference distribution of hcp-Ru/C_3N_4 and fcc-Ru/C_3N_4; Figure S7: (a) Steady-state PL spectra and (b–d) N_2 sorption isotherms of hcp-Ru/C_3N_4, fcc-Ru/C_3N_4 and C_3N_4; Table S1: The ICP-MS results of hcp-Ru/C_3N_4 and fcc-Ru/C_3N_4; Table S2: *EQE* Parameters and Corresponding Results; Table S3: Comparison of the photocatalytic activity and quantum efficiency over g-C_3N_4-based photocatalysts loaded with other materials. Refs. [32,45–52] can be found in Supplementary Materials.

Author Contributions: Methodology, X.C. and J.Y.; Software, G.Z.; Formal analysis, X.S. and Y.Z.; Investigation, X.S., X.C., T.L. and J.X.; Resources, Y.Z.; Writing—original draft, X.S.; Writing—review & editing, J.X. and Z.W.; Supervision, Z.W.; Project administration, J.Y.; Funding acquisition, J.Y. All authors have read and agreed to the published version of the manuscript.

Funding: The authors acknowledge the financial support from Natural Science Foundation of Jiangsu Province (BK20210827), China Postdoctoral Science Foundation (2021M700117), and Postgraduate Research and Practice Innovation Program of Jiangsu Province (SJCX22_1737).

Institutional Review Board Statement: Not applicable.

Informed Consent Statement: Not applicable.

Data Availability Statement: The original contributions presented in this study are included in the article/Supplementary Materials. Further inquiries can be directed to the corresponding author.

Conflicts of Interest: Authors Yubo Zhou and Jianjian Yi were employed by the company Ningbo Solartron Technology Co., Ltd. The remaining authors declare that the research was conducted in the absence of any commercial or financial relationships that could be construed as a potential conflict of interest.

References

1. Ben-Shahar, Y.; Scotognella, F.; Kriegel, I.; Moretti, L.; Cerullo, G.; Rabani, E.; Banin, U. Optimal metal domain size for photocatalysis with hybrid semiconductor-metal nanorods. *Nat. Commun.* **2016**, *7*, 10413. [CrossRef] [PubMed]
2. Ham, R.; Nielsen, C.J.; Pullen, S.; Reek, J.N.H. Supramolecular Coordination Cages for Artificial Photosynthesis and Synthetic Photocatalysis. *Chem. Rev.* **2023**, *123*, 5225–5261. [CrossRef] [PubMed]
3. Zhou, M.; Li, C.; Fang, J. Noble-Metal Based Random Alloy and Intermetallic Nanocrystals: Syntheses and Applications. *Chem. Rev.* **2021**, *121*, 736–795. [CrossRef] [PubMed]
4. Yang, X.; Zhang, Y.; Wang, Y.; Xin, C.; Zhang, P.; Liu, D.; Mamba, B.B.; Kefeni, K.K.; Kuvarega, A.T.; Gui, J. Hollow β-Bi_2O_3@CeO_2 heterostructure microsphere with controllable crystal phase for efficient photocatalysis. *Chem. Eng. J.* **2020**, *387*, 124100. [CrossRef]
5. Wang, Y.; Yu, W.; Wang, C.; Chen, F.; Ma, T.; Huang, H. Defects in photoreduction reactions: Fundamentals, classification, and catalytic energy conversion. *eScience* **2024**, *4*, 100228. [CrossRef]
6. Zhao, M.; Xia, Y. Crystal-phase and surface-structure engineering of ruthenium nanocrystals. *Nat. Rev. Mater.* **2020**, *5*, 440–459. [CrossRef]
7. Yao, Y.; Wang, S.; Li, Z.; Wu, Y. Atomic level engineering of noble metal nanocrystals for energy conversion catalysis. *J. Energy Chem.* **2021**, *63*, 604–624. [CrossRef]
8. Zhai, W.; Li, Z.; Wang, Y.; Zhai, L.; Yao, Y.; Li, S.; Wang, L.; Yang, H.; Chi, B.; Liang, J.; et al. Phase Engineering of Nanomaterials: Transition Metal Dichalcogenides. *Chem. Rev.* **2024**, *124*, 4479–4539. [CrossRef]
9. Yi, J.; Zhu, X.; Zhou, M.; Zhang, S.; Li, L.; Song, Y.; Chen, H.; Chen, Z.; Li, H.; Xu, H. Crystal phase dependent solar driven hydrogen evolution catalysis over cobalt diselenide. *Chem. Eng. J.* **2020**, *396*, 125244. [CrossRef]
10. Chen, Y.; Lai, Z.; Zhang, X.; Fan, Z.; He, Q.; Tan, C.; Zhang, H. Phase engineering of nanomaterials. *Nat. Rev. Chem.* **2020**, *4*, 243–256. [CrossRef]

11. Li, H.; Zhou, X.; Zhai, W.; Lu, S.; Liang, J.; He, Z.; Long, H.; Xiong, T.; Sun, H.; He, Q.; et al. Phase Engineering of Nanomaterials for Clean Energy and Catalytic Applications. *Adv. Energy Mater.* **2020**, *10*, 2002019. [CrossRef]
12. Yun, Q.; Ge, Y.; Huang, B.; Wa, Q.; Zhang, H. Ligand-Assisted Phase Engineering of Nanomaterials. *ACC Chem. Res.* **2023**, *56*, 1780–1790. [CrossRef] [PubMed]
13. Zhang, Z.; Liu, G.; Cui, X.; Chen, B.; Zhu, Y.; Gong, Y.; Saleem, F.; Xi, S.; Du, Y.; Borgna, A.; et al. Crystal Phase and Architecture Engineering of Lotus-Thalamus-Shaped Pt-Ni Anisotropic Superstructures for Highly Efficient Electrochemical Hydrogen Evolution. *Adv. Mater.* **2018**, *30*, e1801741. [CrossRef] [PubMed]
14. Li, P.; Li, W.; Huang, Y.; Huang, Q.; Li, J.; Zhao, S.; Tian, S. Unconventional Phase Synergies with Doping Engineering over Ni Electrocatalyst Featuring Regulated Electronic State for Accelerated Urea Oxidation. *ChemSusChem* **2023**, *16*, e202201921. [CrossRef]
15. Yi, J.; Zhang, G.; Cao, X.; Zhu, X.; Li, L.; Wang, X.; Zhu, X.; Song, Y.; Xu, H.; Wang, X. Structurally disordered $MoSe_2$ with rich 1T phase as a universal platform for enhanced photocatalytic hydrogen production. *J. Colloid Interface Sci.* **2024**, *668*, 492–501. [CrossRef]
16. Pelicano, C.M.; Saruyama, M.; Takahata, R.; Sato, R.; Kitahama, Y.; Matsuzaki, H.; Yamada, T.; Hisatomi, T.; Domen, K.; Teranishi, T. Bimetallic Synergy in Ultrafine Cocatalyst Alloy Nanoparticles for Efficient Photocatalytic Water Splitting. *Adv. Funct. Mater.* **2022**, *32*, 2202987. [CrossRef]
17. She, P.; Qin, J.S.; Sheng, J.; Qi, Y.; Rui, H.; Zhang, W.; Ge, X.; Lu, G.; Song, X.; Rao, H. Dual-Functional Photocatalysis for Cooperative Hydrogen Evolution and Benzylamine Oxidation Coupling over Sandwiched-Like Pd@TiO_2@$ZnIn_2S_4$ Nanobox. *Small* **2022**, *18*, 2105114. [CrossRef]
18. Li, R.; Wu, D.; Rao, P.; Deng, P.; Li, J.; Luo, J.; Huang, W.; Chen, Q.; Kang, Z.; Shen, Y.; et al. General approach for atomically dispersed precious metal catalysts toward hydrogen reaction. *Carbon Energy* **2023**, *5*, e294. [CrossRef]
19. Zhang, D.; Zhang, C.; Ye, B.; Ma, H.; Wang, C.; Zhuang, T.; Lv, Z. Droplet microreactor continuous synthesis of hierarchical Rh on CdZnS snowflakes for enhanced photocatalytic hydrogen evolution. *Chem. Eng. J.* **2024**, *499*, 156123. [CrossRef]
20. Su, K.; Wang, Y.; Zhang, C.; Gao, Z.; Han, J.; Wang, F. Tuning the Pt species on Nb_2O_5 by support-induced modification in the photocatalytic transfer hydrogenation of phenylacetylene. *Appl. Catal. B Environ. Energy* **2021**, *298*, 120554. [CrossRef]
21. Ding, Z.; Li, X.; Kang, C.; Yan, S.; Zhao, D.; Cai, H.; Zhang, S.-Y.; Zeng, Y.-J. Single Ru atoms confined into MOF/C_3N_4 for dual improved photocatalytic carbon dioxide reduction and nitrogen fixation. *Chem. Eng. J.* **2023**, *473*, 145256. [CrossRef]
22. Saito, D.; Yamazaki, Y.; Tamaki, Y.; Ishitani, O. Photocatalysis of a Dinuclear Ru(II)-Re(I) Complex for CO_2 Reduction on a Solid Surface. *J. Am. Chem. Soc.* **2020**, *142*, 19249–19258. [CrossRef] [PubMed]
23. Hamza, M.A.; Evans, J.D.; Andersson, G.G.; Metha, G.F.; Shearer, C.J. Ultrathin Ru-$CdIn_2S_4$ nanosheets for simultaneous photocatalytic green hydrogen production and selective oxidation of furfuryl alcohol to furfural. *Chem. Eng. J.* **2024**, *493*, 152603. [CrossRef]
24. Zhu, S.; Li, Z.; Hou, L.; Kim, M.G.; Jang, H.; Liu, S.; Liu, X. Revealing The Role of Electronic Asymmetricity on Supported Ru Nanoclusters for Alkaline Hydrogen Evolution Reaction. *Adv. Funct. Mater.* **2023**, *34*, 2314899. [CrossRef]
25. Dong, Y.; Zhang, W.; Hu, Z.; Ng, Y.H.; Wei, Z.; Liu, Y.; Deng, J.; Dai, H.; Jing, L. Advancing CO_2 to CH_4 conversion: The pivotal role of RuCu alloy in crystalline red phosphorus photocatalysis. *Appl. Catal. B Environ. Energy* **2024**, *357*, 124347. [CrossRef]
26. Nguyen, Q.N.; Kim, E.M.; Ding, Y.; Janssen, A.; Wang, C.; Li, K.K.; Kim, J.; Fichthorn, K.A.; Xia, Y. Elucidating the Role of Reduction Kinetics in the Phase-Controlled Growth on Preformed Nanocrystal Seeds: A Case Study of Ru. *J. Am. Chem. Soc.* **2024**, *146*, 12040–12052. [CrossRef]
27. Lin, J.-T.; Liu, Y.-H.; Tsao, C.-Y.; Wu, C.-Y.; Hsieh, C.-J.; Chen, M.-Z.; Chang, C.-W.; Hsiao, Y.-C.; Chen, H.-L.; Yang, T.-H. Toward a Quantitative Understanding of Crystal-Phase Engineering of Ru Nanocrystals. *Chem. Mater.* **2023**, *35*, 4276–4285. [CrossRef]
28. Fan, Z.; Guo, X.; Yang, M.; Jin, Z. Mechanochemical preparation and application of graphdiyne coupled with CdSe nanoparticles for efficient photocatalytic hydrogen production. *Chin. J. Catal.* **2022**, *43*, 2708–2719. [CrossRef]
29. Yu, B.; Li, H.; White, J.; Donne, S.; Yi, J.B.; Xi, S.B.; Fu, Y.; Henkelman, G.; Yu, H.; Chen, Z.L.; et al. Tuning the Catalytic Preference of Ruthenium Catalysts for Nitrogen Reduction by Atomic Dispersion. *Adv. Funct. Mater.* **2020**, *30*, 1905665. [CrossRef]
30. Liu, D.; Jiang, L.; Chen, D.; Hao, Z.; Deng, B.; Sun, Y.; Liu, X.; Jia, B.; Chen, L.; Liu, H. Twin S-Scheme g-C_3N_4/$CuFe_2O_4$/$ZnIn_2S_4$ Heterojunction with a Self-Supporting Three-Phase System for Photocatalytic CO_2 Reduction: Mechanism Insight and DFT Calculations. *ACS Catal.* **2024**, *14*, 5326–5343. [CrossRef]
31. Zhou, G.; Zhang, L.; Xia, Y.; Yin, W.; Zhu, X.; Hou, J.; Wang, S.; Ning, X.; Wang, X. Remarkably enhanced hydrogen evolution of g-C_3N_4 nanosheet under simulated sunlight via AgPt alloy co-catalyst with low amount of Pt. *J. Clean. Prod.* **2024**, *434*, 139950. [CrossRef]
32. Chen, Z.; Bu, Y.; Wang, L.; Wang, X.; Ao, J.-P. Single-sites Rh-phosphide modified carbon nitride photocatalyst for boosting hydrogen evolution under visible light. *Appl. Catal. B Environ. Energy* **2020**, *274*, 119117. [CrossRef]

33. Kusada, K.; Kobayashi, H.; Yamamoto, T.; Matsumura, S.; Sumi, N.; Sato, K.; Nagaoka, K.; Kubota, Y.; Kitagawa, H. Discovery of Face-Centered-Cubic Ruthenium Nanoparticles: Facile Size-Controlled Synthesis Using the Chemical Reduction Method. *J. Am. Chem. Soc.* **2013**, *135*, 5493–5496. [CrossRef]
34. Saruyama, M.; Pelicano, C.M.; Teranishi, T. Bridging electrocatalyst and cocatalyst studies for solar hydrogen production via water splitting. *Chem. Sci.* **2022**, *13*, 2824–2840. [CrossRef]
35. Pelicano, C.M.; Tong, H. Recent advances in cocatalyst engineering for solar-driven overall water splitting. *Appl. Res.* **2023**, *3*, e202300080. [CrossRef]
36. Zhang, S.; Yi, J.; Liu, M.; Shi, L.; Chen, M.; Wu, L. High-Density Atomically Dispersed Metals Activate Adjacent Nitrogen/Carbon Sites for Efficient Ammonia Electrosynthesis from Nitrate. *ACS Nano* **2024**, *18*, 26722–26732. [CrossRef]
37. Wang, H.; Xu, C.; Chen, Q.; Ming, M.; Wang, Y.; Sun, T.; Zhang, Y.; Gao, D.; Bi, J.; Fan, G. Nitrogen-Doped Carbon-Stabilized Ru Nanoclusters as Excellent Catalysts for Hydrogen Production. *ACS Sustain. Chem. Eng.* **2019**, *7*, 1178–1184. [CrossRef]
38. Wang, X.; Li, N.; Wang, G.C.; Liu, M.; Zhang, C.; Liu, S. Ultrafine Nanoclusters Unlocked 3d-4f Electronic Ladders for Efficient Electrocatalytic Water Oxidation. *ACS Nano* **2024**, *18*, 20518–20529. [CrossRef]
39. Wu, Y.; Liu, Y.; Emrick, T.; Russell, T.P. Polymer design to promote low work function surfaces in organic electronics. *Prog. Polym. Sci.* **2020**, *103*, 101222. [CrossRef]
40. Hong, S.J.; Lee, S.; Jang, J.S.; Lee, J.S. Heterojunction $BiVO_4/WO_3$ electrodes for enhanced photoactivity of water oxidation. *Energy Environ. Sci.* **2011**, *4*, 1781–1787. [CrossRef]
41. Guan, Q.; Ran, W.; Zhang, D.; Li, W.; Li, N.; Huang, B.; Yan, T. Non-Metal Sulfur Doping of Indium Hydroxide Nanocube for Selectively Photocatalytic Reduction of CO_2 to CH_4: A "One Stone Three Birds" Strategy. *Adv. Sci.* **2024**, *11*, 2401990. [CrossRef] [PubMed]
42. Yi, J.; Fei, T.; Li, L.; Yu, Q.; Zhang, S.; Song, Y.; Lian, J.; Zhu, X.; Deng, J.; Xu, H.; et al. Large-scale production of ultrathin carbon nitride-based photocatalysts for high-yield hydrogen evolution. *Appl. Catal. B Environ. Energy* **2021**, *281*, 119475. [CrossRef]
43. Zhou, Z.; Yang, Y.; Hu, L.; Zhou, G.; Xia, Y.; Hu, Q.; Yin, W.; Zhu, X.; Yi, J.; Wang, X. Phase Control of Cobalt Selenide: Unraveling the Relationship Between Phase Property and Hydrogen Evolution Catalysis. *Adv. Mater. Interfaces* **2022**, *9*, 2201473. [CrossRef]
44. Wang, Y.; Qian, W.; Zhou, G.; Zhang, S.; Zhu, X.; Li, L.; Zhu, X.; Wang, X.; Han, X.; Yi, J. MOF-derived phase-selective synthesis of In_2O_3 with appropriate surface atomic arrangement for CO_2 photoreduction. *Chem. Eng. J.* **2024**, *501*, 157513. [CrossRef]
45. Liu, Y.; Xu, X.; Li, A.; Si, Z.; Wu, X.; Ran, R.; Weng, D. A strategy to construct (reduced graphene oxide, γ-Fe_2O_3)/C_3N_4 step-scheme photocatalyst for visible-light water splitting. *Catal. Commun.* **2021**, *157*, 106327. [CrossRef]
46. Shen, J.; Luo, C.; Qiao, S.; Chen, Y.; Tang, Y.; Xu, J.; Fu, K.; Yuan, D.; Tang, H.; Zhang, H.; et al. Single-Atom Cu Channel and N-Vacancy Engineering Enables Efficient Charge Separation and Transfer between C_3N_4 Interlayers for Boosting Photocatalytic Hydrogen Production. *ACS Catal.* **2023**, *13*, 6280–6288. [CrossRef]
47. Wang, W.; Bai, X.; Ci, Q.; Du, L.; Ren, X.; Phillips, D.L. Near-Field Drives Long-Lived Shallow Trapping of Polymeric C_3N_4 for Efficient Photocatalytic Hydrogen Evolution. *Adv. Funct. Mater.* **2021**, *31*, 2103978. [CrossRef]
48. Xing, F.; Wang, C.; Liu, S.; Jin, S.; Jin, H.; Li, J. Interfacial Chemical Bond Engineering in a Direct Z-Scheme g-C_3N_4/MoS_2 Heterojunction. *ACS Appl. Mater. Interfaces* **2023**, *15*, 11731–117340. [CrossRef]
49. Wang, S.; Chen, L.; Zhao, X.; Zhang, J.; Ao, Z.; Liu, W.; Wu, H.; Shi, L.; Yin, Y.; Xu, X.; et al. Efficient photocatalytic overall water splitting on metal-free 1D SWCNT/2D ultrathin C_3N_4 heterojunctions via novel non-resonant plasmonic effect. *Appl. Catal. B Environ. Energy* **2020**, *278*, 119312. [CrossRef]
50. Shao, M.; Chen, W.; Ding, S.; Lo, K.H.; Zhong, X.; Yao, L.; Ip, W.F.; Xu, B.; Wang, X.; Pan, H. WX_y/g-C_3N_4 (WX_y = W_2C, WS_2, or W_2N) Composites for Highly Efficient Photocatalytic Water Splitting. *ChemSusChem* **2019**, *12*, 3355–3362. [CrossRef]
51. Tahir, B.; Tahir, M.; Alraeesi, A.; Kumar, N.; Al-Marzouqi, M. Synergistic effect of bimetallic RuCo loaded N-defective g-C_3N_4 nanosheets with cleavage of metal-hydrogen bonds for H_2 production in a continuous flow photoreactor. *Int. J. Hydrogen Energy* **2024**, *95*, 402–416. [CrossRef]
52. Zhou, X.; Yu, X.; Peng, L.; Luo, J.; Ning, X.; Fan, X.; Zhou, X.; Zhou, X. Pd(II) coordination molecule modified g-C_3N_4 for boosting photocatalytic hydrogen production. *J. Colloid Interface Sci.* **2024**, *671*, 134–144. [CrossRef] [PubMed]

materials

MDPI

Article

Preparation and Photocatalytic Degradation Performance of $C@Cd_xMn_{1-x}S$ to Tetracycline Hydrochloride

Yabin Su, Zedong Zeng, Haowen Chen, Zuosheng Lv, Cen Tan and Congjin Chen *

Guangxi Key Laboratory of Petrochemical Resources Processing and Process Intensification Technology, School of Chemistry and Chemical Engineering, Guangxi University, Nanning 530004, China; 18934709339@163.com (Y.S.); 15586229137@163.com (Z.Z.); 19976192847@163.com (H.C.); sense1ess030503@163.com (Z.L.); tancen23@163.com (C.T.)

* Correspondence: gxdxccj@163.com or chencongjin@gxu.edu.cn

Abstract: $Cd_xMn_{1-x}S$ solid solutions were synthesized by incorporating Mn^{2+} into CdS and the optimal ratio of Mn^{2+} to Cd^{2+} was explored via photocatalytic degradation performance for tetracycline (TC). Subsequently, the composite catalyst $C@Cd_xMn_{1-x}S$ was prepared by loading $Cd_xMn_{1-x}S$ onto the biomass gasification carbon residue (C) by hydrothermal method and characterized by various characterization tests. The optimal TC photodegradation condition and degradation mechanism catalyzed by $C@Cd_xMn_{1-x}S$ was investigated. The results showed $Cd_{0.6}Mn_{0.4}S$ had the optimal photocatalytic degradation efficiency, which is about 1.3 times that of CdS. The TC photodegradation efficiency by $C@Cd_{0.6}Mn_{0.4}S$ prepared at the mass ratio of C to $Cd_{0.6}Mn_{0.4}S$ of 1:2 was the best, which was 1.24 times that of $Cd_{0.6}Mn_{0.4}S$ and 1.61 times that of CdS. Under the optimal conditions (visible light irradiation for 60 min, $C@Cd_{0.6}Mn_{0.4}S$ of 20 mg, 40 mL TC solution of 40 mg/L), the TC degradation efficiency was 90.35%. The degradation efficiencies of 20 mg/L levofloxacin, ciprofloxacin, and 40 mg/L oxytetracycline catalyzed by $C@Cd_{0.6}Mn_{0.4}S$ range from 89.88% to 98.69%. In the photocatalytic reaction system, $\bullet O_2^-$ and h^+ are the dominant active species, which directly participate in the photocatalytic degradation reaction of TC, and •OH contributes little. The work provides a strategy to improve the photocatalytic performance of CdS for photocatalytic degradation antibiotics, and opens an interesting insight to deal with solid waste from biomass gasification.

Keywords: biomass gasification carbon residue; $C@Cd_xMn_{1-x}S$; composite catalyst; photocatalytic degradation; tetracycline hydrochloride

1. Introduction

Among the organic pollutants existing in the aquatic environment, antibiotics have received special attention. Antibiotics are secondary metabolites or synthetic analogues produced by bacteria, mold, or other microorganisms. They are widely used in the treatment of infectious diseases in humans and animals, and are also widely used in aquaculture and livestock breeding. Antibiotics are not fully absorbed in humans and animals, and most of them enter waste treatment plants or enter the environment directly in the form of raw or active metabolites. Although the half-life of most antibiotics is short, due to their frequent use and entry into the environment, they have posed potential ecological risks to the ecological environment and human health [1,2]. Antibiotics and their metabolites have been detected in surface water, groundwater, sewage, and drinking water, with concentrations ranging from mg/L to μg/L [1,2]. Tetracycline is one of the antibiotics that widely exists in the aquatic environment [3]. However, tetracycline has a stable structure. Removal methods of tetracycline from water usually involve in biodegradation, adsorption, Fenton or Fenton-like reaction, persulfate advanced oxidation process, photocatalytic oxidative degradation, etc. Among these methods, biodegradation requires specific microbes and takes a long time, adsorption may cause secondary pollution, and Fenton or Fenton-like reaction and persulfate advanced oxidation process all require additional chemicals. Therefore, the green, harmless, and simple-to-operate semiconductor photocatalytic technology is often used to degrade tetracycline [3–5].

Many semiconductor photocatalysts can perform efficient photocatalysis reaction under visible light, such as TiO_2, ZnO, Bi_2WO_6, Bi_2MoO_6, g-C_3N_4, CdS, etc. [6–9]. However, most semiconductors have shortcomings such as high price, wide band gap, and low efficiency [10]. Among existing photocatalysts, CdS is a metal semiconductor with a narrow band gap (2.4 eV). However, the photogenerated charge carrier recombination rate of CdS alone is high, there is the photocorrosion, and CdS is prone to agglomeration, resulting in low efficiency, which greatly restricts its practical application [7]. From the existing reports, it can be summarized that the regulation of physics factors such as morphology, crystal structure, crystallinity, and particle size have a significant impact on the performance of CdS [11]. CdS can be prepared into nanorods [12], nanosheets [13], nanotubes [14], hollow structures [15], etc. Constructing a heterojunction by combining CdS catalysts with other materials [8,9,16,17] or loading CdS catalysts on a carrier with a large specific surface area [18,19] are common methods to solve the problem of CdS photocorrosion. Among carriers with a large specific surface area, the biomass gasification carbon residue (C) contains some hydrophilic/lipophilic functional groups and aromatic carbon structures, as well as a high carbon content and developed porous structure. It has been reported to be used for the catalyst's carrier [20]. It is speculated that the biomass gasification carbon residues can be used as the loading material of CdS-based photocatalysts to increase the contact chance between photocatalysts and contaminants, and increases the exposure of the active sites of photocatalysts, realizing the effective separation of photo-induced carriers, hindering the rapid recombination of photo-induced carriers, and thereby improving the photocatalytic activity.

Solid solution strategy is also a feasible method to modify CdS catalysts [21]. A solid solution is performed by selecting another metallic element of atom size and properties similar to the Cd atom for ion exchange at a specific location in the crystal structure without changing the overall crystal structure or symmetry. The photocatalytic activity and stability of solid solution materials can be improved by adjusting the ratio of metal ions to change the band position and band gap width. Researchers often construct solid solutions with CdS by introducing transition metals such as Zn, Mn, and Cu [22–28]. Wang et al. [29] prepared a series of 3D dendritic $Mn_xCd_{1-x}S$ (LMCS-x) photocatalysts using a simple hydrothermal method without the addition of template agent and surfactant. With Pt as the co-catalyst, the photocatalytic cracking performance of aquatic hydrogen was tested. The hydrogen production efficiency of LMCS-2 is three times that of CdS.

Aimed to make up for the limitations of the single-component CdS and exert the synergy effect of CdS and MnS in $Cd_xMn_{1-x}S$ and thereby improve the photocatalytic performance and stability of the CdS photocatalyst, in the work, $Cd_xMn_{1-x}S$ solid solutions were synthesized by incorporating Mn^{2+} into CdS using a hydrothermal method. The optimal ratio of Mn^{2+} to Cd^{2+} was explored through various characterization tests and photocatalytic degradation performance evaluation. Subsequently, the composite catalyst C@$Cd_xMn_{1-x}S$ was prepared by loading $Cd_xMn_{1-x}S$ onto the biomass gasification carbon residue (C) by hydrothermal method, characterized by scanning electron microscope–Energy-dispersive X-ray spectroscopy (SEM-EDS), Fourier transform infrared spectroscopy (FTIR), X-ray diffraction (XRD), N_2 adsorption–desorption isotherm, X-ray photoelectron spectroscopy (XPS), and other characterization methods. Taking TC degradation efficiency as the main index, the catalytic activity of C@$Cd_xMn_{1-x}S$ under different pH values, catalyst dosages, different TC concentrations, and reaction systems was explored; the reusability, stability, and universality of the composite catalyst were determined; finally, the mechanism of photocatalysis was inferred. The work will provide a strategy to improve the photocatalytic performance of CdS for photocatalytic degradation antibiotics, and opens an interesting insight to deal with solid waste from biomass gasification.

2. Materials and Methods

2.1. Reagents and Instruments

Phosphoric acid (H_3PO_4), thioacetamide CH_3CSNH_2 (TAA), ethanol (EtOH), sodium chloride (NaCl), sodium nitrate ($NaNO_3$), sodium sulfate (Na_2SO_4), magnesium chloride ($MgCl_2$), and hydrochloric acid (HCl) were purchased from Guangdong Guanghua Technology Co., Ltd. (Guangzhou, China). Cadmium nitrate tetrahydrate ($Cd(NO_3)_2 \cdot 4H_2O$), manganese acetate tetrahydrate ($Mn(OAc)_2 \cdot 4H_2O$), humic acid (HA), ciprofloxacin (CIP), levofloxacin (LVFX), and oxytetracycline (OTC) were purchased from Shanghai Mclean Biochemical Technology Co., Ltd. (Shanghai, China) *p*-benzoquin (p-BQ) and tetracycline hydrochloride (TC) were purchased from Arald Inc. (Shanghai, China) Co., Ltd. Sodium hydroxide (NaOH), isopropyl alcohol (IPA), disodium EDTA (EDTA-2Na), sodium bicarbonate ($NaHCO_3$), disodium hydrogen phosphate dehydrates ($Na_2HPO_4 \cdot 2H_2O$), and sodium carbonate (Na_2CO_3) were purchased from Sinopharm Group Chemical Reagent Co., Ltd., (Beijing, China). All reagents are analytical purity level. Biomass gasification carbon residues were provided by the Demonstration Project of Multi-Product Comprehensive Utilization Technology of Biomass Gasification. The water used in the experiment is deionized water.

The following instruments were used in this study: MiniFlex600 X-ray diffractometer (XRD), Rigaku, Tokyo, Japan; Niolet iN10 Fourier Transform Infrared Spectroscopy (FT-IR) microscope, K-Alpha X-ray photoelectron spectrometer (XPS), Thermo Fisher Scientific, Waltham, MA, USA; Zeiss Gemini 300 Field Emission Scanning Electron Microscope (SEM), Carl Zeiss, Jena, Germany; ASAP2460 Fully Automatic BET Surface Area and Porosity Analyzer, Micromeritics, Norcross, GA, USA; FLS1000 Steady-state/Transient Fluorescence Spectrometer (PL), Edinburgh, UK; CHI 650E Electrochemical Workstation, Shanghai Chenhua Instrument Co., Ltd., Shanghai, China; DHG-9070 Electric Heating Constant Temperature Blower Drying Oven, Shanghai Jinghong Laboratory Equipment Co., Ltd., Shanghai, China; UV-2802 Ultraviolet-Visible Spectrophotometer, Shanghai UNIC Instruments Co., Ltd., Shanghai, China.

2.2. Preparation Method

2.2.1. Preparation of CdS and MnS

CdS (MnS) nanomaterials were synthesized by hydrothermal methods. 2 mmol of Cd $(NO_3)_2 \cdot 4H_2O$ (Mn $(OAc)_2 \cdot 4H_2O$) was dissolved in 20 mL of deionized water. After complete dissolution, 2.5 mmol of TAA was added and dissolved, followed by the addition of 10 mL of 2M NaOH. After continuous stirring for 30 min, the mixture was put into an autoclave and reacted at 180 °C for 12 h. After cooling, the products were filtered, washed alternately with deionized water and ethanol several times and dried for 6 h at 70 °C under vacuum; CdS (MnS) was obtained, then ground and stored for later use.

2.2.2. Preparation of $Cd_xMn_{1-x}S$

The $Cd_xMn_{1-x}S$ solid solution was synthesized by hydrothermal methods. An appropriate amount of Mn $(OAc)_2 \cdot 4H_2O$ and Cd $(NO_3)_2 \cdot 4H_2O$ were dissolved in 20 mL of deionized water. The molar ratios of Mn^{2+} to Cd^{2+} were controlled as 5:0, 4:1, 3:2, 2:3, 1:4, and 0:5, respectively (Cd^{2+} + Mn^{2+} = 2 mmol). After complete dissolution, 2.5 mmol TAA was added and dissolved, followed by the addition of 10 mL of 2 mol/L NaOH. After continuous stirring for 30 min, the mixture was put into an autoclave and reacted at 180 °C for 12 h. After cooling, the sample was filtered, washed alternately with deionized water and ethanol several times, and dried for 6 h at 70 °C under vacuum; $Cd_xMn_{1-x}S$ was obtained and labeled as $Cd_xMn_{1-x}S$ (x = 0, 0.2, 0.4, 0.6, 0.8, 1). Then each sample was ground and stored for later use.

2.2.3. Treatment of Biomass Gasification Carbon Residue

Firstly, the original biomass gasification residues were ground through a 100-mesh sieve; 10.000 g of the sieved biomass gasification residue was poured into a 200 mL NaOH (1 mol/L) aqueous solution and stirred for 12 h at 80 °C to remove silicates. After filtration and washing, the washed biomass gasification residue was poured into a 200 mL H_3PO_4 (1 mol/L) aqueous solution and stirred for 12 h at room temperature to remove inorganic metal ions. Then, the carbon powder after acid washing was washed alternately with deionized water and absolute ethanol until neutral, suction filtered, dried at 70 °C for 12 h; the sample was then ground to obtain the biochar (C), and preserved for the subsequent use.

2.2.4. Preparation of C@$Cd_{0.6}Mn_{0.4}S$ Composite Catalyst

The C@$Cd_{0.6}Mn_{0.4}S$ composite materials with different mass ratios of C to C@$Cd_{0.6}Mn_{0.4}S$ (labeled as C@$Cd_{0.6}Mn_{0.4}S$ (1:X), where X = 1, 2, 3, 4) were obtained by introducing biochar (C) as an auxiliary catalyst by chemical deposition. Quantities of 0.8 mmol Mn $(OAc)_2 \cdot 4H_2O$ and 1.2 mmol Cd $(NO_3)_2 \cdot 4H_2O$ were put into 20 mL deionized water and completely dissolved. Then, C (C: $Cd_{0.6}Mn_{0.4}S$ with mass ratio of 1:1, 1:2, 1:3, and 1:4) was added. After ultrasonic treatment for 30 min, 2.5 mmol TAA was added, dissolved, and then 10 mL of 2 mol/L NaOH was added. After continuous stirring for 30 min, the mixture was put into an autoclave and reacted at 180 °C for 12 h. After cooling, the products were filtered, washed alternately with deionized water and ethanol several times, and dried for 6 h at 70 °C under vacuum. The preparation process of C@$Cd_xMn_{1-x}S$ is shown in Scheme 1.

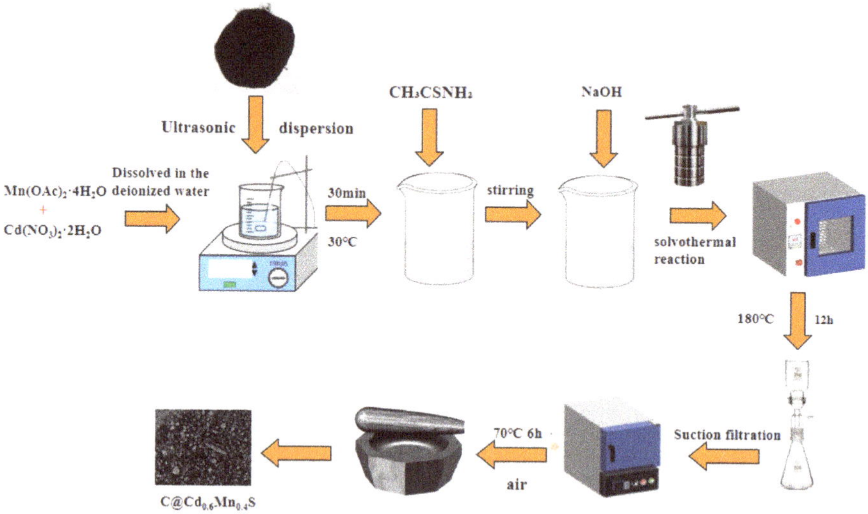

Scheme 1. The preparation process of C@$Cd_xMn_{1-x}S$.

2.3. Structural Characterization and Performance Testing

XRD test: the crystal form of the sample was detected by X-ray diffractometer at room temperature; Cu target; accelerating voltage: 40 kV; tube current: 40 mA; scanning range: 10° to 80°; scanning rate: 5 °/min. FTIR test: potassium bromide pellet method; scanning wavenumber range: 4000 to 400 cm^{-1}; scanning times: 32 times, resolution: 4 cm^{-1}. N_2 adsorption–desorption test: N_2 adsorption–desorption isotherms of the sample were obtained using an automatic specific surface and porosity analyzer; the degassing temperature and degassing time were 200 °C and 8 h, respectively. SEM test: accelerating voltage: 3.0 kV; gold-plating treatment. XPS test: Al Kα excitation source; energy: 1486.6 eV; the spectrometer calibration and charge correction were performed at the binding energy of C 1s of 284.8 eV. Photoluminescence (PL) test: utilized a steady-state/transient fluorescence spectrometer, with xenon lamp as the excitation light source; excitation wavelength: 380 nm; scanning range: 400–800 nm. Ultraviolet-visible diffuse reflectance (UV-Vis DRS, UV3600, Tokyo, Japan) test: with $BaSO_4$ as the reference material, the measurement range is 200–800 nm.

2.4. Photodegradation of TC Catalyzed by C@$Cd_{0.6}Mn_{0.4}S$

A certain amount of C@$Cd_{0.6}Mn_{0.4}S$ was poured into the photocatalytic test tube containing 40 mL of TC standard solution with a mass concentration of 40 mg/L. The mixture was stirred for 30 min under dark conditions to achieve adsorption–desorption equilibrium. Then, a photodegradation experiment was carried out in the photocatalytic reaction system equipped with a 300 W xenon lamp with a 420 nm cut-off filter. Then, 2 mL of the reaction solution was taken out and filtered through a 0.22 μm water system filter membrane at 15-min intervals; the absorbance (A) of filtrate of the reaction solution was determined at the maximum absorption wavelength (357 nm) of TC using an ultraviolet–visible spectrophotometer, and the mass concentration (ρ) of TC was calculated by the calibration curve Equation (1) for the relationship between the absorbance (A) and the mass concentration (ρ) of the TC standard solution. The degradation efficiency (η) of TC is calculated according to Equation (2).

$$A = 0.03159\,\rho + 0.0057\ (R^2 = 0.9988) \tag{1}$$

$$\eta = (1 - \rho_t/\rho_0) \times 100\% \tag{2}$$

where η is the degradation efficiency of TC, %; ρ_t is the mass concentration of TC at the degradation time t, mg/L; and ρ_0 is the initial mass concentration of TC, mg/L.

The effects of different proportions of the composite catalyst, the dosage of the catalyst, different reaction systems, TC concentration, pH value, different anions (HPO_4^{2-}, HCO_3^-, NO_3^-, Cl^-, CO_3^{2-}, SO_4^{2-}) on photocatalytic degradation efficiency of TC were investigated, respectively. Experiments on the stability and reusability of the catalyst were performed as follows: 20 mg of C@$Cd_{0.6}Mn_{0.4}S$ was put into 40 mL of TC aqueous solution with a mass concentration of 40 mg/L at 30 °C to degrade TC. The time gradients were set as 15, 30, 45, and 60 min. Samples were taken out and filtered after each time gradient, and the absorbance of the filtrate was determined using a 2802-PC UV-visible spectrophotometer, Unico (Shanghai) Instrument Co., Ltd., Shanghai, China. The concentration of TC in the degraded solution was calculated using the calibration curve equation, and the degradation efficiency was calculated according to Formula (2). After the degradation experiment, C@$Cd_{0.6}Mn_{0.4}S$ was collected using a magnet after catalytic reaction, washed repeatedly with deionized water and ethanol-water, and then filtered. It was placed in an oven and dried for 12 h at 70 °C for reuse. The catalyst was reused 5 times.

3. Results and Discussion

3.1. Characterization Results

3.1.1. SEM and EDS Analysis

Figure 1 shows the SEM images of biochar (a,b), CdS (c,d), MnS (e,f), C@$Cd_{0.6}Mn_{0.4}S$ (g,h), and C@$Cd_{0.6}Mn_{0.4}S$ (i,j), as well as the EDS images (k–p) of C@$Cd_{0.6}Mn_{0.4}S$ (mass ratio of C:$Cd_{0.6}Mn_{0.4}S$ = 1:2). Figure 1a,b indicate that biochar (C) has a fractured or intact columnar pore structure (the enlarged pores are clearly visible), which can load $Cd_{0.6}Mn_{0.4}S$. Figure 1c–f shows that CdS and MnS are aggregated from multiple spherical nanoparticles. As presented in Figure 1g,h, the as-prepared composite material C@$Cd_{0.6}Mn_{0.4}S$ presents irregular nanoparticles and uneven distribution under different magnification levels, and these nanoparticles aggregate with each other, with an average particle size of less than 200 nm. As shown in Figure 1i,j, C@$Cd_{0.6}Mn_{0.4}S$ nanoparticles grow closely on the surface of C, and the C surface, originally relatively smooth, becomes rough and the aggregation phenomenon of C@$Cd_{0.6}Mn_{0.4}S$ weakens, which means that C@$Cd_{0.6}Mn_{0.4}S$ can provide more active sites. As for the ratio of four elements of C, Cd, S, and Mn, Figure 1k shows that the atom percentage is 86.51%, 3.54%, 5.06%, and 2.66%, and weight percentage is 58.1%, 22.18%, 9.06%, and 8.21%, respectively. The atom ratio of Cd/Mn/S is about 0.70:0.53:1, not 0.6:0.4:1, which resulted for the following reasons: SEM–EDS testing is a powerful material surface analysis technique that can obtain information and chemical composition on the surface of a sample. However, the quantification of the chemical composition via SEM–EDS is a semi-quantitative method and is not accurate enough. In addition, O is originated from biochar. Figure 1l–p show that the elements of Cd, C, S, Mn, and O are uniformly distributed on the surface of the composite material, which also more fully demonstrates the growth and dispersion of $Cd_{0.6}Mn_{0.4}S$ nanoparticles on the surface of C.

3.1.2. XRD Analysis

Figure 2a,b are the XRD patterns and FITR spectra of C, CdS, MnS, $Cd_{0.6}Mn_{0.4}S$, and C@$Cd_{0.6}Mn_{0.4}S$, respectively. In Figure 2a, biochar (C) presents a typical broad peak of amorphous carbon at 2θ = 26.6° ((002) reflection) and 45.3° ((100) reflection), indicating the formation of amorphous carbon [20,30]; MnS shows obvious diffraction peaks at 2θ = 25.8°, 27.7°, 29.3°, 38.2°, 45.5°, 53.0°, 54.1°, 55.0°, and 72.3°, which belong to the (100), (002), (101), (102), (110), (200), (112), (201), and (203) crystal faces of hexagonal γ-MnS (JCPDS No. 89-4089) [31,32], and 2θ = 29.3°, 35°, 62.1° and 75° belong to the (111), (200), (222), (200), (400) crystal faces of cubic α-MnS (JCPDS No. 89-4952) [32,33], indicating the obtained MnS is a hybrid of hexagonal γ-MnS and cubic α-MnS; the as-prepared CdS shows characteristic diffraction peaks: the peaks at 2θ = 24.9°, 26.5°, 44.0°, 52.1°, 70.0°, and 72.6° correspond to the (100), (002), (101), (112), (210), and (114) crystal faces, which are consistent with the standard peaks of hexagonal CdS (JCPDS 77-2306) [32]; the peaks at 2θ = 26.9, 31.0, 44.3, and 52.3° correspond to the (111), (200), (220), and (311) crystal faces of cubic CdS (JCPDS 89-440) [32], indicating the successful preparation of CdS and that the as-synthesized CdS is hybrid of hexagonal CdS and cubic CdS. By comparing the JCPDS standard cards of MnS and CdS, the characteristic peaks of the as-synthesized $Cd_{0.6}Mn_{0.4}S$ solid solution are not a simple superposition of the peak shapes of CdS and MnS, indicating that the composite material is not a simple physical mixture of CdS and MnS, but that a special solid solution structure has been formed. This indicates that the $Cd_{0.6}Mn_{0.4}S$ solid solution has been successfully formed. The diffraction peaks of C@$Cd_{0.6}Mn_{0.4}S$ match those of $Cd_{0.6}Mn_{0.4}S$, and the XRD patterns of the two are almost the same. The as-prepared $Cd_{0.6}Mn_{0.4}S$ and C@$Cd_{0.6}Mn_{0.4}S$ are both hexagonal and cubic mixed crystal phase structures. In addition, XRD of C@$Cd_{0.6}Mn_{0.4}S$, since biochar is amorphous carbon, only has a broad peak response to the weak graphitized microcrystalline structure, and cannot be clearly observed in the composite material.

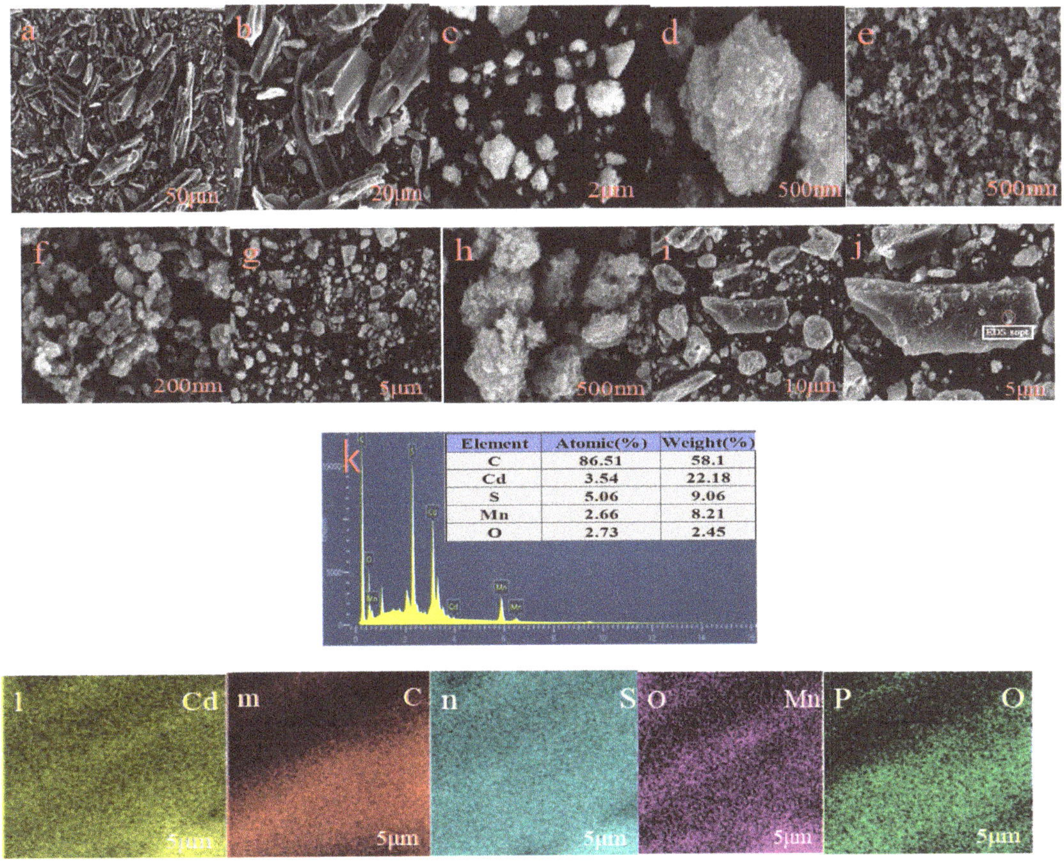

Figure 1. SEM images of C (**a**,**b**), CdS (**c**,**d**), MnS (**e**,**f**), Cd 0.6Mn 0.4S (**g**,**h**), C@$Cd_{0.6}Mn_{0.4}S$ (**i**,**j**), (**e**) EDS spectrum (**k**) and EDS elemental distribution map (**l**–**p**) of C@$Cd_{0.6}Mn_{0.4}S$.

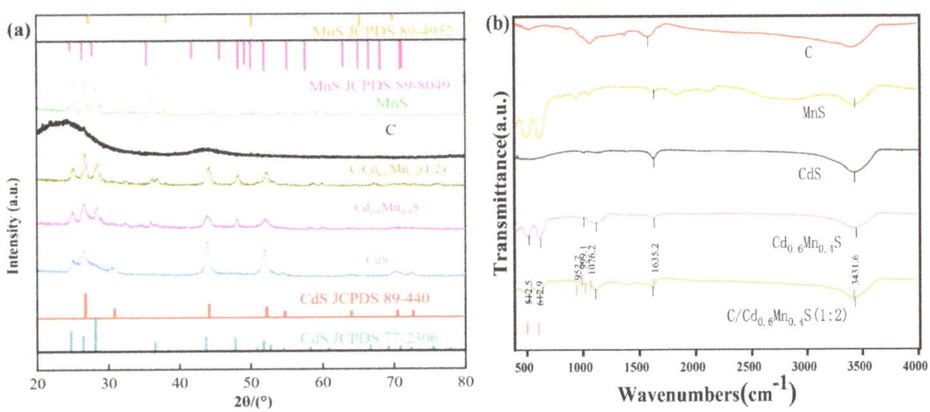

Figure 2. XRD patterns (**a**) and FTIR spectra (**b**) of C, CdS, MnS, $Cd_{0.6}Mn_{0.4}S$, and C@ C@$Cd_{0.6}Mn_{0.4}S$.

Figure 2b presents that the stretching vibration of –OH is at 3431.6 cm^{-1} in all samples; the stretching vibration of C=C/C=O bond is around 1635.2 cm^{-1} [34]; the vibration of C–O/C–C and C–H are overlapped at 1076.2 cm^{-1} [30]; the two peaks at 612.9 and 952.7 cm^{-1} are assigned the stretching vibration of Mn–S bond; and the stretching vibration of Cd–S bond is at 999.1 cm^{-1} [25,26,31,32]. However, the peaks of Mn–S bond and Cd–S bond in C@$Cd_{0.6}Mn_{0.4}S$ are lower than those in pure MnS, CdS, and the $Cd_{0.6}Mn_{0.4}S$ composite material. There are typical functional groups corresponding to C, as well as the characteristic peaks of CdS and MnS in the composite material, indicating that $Cd_{0.6}Mn_{0.4}S$ is successfully loaded on biochar.

3.1.3. XPS Analysis

The composition and elemental valence state of the C@$Cd_{0.6}Mn_{0.4}S$ solid solution was investigated by XPS. Figure 3a, the survey XPS spectra of the C@$Cd_{0.6}Mn_{0.4}S$ solid solution, shows that the C@$Cd_{0.6}Mn_{0.4}S$ is mainly composed of elements: Cd, Mn, S, C, and O; their atomic percentages are 10.01%, 8.61%,10.45%, 70.03%, and 0.9%, respectively. Among them, O is originated from biochar (the treated biomass gasification carbon residue). The results of element composition and content were almost consistent with those of SEM–EDS. Narrow-spectrum scanning of these elements was conducted to determine their valence states. As presented in Figure 3b, the S 2p spectrum can be divided into peaks at the binding energies of 163.25 and 162.03 eV, which matches well with S $2p_{1/2}$ and S $2p_{3/2}$ of S^{2-} in the solid solution [26]. As shown in Figure 3c, the two strong peaks at the binding energies of 655.33 and 642.01 eV correspond to Mn $2p_{1/2}$ and Mn $2p_{3/2}$, respectively, indicating that the Mn element exists in the catalyst in the form of Mn^{2+} [31,32]. As can be seen from Figure 3d, the two peaks at the binding energies of 405.60 and 412.35 eV correspond to Cd $3d_{5/2}$ and Cd $3d_{3/2}$, respectively, confirming that Cd exists in the catalyst in the form of Cd^{2+} [26]. In Figure 3e, the XPS spectrum of C 1s has a total of three peaks, corresponding to C=C/C–C (284.80 eV), C–O (286.37 eV), and C=O (288.79 eV) [35]. The peak of the C=C bond proves that C has a highly aromatic structure. The chemical valence states of Cd, Mn, and S in the solid solution existing in the form of Cd^{2+}, Mn^{2+}, and S^{2-} illustrated that Mn^{2+} has been successfully incorporated into the crystal structure of CdS, verifying the successful synthesis of the C@$Cd_{0.6}Mn_{0.4}S$ solid solution.

3.1.4. N_2 Adsorption–Desorption Isotherm Analysis

Figure 4a shows that the N_2 adsorption–desorption isotherms of the three samples all belong to type IV according to the International Union of Pure and Applied Chemistry (IUPAC) classification of physisorption isotherms [36]. In the low-pressure region, the adsorption capacity slowly increases with the increase of pressure. Subsequently, with the increase of pressure at higher relative pressures, the adsorption amounts of the adsorption materials C (0.6–1), $Cd_{0.6}Mn_{0.4}S$ (0.9–1.0), and C@$Cd_{0.6}Mn_{0.4}S$ (0.8–1) increase suddenly. The desorption curves show obvious hysteresis loops, and the shape of the hysteresis loops belongs to type H3, indicating that the material has a slit pore structure caused by the aggregation of mesoporous particles and the stacking of flaky particles [36]. The specific surface area of the sample was calculated by Brunauer–Emmett–Teller (BET) methods, and the pore volume and pore size of the sample were calculated by the Barrett–Joyner–Halenda (BJH) methods. Figure 4b shows that the pore diameters of the three materials are unevenly distributed within the range of 1–49 nm, mainly concentrated in the range of 1–12 nm, which further proves that the material presents a mesoporous structure. Biochar exhibits a relatively high specific surface area (645.91 m^2/g), and the specific surface area of C@$Cd_{0.6}Mn_{0.4}S$ (201.63 m^2/g) is larger than that of sole $Cd_{0.6}Mn_{0.4}S$ (45.02 m^2/g), indicating that $Cd_{0.6}Mn_{0.4}S$ are loaded on the surface of biochar pores. The loaded

$Cd_{0.6}Mn_{0.4}S$ can significantly improve the dispersibility of $Cd_{0.6}Mn_{0.4}S$ nanoparticles to expose more reaction sites for the catalyst and enhance the photocatalytic activity of the material.

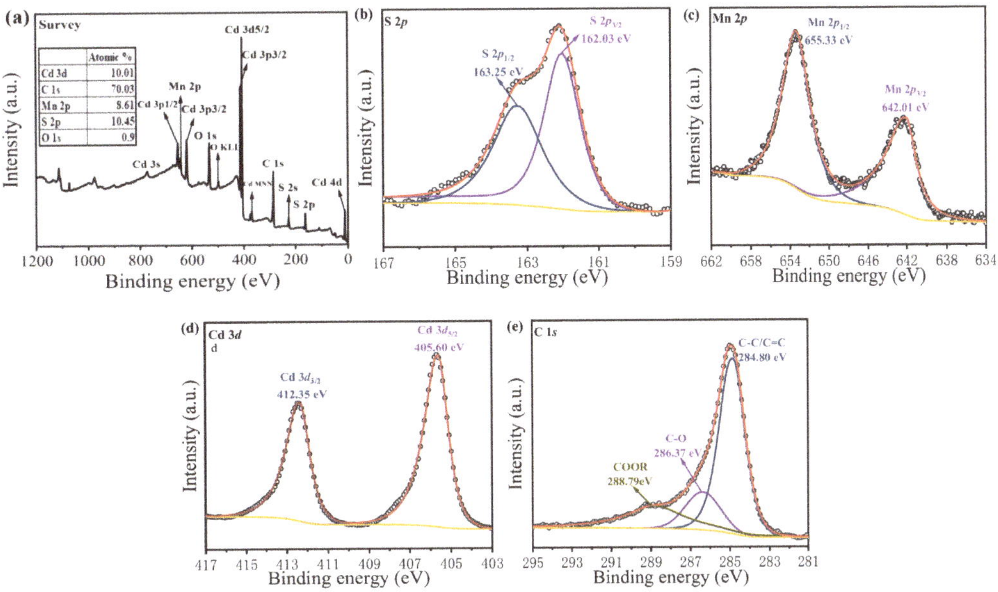

Figure 3. XPS spectra of $C@Cd_{0.6}Mn_{0.4}S$: (**a**) Survey spectrum; (**b**) S2p sub-spectrum; (**c**) Mn 2p sub-spectrum; (**d**) Cd 3p sub-spectrum; (**e**) C 1s sub-spectrum.

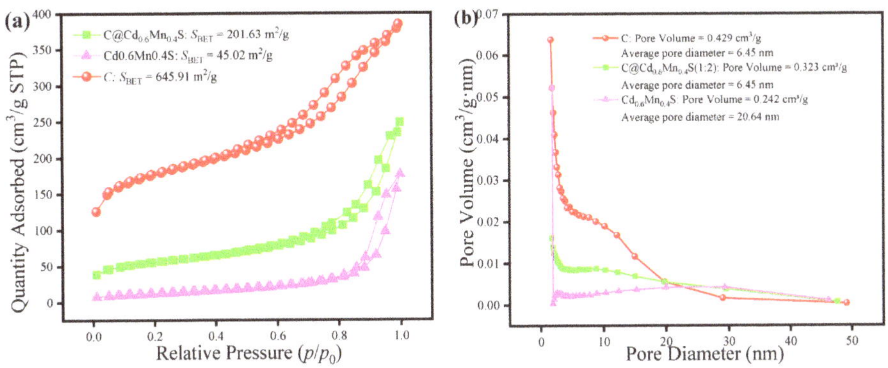

Figure 4. N_2 adsorption–desorption isotherms (**a**) and pore size distribution diagrams (**b**) of C, $C@Cd_{0.6}Mn_{0.4}S$ and $Cd_{0.6}Mn_{0.4}S$.

3.1.5. The Optical Absorption Properties and Energy Band Potential Analysis

The UV–Vis light absorption performance of CdS, MnS, $Cd_{0.6}Mn_{0.4}S$, and $C@Cd_{0.6}Mn_{0.4}S$ was investigated via UV–Vis diffuse reflectance absorption spectroscopy (UV-Vis DRS). As shown in Figure 5a, compared with CdS, the visible light absorption of $Cd_{0.6}Mn_{0.4}S$ has a significant improvement. It can be speculated that the introduction of Mn^{2+} and biochar leads to an enhancement for its visible light absorption. The energy bandgap (E_g) of the materials is estimated by the Kubelta–Munk equation $(\alpha h\upsilon)^2 = A(h\upsilon - E_g)$, where α, υ, h, and A are the absorption coefficient, light frequency, Planck's constant, and a constant, respectively [26,37], and is shown in Figure 5b. E_g of CdS, MnS, $Cd_{0.6}Mn_{0.4}S$, and

$C@Cd_{0.6}Mn_{0.4}S$ are 2.16, 1.64, 2.12, and 1.73 eV, respectively. Due to the introduction of Mn^{2+} and biochar, E_g of $C@Cd_{0.6}Mn_{0.4}S$ decreases and is less than that of $Cd_{0.6}Mn_{0.4}S$ and CdS. Based on the formula $\lambda = 1240/E_g$ [37] (E_g is the energy bandgap, eV; λ is the excitation wavelength, nm), λ increases as E_g decreases, which causes an absorption wavelength of $C@Cd_{0.6}Mn_{0.4}S$ redshift, and enables $C@Cd_{0.6}Mn_{0.4}S$ to have a higher visible light absorption ability. Based on the test data of valence band X-ray photoelectron spectroscopy (VB XPS), the linear part of the obtained graph near 0 eV is extrapolated to intersect with the horizontal extension line, the intersection point is the valence band potential (E_{VB}), and E_{VB} of $Cd_{0.6}Mn_{0.4}S$ as determined by an XPS valence band analysis is 1.32 eV (Figure 5c, measured from the Fermi level). According to the empirical formula $E_{CB} = E_{VB} - E_g$, the conduction band potential (E_{CB}) of $Cd_{0.6}Mn_{0.4}S$ is calculated as –0.80 eV.

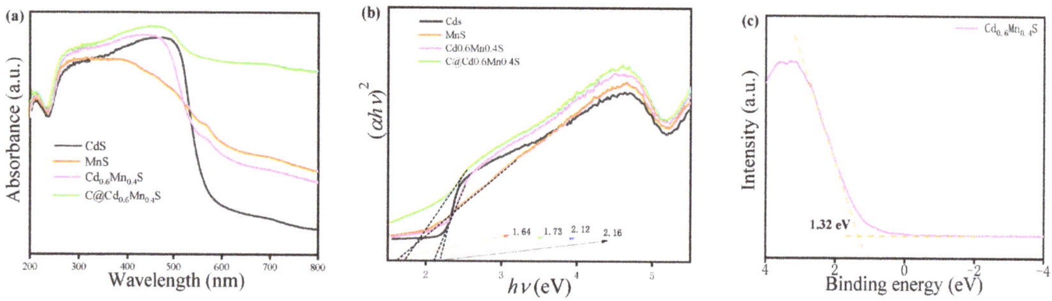

Figure 5. UV–Vis DRS absorbance spectra (**a**), $(\alpha h\upsilon)^2$ vs. $h\upsilon$ plots (**b**) of CdS, $Cd_{0.6}Mn_{0.4}S$, $C@Cd_{0.6}Mn_{0.4}S$ and MnS, and XPS valence band spectrum of $Cd_{0.6}Mn_{0.4}S$ (**c**).

3.2. Selection of $C@Cd_{0.6}Mn_{0.4}S$ and Its Photocatalytic Degradation Performance

3.2.1. Selection of $Cd_{0.6}Mn_{0.4}S$

The photocatalytic degradation of tetracycline hydrochloride (TC) was carried out, catalyzed by the solid solutions $Cd_xMn_{1-x}S$ (x = 0, 0.2, 0.4, 0.6, 0.8, 1). Firstly, each solid solution of $Cd_xMn_{1-x}S$ (x = 0, 0.2, 0.4, 0.6, 0.8, 1) with a mass of 20 mg was put into 40 mL of TC aqueous solution (40 mg/L). After holding for 30 min under non-illumination conditions to reach the adsorption–desorption equilibrium, the reaction lasted for 60 min under visible light conditions, and the absorbance was measured every 15 min. The effects of different solid solutions of $Cd_xMn_{1-x}S$ (x = 0, 0.2, 0.4, 0.6, 0.8, 1) on the degradation efficiency of TC were different. The results are shown in Figure 6a. When the amount of Mn^{2+} was doped with an appropriate ratio of Mn^{2+} to Cd^{2+}, the photocatalytic activity of the synthesized solid solution increased significantly. However, when the amount of the doped Mn^{2+} was too high, the photocatalytic activity decreased instead. When x = 1, $Cd_xMn_{1-x}S$ is CdS, and the catalytic activity of CdS decreased within the photoreaction process due to its susceptibility to photocorrosion. $Cd_{0.8}Mn_{0.2}S$ and $Cd_{0.6}Mn_{0.4}S$ were obtained after doping Mn^{2+} to CdS; their stability was significantly improved compared to CdS, and the catalytic effect was also significantly enhanced. However, because the absorbance of MnS to light is lower than that of CdS, with the increase in the doping amount of Mn^{2+}, the photosensitivity of $Cd_xMn_{1-x}S$ significantly decreases, resulting in a reduction in photocatalytic activity. Therefore, considering the activity comprehensively, stability, and the degradation efficacy of the $Cd_xMn_{1-x}S$ photocatalysis, the degradation efficacy of $Cd_{0.8}Mn_{0.2}S$ and $Cd_{0.6}Mn_{0.4}S$ does not differ much with the extension of degradation time, and $Cd_{0.8}Mn_{0.2}S$ has less Mn^{2+} doped, which cannot clearly explore the influence of Mn^{2+} doping on CdS. Therefore, $Cd_{0.6}Mn_{0.4}S$ was selected as the optimized material for photocatalytic degradation of TC. The photodegradation efficiency of TC catalyzed by $Cd_{0.6}Mn_{0.4}S$ was about 1.3 times that catalyzed by CdS.

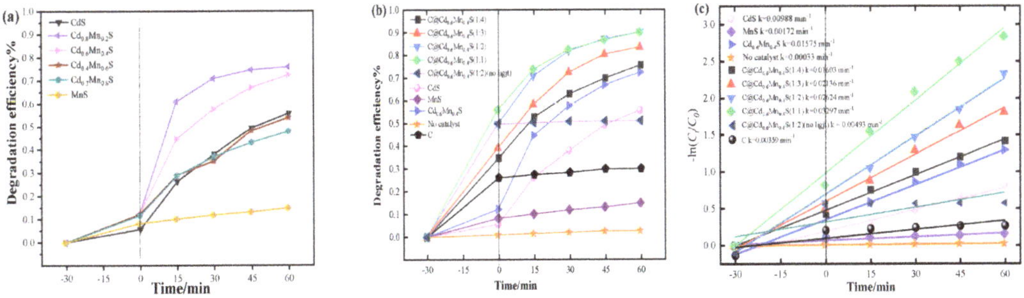

Figure 6. Effects of $Cd_xMn_{1-x}S$ with different proportions on the degradation efficiency of TC (**a**); the influence of catalysts at different mass ratio of biochar to $Cd_{0.6}Mn_{0.4}S$ on the degradation efficiency of TC (**b**); and the quasi-first-order kinetics fitting (**c**) (Reaction conditions: the dosage of each catalyst was 20 mg, the mass concentration of TC solution was 40 mg/L, the solution volume was 40 mL, and the temperature was 30 °C).

3.2.2. Selection of the Mass Ratio of Biochar to $Cd_{0.6}Mn_{0.4}S$ in C@$Cd_{0.6}Mn_{0.4}S$

After combining biochar with $Cd_{0.6}Mn_{0.4}S$ at different mass ratios (during the preparation, the mass ratio of biochar to $Cd_{0.6}Mn_{0.4}S$ = 1:X, X = 1, 2, 3, 4), $Cd_{0.6}Mn_{0.4}S$ was loaded on biochar. The photodegradation efficiency of TC by catalysts at different mass ratios of biochar to $Cd_{0.6}Mn_{0.4}S$ was compared, and the results are shown in Figure 6b. Figure 6b shows that the photodegradation efficiency of TC was only 2.83% without any photocatalyst, and were 55.82% and 14.94% with CdS and MnS alone, respectively, after 60 min of illumination, indicating that a single catalyst of CdS or MnS could realize the photodegradation of TC; yet the degradation efficiency of TC was very low. When MnS was doped into CdS, the photodegradation efficiency of TC reached 72.66% catalyzed by $Cd_{0.6}Mn_{0.4}S$, which was greater than that of TC catalyzed by either CdS or MnS, yet was still not satisfactory. However, after $Cd_{0.6}Mn_{0.4}S$ was loaded on biochar, since biochar could adsorb TC effectively on its surface and then facilitate the reaction catalyzed by $Cd_{0.6}Mn_{0.4}S$, the photodegradation efficiency of TC catalyzed by C@$Cd_{0.6}Mn_{0.4}S$ (1:1) reached 90.01%. When the mass ratio of biochar to $Cd_{0.6}Mn_{0.4}S$ was 1:2 (C@$Cd_{0.6}Mn_{0.4}S$ (1:2)), the degradation efficiency of TC was slightly better within 30 min, and then gradually approached that catalyzed by C@$Cd_{0.6}Mn_{0.4}S$ (1:1), and was even higher, reaching 90.05%, which is 1.24 times that catalyzed by $Cd_{0.6}Mn_{0.4}S$ and 1.61 times that catalyzed by CdS. The degradation efficiency data of TC catalyzed by the catalysis material was fitted using the first-order reaction kinetics model to obtain the reaction rate constant (*K*), and the results are shown in Figure 6c. Figure 6c shows that the TC degradation rate constant *K* value catalyzed by $Cd_{0.6}Mn_{0.4}S$ is 0.01575 min^{-1}, which is greater than that catalyzed by CdS or MnS, but it is less than the *K* value catalyzed by C@$Cd_{0.6}Mn_{0.4}S$. The order of each *K* value in terms of magnitude is as follows: *K* C@$Cd_{0.6}Mn_{0.4}S$ (1:1) > *K* C@$Cd_{0.6}Mn_{0.4}S$ (1:2) > *K* C@$Cd_{0.6}Mn_{0.4}S$ (1:3) > *K* C@$Cd_{0.6}Mn_{0.4}S$ (1:4) > *K* $Cd_{0.6}Mn_{0.4}S$ > *K* CdS > *K* MnS > *K* C > *K* blank. Loading $Cd_{0.6}Mn_{0.4}S$ on biochar, TC can be photocatalyzed to degrade more effectively, and as more biochar added, the *K* value increases. However, it was found in the experiment that the more biochar is present in the composite catalyst, the stronger the adsorption effect, and the recycling effect of the catalyst is relatively poor. Considering all factors, the C@$Cd_{0.6}Mn_{0.4}S$ (1:2) catalyst was selected for the subsequent experiments (the photodegradation rate constant *K* value of TC catalyzed by C@$Cd_{0.6}Mn_{0.4}S$ (1:2) was 0.02624 min^{-1}). To illustrate the role of light, a control experiment with C@$Cd_{0.6}Mn_{0.4}S$ (1:2) in the absence of light was conducted, and the result (Figure 6b) illustrates the condition of the absence of light: TC removal depends on adsorption and TC is hardly degraded.

It may be that TC degradation requires light to provide the activation energy needed for the reaction.

3.3. *Effects of Reaction Conditions on TC Photodegradation Efficiency Catalyzed by C@Cd$_{0.6}$Mn$_{0.4}$S*

When exploring the influence of reaction conditions on TC degradation efficiency catalyzed by C@Cd$_{0.6}$Mn$_{0.4}$S, C@Cd$_{0.6}$Mn$_{0.4}$S was prepared at the theoretical mass ratio of biochar to Cd$_{0.6}$Mn$_{0.4}$S, i.e., 1:2. The preparation process was described in detail in Section 2.2.4.

The influence of catalyst dosage on TC photodegradation is shown in Figure 7a. Figure 7a illustrates that when the catalyst dosage is 5, 10, 20, 30, 40, and 50 mg, and the photodegradation of 40 mL TC aqueous solution with a mass concentration of 40 mg/L lasted for 60 min under the conditions of a temperature of 30 °C and visible light illumination, the photodegradation efficiency of TC increased with the catalyst dosage increase. The corresponding degradation efficiencies are 53.95%, 76.94%, 90.17%, 96.36%, 97.79%, and 98.01%, respectively. The degradation efficiency reached the maximum when the catalyst dosage was 50 mg. However, after the catalyst dosage increased from 40 mg to 50 mg, the degradation efficiency did not increase significantly. Therefore, choosing a catalyst dosage of 40 mg is appropriate.

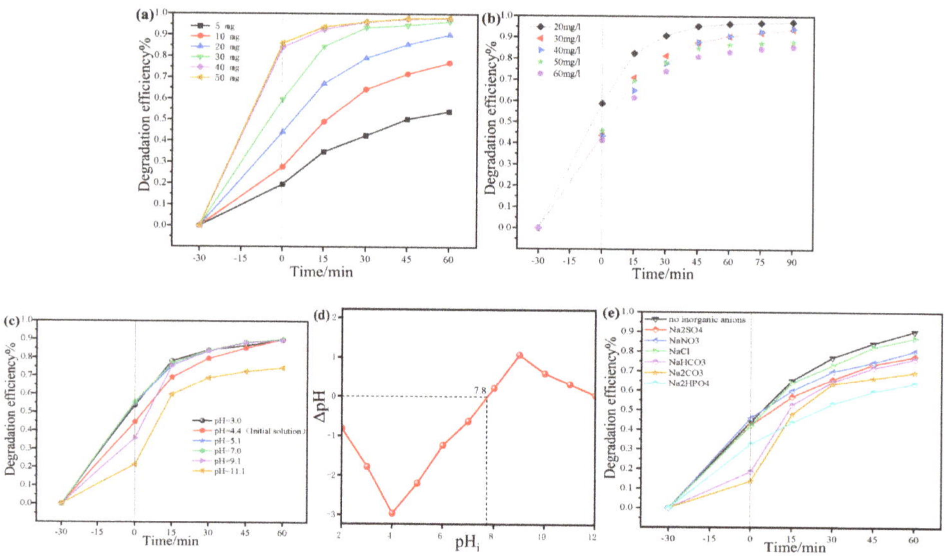

Figure 7. Effects of the dosage of C@Cd$_{0.6}$Mn$_{0.4}$S (**a**), the initial mass concentration of TC (**b**), the initial solution pH (**c**), the pHzpc of C@Cd$_{0.6}$Mn$_{0.4}$S (**d**), and inorganic anions and humic acids (**e**) on photocatalytic degradation of TC catalyzed by C@Cd$_{0.6}$Mn$_{0.4}$S (reaction conditions: C@Cd$_{0.6}$Mn$_{0.4}$S dosage of 20 mg, temperature of 30 °C, TC aqueous solution mass concentration of 40 mg/L and volume of 40 mL).

The influence of TC concentrations on the photocatalytic degradation efficiency of TC is presented in Figure 7b. With a dosage of C@Cd$_{0.6}$Mn$_{0.4}$S of 20 mg, a solution volume of 40 mL, a temperature of 30 °C, and a photocatalytic degradation time of 60 min, when the TC aqueous solution concentrations were 20, 30, 40, 50, and 60 mg/L, the degradation efficiency of TC gradually decreased with TC concentration increase, at 96.45%, 90.60%, 90.48%, 86.61%, and 83.14%, respectively. The results are due to the dosage of the catalyst remaining unchanged, and the amount of active substances produced by illumination remaining unchanged. When the volume of the TC solution to be degraded remained unchanged and the concentration increased, the number of TC molecules to be degraded increased, making the same amount of active substances insufficient to degrade all TC molecules, resulting in a decrease in the degradation efficiency. In addition, a higher mass concentration of TC may prevent the catalyst from absorbing light through shielding, further reducing the degradation efficiency. When the TC concentration was 30 and 40 mg/L, the degradation efficiencies did not differ much. Therefore, a TC concentration of 40 mg/L was chosen for the subsequent experiments.

The pH value of solutions dominates the existing forms of TC molecules and the surface charge of the catalysts, thus affecting TC degradation efficiency. Under different pH conditions, TC molecules exist in different forms (pK_1 = 3.30, pK_2 = 7.68, pK_3 = 9.68). Depending on the pH value of the solution, TC molecules can be dissociated into cations, zwitter-ions, and anions. When pH < 3.30, the main form of TC is TCH_3^+ (cation). When 3.30 < pH < 7.68, the main form of TC is TCH_2^0 (ampholyte, neutral). When 7.68 < pH < 9.68, the main form of TC is TCH^- (anion). When pH > 9.68, the main form of TC is TC^{2-} (anion). In general, TC mainly exists in the forms of cation, ampholyte, and anion at pH = 3.3, pH = 3.3–7.8, and pH = 7.8–9.68 [38,39]. The influence of the degradation efficiency of TC catalyzed by $C@Cd_{0.6}Mn_{0.4}S$ at different pH values of 3.0, 4.4 (original aqueous solution pH value), 5.1, 7.0, 9.1, and 11.1 was explored. When the initial pH of the TC aqueous solution with a volume of 40 mL and an initial concentration of 40 mg/L was adjusted to 3.0, 4.4 (original solution pH value), 5.1, 7.0, 9.1, and 11.1, respectively, by using 1 mol/L NaOH or 1 mol/L HCl solution, and $C@Cd_{0.6}Mn_{0.4}S$ of 20 mg was added to catalyze photodegradation reaction of TC at 30 °C for 60 min, the degradation efficiencies of TC were 89.60%, 89.99%, 90.02%, 90.07%, 89.56%, and 74.89%, respectively (Figure 7c). When pH is between 3 and 9, the degradation efficiencies of TC do not differ much. The zero potential of $C@Cd_{0.6}Mn_{0.4}S$ measured by the charge titration method is approximately 7.8 (Figure 7d). When the solution pH is less than 7.8, the surface of $C@Cd_{0.6}Mn_{0.4}S$ is positively charged and substances with negative charges are more easily adsorbed on the surface of the catalyst; while when the solution pH is greater than 7.8, the surface of $C@Cd_{0.6}Mn_{0.4}S$ is negatively charged, making it easier to adsorb substances with positive charges. When the solution pH is 3 to 9, TC mainly exists in the form of ampholyte, reducing the electrostatic repulsion between TC and $C@Cd_{0.6}Mn_{0.4}S$. However, when pH is 11.1, the degradation efficiency is the lowest, probably because TC mainly exists in the form of anion (TC^{2-}) at a pH of 11.1, generating an electrostatic repulsion with the catalyst, thereby hindering TC from being adsorbed on the surface of the photocatalyst and ultimately leading to a decrease in the degradation efficiency. Based on the abovementioned dynamics, it can be concluded that when pH is between 3 and 9, the influence of pH on the photodegradation of TC by $C@Cd_{0.6}Mn_{0.4}S$ is not significant, and there is no need to adjust the pH of TC aqueous solution.

$C@Cd_{0.6}Mn_{0.4}$ may have complex interactions with inorganic ions and organics in water environments, and whether ions and organics commonly found in natural water would influence the photocatalytic degradation of TC by $C@Cd_{0.6}Mn_{0.4}$ should be investigated. Common inorganic ions and organics, namely HPO_4^{2-}, HCO_3^-, NO_3^-, Cl^-, CO_3^{2-}, SO_4^{2-}, and humic acid (HA), were chosen to study their effects on TC photocatalytic degradation by $C@Cd_{0.6}Mn_{0.4}$. The results are shown in Figure 7e. When 10 mmol/L of HPO_4^{2-}, CO_3^{2-}, HCO_3^-, SO_4^{2-}, NO_3^-, Cl^-, and 10 mg/L of HA, respectively, was added to the TC solution, the corresponding degradation efficiencies were 63.61%, 69.30%, 76.12%, 77.41%, 80.25%, 86.88%, and 83.33%, respectively. The presence of NO_3^-, Cl^-, and HA has little influence on the photocatalytic degradation efficiency of TC by $C@Cd_{0.6}Mn_{0.4}$, which may be because NO_3^- and Cl^- do not shield the photocatalytic active sites. HA is a macromolecular polymer with aromatic nuclei as the main body and various functional groups attached; the main functional groups are phenolic hydroxyl groups, hydroxyl groups, methoxy groups, and cyclic compounds containing nitrogen. HA is an amphoteric colloid, both negatively and positively charged on its surface, usually mainly negatively charged, and the source of electricity is mainly the oxygen-containing group on the molecular surface, such as the dissociation of carboxyl, hydroxyl, methyl, acyl, and other groups and the protonation of the amine group. These HA' characters may cause it to interact with TC and make HA have little influence on the photocatalytic degradation efficiency of TC. HCO_3^- and SO_4^{2-} have slight influence on the degradation TC, which is likely due to the capture of active substances by HCO_3^- and SO_4^{2-} during photocatalytic degradation [28]. The influence of CO_3^{2-} and HPO_4^{2-} on the degradation efficiency of TC is relatively high; possibly CO_3^{2-} and HPO_4^{2-} show alkalinity in the solution, causing the photocatalytic degradation efficiency of TC to be reduced, as shown in the results of Figure 7e.

3.4. *Universality and Reusability of Photocatalytic Properties of $C@Cd_{0.6}Mn0_{.4}S$*

The photocatalytic degradation performance of $C@Cd_{0.6}Mn_{0.4}S$ on the other antibiotics under the same conditions is shown in Figure 8a. As illustrated in Figure 8a, $C@Cd_{0.6}Mn_{0.4}S$ has a good photocatalytic degradation efficacy not only on TC, but also on the other antibiotics such as oxytetracycline (OTC), levofloxacin (LEV), and ciprofloxacin (CIP). Their photocatalytic degradation efficiencies within 60 min are 90.21%, 95.09%, 96.19%, and 95.36%, respectively. To verify the reusability of $C@Cd_{0.6}Mn_{0.4}S$ for photocatalytic degradation of TC, five cycling experiments were carried out on the photocatalytic degradation. The experimental conditions were 30 min of dark adsorption and 60 min of photocatalytic degradation. The results are shown in Figure 8b. After five cycling experiments, the photocatalytic degradation efficiency of TC by $C@Cd_{0.6}Mn_{0.4}S$ gradually decreased from 90.05% to 79.23%. Comparing TC degradation efficiency by $C@Cd_{0.6}Mn_{0.4}S$ with those by the reported photocatalysts, and the comparison results of TC photocatalytic degradation efficiency and the reaction conditions by $C@Cd_{0.6}Mn_{0.4}S$ with those in the reported literatures by various photocatalysts, are shown in Table 1, and it is concluded that the photocatalytic degradation performance of $C@Cd_{0.6}Mn_{0.4}S$ is comparable to those of the reported catalysts in the previous study in terms of the degradation efficiency of TC. In addition, the main reason for the decrease in photocatalytic efficiency observed in the reusability test is that the catalyst $C@Cd_{0.6}Mn_{0.4}S$ experiences a slight mass loss in the process of recycling, and TC or its degradation intermediates would occupy little of the active sites of the catalyst. The XRD pattern of $C@Cd_{0.6}Mn_{0.4}S$ before and after photocatalytic degradation of TC is shown in Figure 8c. Compared with the catalyst before the reaction, the main peaks of the catalyst after five reactions exhibit almost no change, except that XRD peaks of the used $C@Cd_{0.6}Mn_{0.4}S$ sample narrowed, indicating that the crystal phase of $C@Cd_{0.6}Mn_{0.4}S$ remained intact after the reaction, confirming that $C@Cd_{0.6}Mn_{0.4}S$ has good stability. XRD peaks of the used $C@Cd_{0.6}Mn_{0.4}S$ sample narrowed, which is supposedly evidence of recrystallization, which took place after the sample dried for 12 h at 70 °C.

Table 1. Comparison of TC photocatalytic degradation efficiency by $C@Cd_{0.6}Mn_{0.4}S$ with those by the reported photocatalysts.

Catalyst	TC	Reaction Conditions	Degradation Efficiency	Degradation Efficiency After Cycles	Ref.
$Bi_{12}O_{17}Cl_2$	0.02 g/L,40 mL	catalyst = 0.025 g/L	60 min, 92.2%	60 min, 80.5%	[40]
Zn-HT-CN	0.02 g/L,10 mL	catalyst = 0.3g/L	40 min, 94.41%	40 min, 78.78%	[41]

Table 1. *Cont.*

Catalyst	TC	Reaction Conditions	Degradation Efficiency	Degradation Efficiency After Cycles	Ref.
In_2O_3Nanoparticl/$Bi_4O_5Br_2$	0.035 g/L,100 mL	catalyst = 0.02 g/L	100 min, 92.4%	100 min, 82.4%	[42]
$Zn_{0.2}Ti_{0.8}$-NML	0.02g/L,100 mL	catalyst = 0.02 g/L	40 min, 95.6%	40 min, 85%	[43]
MCN@CdS	0.02 g/L,100 mL	catalyst = 0.02 g/L in presence of O_2	80 min, 99.5%	80 min, 90%	[44]
Ce−Bi_2WO_6	0.001 g/L,50 mL	catalyst = 0.6 g/L, pH = 6.7	40 min, 90.4%	40 min, 89%	[45]
C@$Cd_{0.6}Mn_{0.4}S$	0.04 g/L40 mL	catalyst = 0.02 g/L, pH = 5.1	60 min, 90.05%	60 min, 79.23%	This work

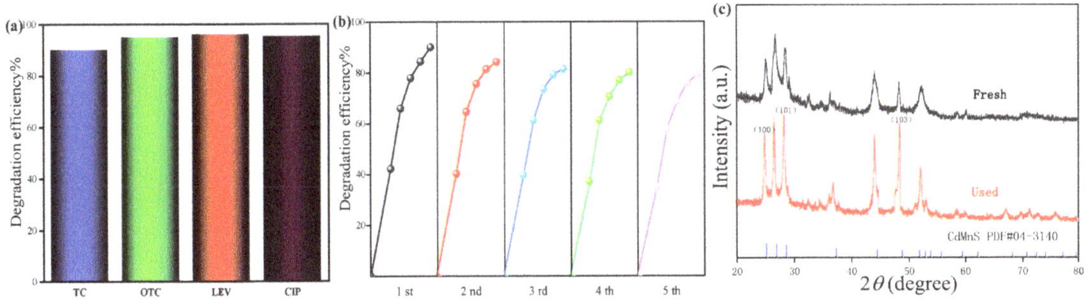

Figure 8. The photodegradation efficiencies of antibiotics catalyzed by C@$Cd_{0.6}Mn_{0.4}S$ (**a**); cyclic stability during degrading TC (**b**), and XRD pattern before and after five cycle reactions for degrading TC (**c**). (Reaction conditions: C@Cd $_{0.6}Mn_{0.4}S$ dosage of 20 mg, temperature of 30 °C, TC solution and OTC solution mass concentration of 40 mg/L, LEV solution and CIP solution of mass concentration of 20 mg/L, and the volume of each solution of 40 mL).

3.5. Photocatalytic Degradation Mechanism

In the C@$Cd_{0.6}Mn_{0.4}S$ photocatalytic degradation of the TC system, 2 mmol/L disodium ethylenediaminetetraacetate (EDTA-2Na), *p*-quinone (*p*-BQ), and isopropyl alcohol (IPA) were added as the scavengers for photogenerated holes (h^+), superoxide radicals (•O_2^-), and hydroxyl radicals (•OH), respectively. The influence of the scavengers on TC degradation efficiencies was analyzed. The results are shown in Figure 9a. EDTA-2Na (h^+ scavenger) and *p*-BQ (•O_2^- scavenger) have inhibitory effects on the degradation of TC, indicating that in the reaction system, both •O_2^- and h^+ are active substances of the reaction and directly participate in the degradation of TC molecules. However, the addition of IPA (•OH scavenger) has little influence on TC degradation efficiency, indicating that in this reaction system, •OH contributes little to the degradation of TC.

The photoluminescence (fluorescence) intensity can reflect the recombination degree of charge carriers produced by the material exposed to light. The higher the intensity of photoluminescence, the higher the recombination rate of electrons and holes, and the lower the photocatalytic performance. Figure 9b illustrated that the luminescence spectrum shape of C@$Cd_{0.6}Mn_{0.4}S$ is almost the same as that of the $Cd_{0.6}Mn_{0.4}S$, and the luminescence intensity of C@$Cd_{0.6}Mn_{0.4}S$ and $Cd_{0.6}Mn_{0.4}S$ is lower than that of the CdS monomer. The order of luminescence intensity from strong to weak is: CdS > $Cd_{0.6}Mn_{0.4}S$ > C@$Cd_{0.6}Mn_{0.4}S$, and there is no fluorescence on the biochar. The results indicate that Mn^{2+} doped and biochar support strategies can effectively inhibit the recombination rate of photoexcited electrons and holes, which is consistent with the changing trend of the photocatalytic activity of the obtained composite materials. These results may be due to the loose and porous structure of biochar from biomass gasification carbon residue, which makes it easier for Cd^{2+} and Mn^{2+} to disperse and anchor on the surface of biochar. Furthermore, biochar has excellent electrical conductivity; the two factors significantly increase the binding rate of Cd^{2+} and Mn^{2+}, the dispersity of $Cd_{0.6}Mn_{0.4}S$, and the charge transfer of the material, thereby enhancing the photocatalytic performance of $Cd_{0.6}Mn_{0.4}S$. Based on the above analysis, the possible mechanism of photocatalytic degradation of TC by C@$Cd_{0.6}Mn_{0.4}S$ is speculated as shown in Scheme 2.

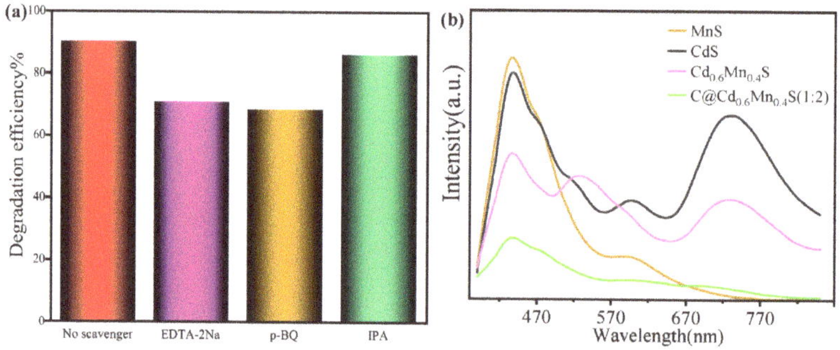

Figure 9. Effects of different quenchers on degradation efficiencies of TC (**a**) (Reaction conditions: C@$Cd_{0.6}Mn_{0.4}S$ dosage of 20 mg, TC solution mass concentration of 40 mg/L, solution volume of 40 mL and temperature of 30 °C), and fluorescence intensity of the prepared material (**b**).

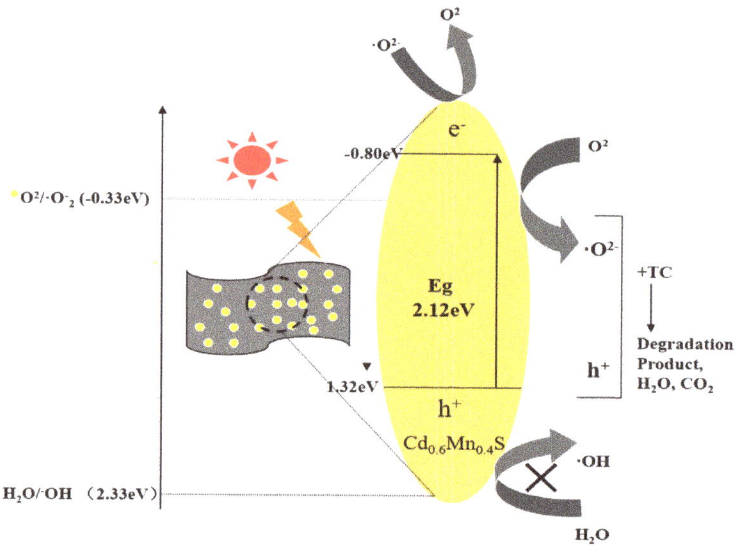

Scheme 2. Schematic diagram of the photocatalytic degradation mechanism of TC by C@$Cd_{0.6}Mn_{0.4}S$ under visible light irradiation.

Under the irradiation of visible light whose energy is greater than or equal to the bandgap energy, C@$Cd_{0.6}Mn_{0.4}S$ can generate photo-generated electrons (e^-) and holes (h^+), and electrons (e^-) in the valence band (VB) will be excited and transited to the conduction band (CB). At the same time, holes (h^+) are left on the VB as active species. Since $Cd_{0.}6Mn_{0.4}S$ is loaded on biochar, more active sites are provided, which increases the contact chance between $Cd_{0.6}Mn_{0.4}S$ and TC, and increases the exposure of the active sites of $Cd_{0.6}Mn_{0.4}S$, realizing the effective separation of electrons and holes, hindering the rapid recombination of photogenerated electrons and holes, and thereby improving the photocatalytic activity. Since the CB potential of $Cd_{0.6}Mn_{0.4}S$ is more negative than that of E ($O_2/\bullet O_2^-$ = −0.33 eV) [37,46,47], the electrons (e^-) on CB of $Cd_{0.6}Mn_{0.4}S$ will quickly migrate to the surface of biochar and react with O_2 adsorbed on the active site to form $\bullet O_2^-$, and then react with TC molecules to decompose into small molecules. With the continuous transfer of electrons, holes (h^+) are left on VB of $Cd_{0.6}Mn_{0.4}S$. These remaining h^+, due to their strong oxidation ability, will directly participate in the oxidation decomposing process of TC and play a major role in photocatalytic degradation. Eventually, TC is oxidized, degraded into small molecules, substances, and even into CO_2 and H_2O, etc. The mechanism is shown in Formulas (3)–(5).

$$C@Cd_{0.6}Mn_{0.4}S + h\upsilon \rightarrow C@Cd_{0.6}Mn_{0.4}S(e^-(CB) + h^+(VB)) \tag{3}$$

$$e^- + O_2 \rightarrow \bullet O_2^- \tag{4}$$

$$TC + (\bullet O_2^- + h^+) \rightarrow \text{small organic molecules} \rightarrow CO_2 + H_2O. \tag{5}$$

4. Conclusions

1. The $C@Cd_xMn_{1-x}S$ composite catalyst was prepared by hydrothermal method using biomass gasification residues, Cd $(OAc)_2{\cdot}2H_2O$, Mn $(OAc)_2{\cdot}4H_2O$, and TAA as raw materials.
2. The $Cd_xMn_{1-x}S$ prepared with a mole ratio of Cd to Mn at 6:4 has the good degradation efficacy for TC in response to light reaction. After 60 min of visible light irradiation, the degradation efficiency of TC reaches 72.66%, but it is still not ideal. The $C@Cd_{0.6}Mn_{0.4}S$ prepared after $Cd_{0.6}Mn_{0.4}S$ loaded on biochar has the highest degradation efficiency of 90.05% after light irradiation of 60 min when the mass ratio m(biochar):m($Cd_{0.6}Mn_{0.4}S$) is 1:2. After being recycled 5 times, the degradation efficiency of TC reaches 79.23%, which is still higher than the catalytic effect of the catalyst $Cd_{0.6}Mn_{0.4}S$. The composite catalyst has good cycling stability.
3. $C@Cd_{0.6}Mn_{0.4}S$ is applicable to a wide pH environment (3–9), and NO_3^-, Cl^-, and HA had little effect; HCO_3^- and SO_4^{2-} had a slight effect; and CO_3^{2-} and HPO_4^{2-} had a relative high effect on the degradation of TC.
4. Both $\bullet O_2^-$ and h^+ directly participate in the photodegradation of TC molecules, and •OH contributes little to the photo degradation of TC catalyzed by $C@Cd_{0.6}Mn_{0.4}S$.
5. $C@Cd_{0.6}Mn_{0.4}S$ has universality for different antibiotics. The degradation efficiencies of 20 mg/L levofloxacin, ciprofloxacin, and 40 mg/L oxytetracycline catalyzed by $C@Cd_{0.6}Mn_{0.4}S$ ranged from 89.88% to 98.69%.
6. Biomass gasification residues, as the carrier of the catalyst, can significantly enhance the dispersibility of $Cd_{0.6}Mn_{0.4}S$ nanoparticles, expose more reaction sites of the catalyst, and improve the photocatalytic activity.

Author Contributions: Y.S.: Writing—original draft, Validation, Conceptualization, Data curation, Methodology, Investigation, Formal analysis. Z.Z.: Writing—original draft, Validation, Conceptualization, Data curation, Methodology, Investigation, Formal analysis. H.C.: Writing—original draft, Validation, Conceptualization, Data curation, Methodology, Investigation, Formal analysis. Z.L.: Writing—original draft, Validation, Conceptualization, Data curation, Methodology, Investigation, Formal analysis. C.T.: Conceptualization, Methodology, Validation. C.C.: Writing—review & editing, Funding acquisition, Supervision, Resources, Project administration, Methodology, Conceptualization. All authors have read and agreed to the published version of the manuscript.

Funding: This work was supported by National Natural Science Foundation of China (Grant No. 31660183), Guangxi Natural Science Foundation (2025GXNSF069247), the Dean/Opening Project of Guangxi Key Laboratory of Petrochemical Resource Processing and Process Intensification Technology (No. 2024K002), and Guangxi University 2023 undergraduate Innovation Project Doubling Plan Guangxi Zhuang Autonomous Region level "College students Innovation and Entrepreneurship Training Plan" (No. 43).

Institutional Review Board Statement: Not applicable.

Informed Consent Statement: Not applicable.

Data Availability Statement: The original contributions presented in the study are included in the article, further inquiries can be directed to the corresponding author.

Conflicts of Interest: The authors declare no conflict of interest.

References

1. Ben, Y.; Fu, C.; Hu, M.; Liu, L.; Wong, M.H.; Zheng, C. Human Health Risk Assessment of Antibiotic Resistance Associated with Antibiotic Residues in the Environment: A Review. *Environ. Res.* **2019**, *169*, 483–493. [CrossRef] [PubMed]
2. Singh, R.; Singh, A.P.; Kumar, S.; Giri, B.S.; Kim, K. Antibiotic Resistance in Major Rivers in the World: A Systematic Review on Occurrence, Emergence, and Management Strategies. *J. Clean. Prod.* **2019**, *234*, 1484–1505. [CrossRef]
3. Amangelsin, Y.; Semenova, Y.; Dadar, M.; Aljofan, M.; Bjørklund, G. The Impact of Tetracycline Pollution on the Aquatic Environment and Removal Strategies. *Antibiotics* **2023**, *12*, 440. [CrossRef] [PubMed]
4. He, X.; Kai, T.; Ding, P. Heterojunction Photocatalysts for Degradation of the Tetracycline Antibiotic: A Review. *Environ. Chem. Lett.* **2021**, *19*, 4563–4601. [CrossRef] [PubMed]
5. Zhou, Q.Q.; Zou, Y.Q.; Lu, L.Q.; Xiao, W.J. Mit Sichtbarem Licht Induzierte, Organische Photochemische Reaktionen Über Energietransferrouten. *Angew. Chem.* **2019**, *131*, 1600–1619. [CrossRef]
6. Ahtasham Iqbal, M.; Akram, S.; Khalid, S.; Lal, B.; Hassan, S.U.; Ashraf, R.; Kezembayeva, G.; Mushtaq, M.; Chinibayeva, N.; Hosseini-Bandegharaei, A. Advanced Photocatalysis as a Viable and Sustainable Wastewater Treatment Process: A Comprehensive Review. *Environ. Res.* **2024**, *253*, 118947. [CrossRef]
7. Cheng, L.; Xiang, Q.; Liao, Y.; Zhang, H. Cds-Based Photocatalysts. *Energy Environ. Sci.* **2018**, *11*, 1362–1391. [CrossRef]
8. Meng, A.; Zhu, B.; Zhong, B.; Zhang, L.; Cheng, B. Direct Z-Scheme TiO_2/CdS Hierarchical Photocatalyst for Enhanced Photocatalytic H_2-Production Activity. *Appl. Surf. Sci.* **2017**, *422*, 518–527. [CrossRef]
9. Wang, S.; Zhu, B.; Liu, M.; Zhang, L.; Yu, J.; Zhou, M. Direct Z-Scheme ZnO/CdS Hierarchical Photocatalyst for Enhanced Photocatalytic H_2-Production Activity. *Appl. Catal. B Environ.* **2019**, *243*, 19–26. [CrossRef]
10. Kim, D.; Yong, K. Boron Doping Induced Charge Transfer Switching of a C_3N_4/ZnO Photocatalyst from Z-Scheme to Type II to Enhance Photocatalytic Hydrogen Production. *Appl. Catal. B Environ.* **2021**, *282*, 119538. [CrossRef]
11. Bai, J.; Chen, W.; Shen, R.; Jiang, Z.; Zhang, P.; Liu, W.; Li, X. Regulating Interfacial Morphology and Charge-Carrier Utilization of Ti_3C_2 Modified All-Sulfide CdS/$ZnIn_2S_4$ S-Scheme Heterojunctions for Effective Photocatalytic H_2 Evolution. *J. Mater. Sci. Technol.* **2022**, *112*, 85–95. [CrossRef]
12. Wang, X.; Liow, C.; Qi, D.; Zhu, B.; Leow, W.R.; Wang, H.; Xue, C.; Chen, X.; Li, S. Programmable Photo-Electrochemical Hydrogen Evolution Based on Multi-Segmented CdS-Au Nanorod Arrays. *Adv. Mater.* **2014**, *26*, 3506–3512. [CrossRef]
13. Mahdi, M.A.; Hassan, J.J.; Ng, S.S.; Hassan, Z. Growth of CdS Nanosheets and Nanowires through the Solvothermal Method. *J. Cryst. Growth* **2012**, *359*, 43–48. [CrossRef]
14. Mahapatra, A.K. Synthesis of Quantum-Confined CdS Nanotubes. *J. Nanopart. Res.* **2009**, *11*, 467–475. [CrossRef]
15. Lin, G.; Zheng, J.; Xu, R. Template-Free Synthesis of Uniform CdS Hollow Nanospheres and their Photocatalytic Activities. *J. Phys. Chem. C* **2008**, *112*, 7363–7370. [CrossRef]
16. Ghoreishian, S.M.; Ranjith, K.S.; Park, B.; Hwang, S.; Hosseini, R.; Behjatmanesh-Ardakani, R.; Pourmortazavi, S.M.; Lee, H.U.; Son, B.; Mirsadeghi, S.; et al. Full-Spectrum-Responsive Bi_2S_3@CdS S-scheme Heterostructure with Intimated Ultrathin rGO Toward Photocatalytic Cr(VI) Reduction and H_2O_2 Production: Experimental and DFT Studies. *Chem. Eng. J.* **2021**, *419*, 129530. [CrossRef]
17. Wu, X.; Zhao, J.; Wang, L.; Han, M.; Zhang, M.; Wang, H.; Huang, H.; Liu, Y.; Kang, Z. Carbon Dots as Solid-State Electron Mediator for $BiVO_4$/CDs/CdS Z-Scheme Photocatalyst Working Under Visible Light. *Appl. Catal. B Environ.* **2017**, *206*, 501–509. [CrossRef]
18. Jin, Z.; Gong, H.; Li, H. Visible-Light-Driven Two-Dimensional Metal-Organic Framework Modified Manganese Cadmium Sulfide for Efficient Photocatalytic Hydrogen Evolution. *J. Colloid Interface Sci.* **2021**, *603*, 344–355. [CrossRef]

19. Xu, H.; Yang, S.; Ma, X.; Huang, J.; Jiang, H. Unveiling Charge-Separation Dynamics in CdS/Metal–Organic Framework Composites for Enhanced Photocatalysis. *ACS Catal.* **2018**, *8*, 11615–11621. [CrossRef]
20. Fan, H.; Chen, C.; Lu, J.; Tan, C.; Wang, P.; Hu, J.; Lu, X.; Huang, X. A Novel Strategy for Efficient Utilization of Carbon Residue from Biomass Gasification Loaded $NiCo_2O_4$ to Construct Composite Catalysts for Activation of Peroxymonosulfate to Effectively Remove Antibiotics. *Sep. Purif. Technol.* **2024**, *335*, 126196. [CrossRef]
21. Yuan, Y.; Chen, D.; Yu, Z.; Zou, Z. Cadmium Sulfide-Based Nanomaterials for Photocatalytic Hydrogen Production. *J. Mater. Chem. A Mater. Energy Sustain.* **2018**, *6*, 1166. [CrossRef]
22. Chen, Y.; Zhao, S.; Wang, X.; Peng, Q.; Lin, R.; Wang, Y.; Shen, R.; Cao, X.; Zhang, L.; Zhou, G.; et al. Synergetic Integration of $Cu_{1.94}$ S–$Zn_XCd_{1-X}S$ Heteronanorods for Enhanced Visible-Light-Driven Photocatalytic Hydrogen Production. *J. Am. Chem. Soc.* **2016**, *138*, 4286–4289. [CrossRef]
23. Hamdan, S.; Al-Ali, K.; Vega, L.F.; Muscetta, M.; Yusuf, A.O.; Palmisano, G. Zinc Cadmium Sulphide ($Zn_XCd_{1-X}S$)-Based Photocatalysts for Hydrogen Production from Seawater and Wastewater. *J. Environ. Chem. Eng.* **2024**, *12*, 113937. [CrossRef]
24. Huang, D.; Wen, M.; Zhou, C.; Li, Z.; Cheng, M.; Chen, S.; Xue, W.; Lei, L.; Yang, Y.; Xiong, W.; et al. $Zn_XCd_{1-X}S$ Based Materials for Photocatalytic Hydrogen Evolution, Pollutants Degradation and Carbon Dioxide Reduction. *Appl. Catal. B Environ.* **2020**, *267*, 118651. [CrossRef]
25. Li, L.; Liu, G.; Qi, S.; Liu, X.; Gu, L.; Lou, Y.; Chen, J.; Zhao, Y. Highly Efficient Colloidal $Mn_XCd_{1-X}S$ Nanorod Solid Solution for Photocatalytic Hydrogen Generation. *J. Mater. Chem. A* **2018**, *6*, 23683–23689. [CrossRef]
26. Xiong, M.; Qin, Y.; Chai, B.; Yan, J.; Fan, G.; Xu, F.; Wang, C.; Song, G. Unveiling the Role of Mn-Cd-S Solid Solution and MnS in $Mn_XCd_{1-X}S$ Photocatalysts and Decorating with CoP Nanoplates for Enhanced Photocatalytic H_2 Evolution. *Chem. Eng. J.* **2022**, *428*, 131069. [CrossRef]
27. Zhang, W.; Zhong, Z.; Wang, Y.; Xu, R. Doped Solid Solution: $(Zn_{0.95}Cu_{0.05})_{1-X}Cd_XS$ Nanocrystals with High Activity for H_2 Evolution From Aqueous Solutions Under Visible Light. *J. Phys. Chem. C* **2008**, *112*, 17635–17642. [CrossRef]
28. Li, Q.; Meng, H.; Zhou, P.; Zheng, Y.; Wang, J.; Yu, J.; Gong, J. Zn1–Xcdxs Solid Solutions with Controlled Bandgap and Enhanced Visible-Light Photocatalytic H_2 -Production Activity. *ACS Catal.* **2013**, *3*, 882–889. [CrossRef]
29. Wang, X.; Feng, B.; Shi, L. A Feasible Preparation of Leaf Like 3D Dendrite $Mn_XCd_{1-X}S$ with Photocatalytic Hydrogen Evolution. *Int. J. Hydrogen Energy* **2024**, *51*, 988–997. [CrossRef]
30. Streletskiy, O.A.; Zavidovskiy, I.A.; Nishchak, O.Y.; Khaidarov, A.A.; Savchenko, N.F.; Pavlikov, A.V. Low-Threshold Field Emission Cathode Based on Heat-Treated Dehydrofluorinated Polyvinylidene Fluoride. *J. Exp. Theor. Phys.* **2022**, *135*, 844–852. [CrossRef]
31. He, X.; Zeng, D.; Liu, Y.; Chen, Q.; Yang, J.; Gao, R.; Fujita, T.; Wei, Y. Porous CoxP Nanosheets Decorated $Mn_{0.35}Cd_{0.65}S$ Nanoparticles for Highly Enhanced Noble-Metal-Free Photocatalytic H_2 Generation. *J. Colloid Interface Sci.* **2022**, *625*, 859–870. [CrossRef]
32. Li, H.; Wang, Z.; He, Y.; Meng, S.; Xu, Y.; Chen, S.; Fu, X. Rational Synthesis of $Mn_XCd_{1-X}S$ for Enhanced Photocatalytic H_2 Evolution: Effects of S Precursors and the Feed Ratio of Mn/Cd On its Structure and Performance. *J. Colloid Interface Sci.* **2019**, *535*, 469–480. [CrossRef] [PubMed]
33. Gui, Y.; Qian, L.; Qian, X. Hydrothermal Synthesis of Uniform Rock Salt (A-) MnS Transformation from Wurtzite (Γ-) MnS. *Mater. Chem. Phys.* **2011**, *125*, 698–703. [CrossRef]
34. Admuthe, A.; Kumbhar, S.S.; Chougule, S.K.; Padasare, G.N.; Tonape, M.M. Synthesis and Characterization of MnS Thin Film at Room Temperature for Supercapacitor Application. *Macromol. Symp.* **2020**, *392*, 2000186. [CrossRef]
35. Zhao, N.; Hu, Y.; Du, J.; Liu, G.; Dong, B.; Yang, Y.; Peng, J.; Li, J.; Zhai, M. Ti_3C_2Tx Mxene-Derived Amorphous TiO_2-C Nanosheet Cocatalysts Coupled CdS Nanostructures for Enhanced Photocatalytic Hydrogen Evolution. *Appl. Surf. Sci.* **2020**, *530*, 147247. [CrossRef]
36. Thommes, M.; Kaneko, K.; Neimark, A.V.; Olivier, J.P.; Rodriguez-Reinoso, F.; Rouquerol, J.; Sing, K.S.W. Physisorption of Gases, with Special Reference to the Evaluation of Surface Area and Pore Size Distribution (IUPAC Technical Report). *Pure Appl. Chem.* **2015**, *87*, 1051–1069. [CrossRef]
37. Bu, X.; Chen, C.; Zhao, X.; Huang, Q.; Liao, X.; Fan, H.; Wang, P.; Hu, H.; Zhang, Y.; Huang, Z. Fabrication of Novel Z-Scheme $LaCoO_3$/Activated Biochar/Ag_3PO_4 Heterojunctions for Intensifying Visible-Light-Catalytic Degradation of Bisphenol A. *Appl. Surf. Sci.* **2022**, *588*, 152887. [CrossRef]
38. Wang, P.; Chen, C.; Li, Y.; Wei, J.; Huang, Z. Facile Synthesis of Lignin/Polyethyleneimine Foams Incorporating Zeolitic Imidazole Framework-8 to Efficiently Remove Tetracycline from Aqueous Solution. *Sep. Purif. Technol.* **2024**, *348*, 127769. [CrossRef]
39. Xue, F.; Bi, J.; Chen, L.; Cui, M.; Fei, Z.; Qiao, X. Uniformly Formed Copper-Modulation ZIF-8 Pellets for Efficient Removal of Tetracycline From Wastewater. *Sep. Purif. Technol.* **2024**, *349*, 127849. [CrossRef]
40. Wang, Y.; Xue, S.; Liao, Y.; Lu, Q.; Wang, H.; Zhao, C.; Tang, N.; Du, F. Mof-Derived $Bi_{12}O_{17}Cl_2$ Nanoflakes for the Photocatalytic Degradation of Bisphenol A and Tetracycline Hydrochloride Under Visible Light. *Acs Appl. Nano Mater.* **2024**, *7*, 3188–3198. [CrossRef]

41. Ye, X.; Qi, Y.; Cheng, Y.; Wang, Q.; Han, G. Zinc Chloride-Doped g-C_3N_4 Microtubes for Enhanced Photocatalytic Degradation of Tetracycline Hydrochloride. *Langmuir* **2025**, *41*, 1684–1693. [CrossRef]
42. Zhang, X.; Zha, X.; Luo, Y.; Liu, T.; Chen, G.; He, X. In_2O_3 Nanoparticle/$Bi_4O_5Br_2$ Nanosheet S-Scheme Heterojunctions with Interfacial Oxygen Vacancies for Photocatalytic Degradation of Tetracycline. *ACS Appl. Nano Mater.* **2023**, *6*, 11877–11887. [CrossRef]
43. Tian, Y.; Liu, Y.; Zhu, H.; Nie, D.; Khan, S.; Yang, X. One-Step Construction of Hierarchical Porous and Defect-Rich Zn^{2+} -Doped NH2 -MIL-125(Ti) to Enhance Photocatalytic Degradation of Tetracycline Hydrochloride. *Inorg. Chem.* **2024**, *63*, 19885–19896. [CrossRef] [PubMed]
44. Jing, S.; Zhao, J.; Wang, A.; Ji, Q.; Cheng, R.; Liang, H.; Chen, F.; Kannan, P.; Brouzgou, A.; Tsiakaras, P. Efficient Photocatalytic Production of H_2O_2 and Photodegradation of Tetracycline by CdS/Square Tubular g-C_3N_4 S-Scheme Heterojunction Photocatalyst. *Chem. Eng. J.* **2024**, *479*, 147150. [CrossRef]
45. Zhang, H.; Luo, W.; Sun, H.; Yang, H.; Yang, T.; Liu, D.; Zhu, X.; Zhao, L.; Shu, M.; Yang, F. Ce-Doped Nanosheet as Visible Light Photocatalyst for the Photocatalytic Degradation of Tetracycline Hydrochloride. *ACS Appl. Nano Mater.* **2024**, *7*, 11506–11517. [CrossRef]
46. Huang, Q.; Chen, C.; Zhao, X.; Bu, X.; Liao, X.; Fan, H.; Gao, W.; Hu, H.; Zhang, Y.; Huang, Z. Malachite Green Degradation by Persulfate Activation with $CuFe_2O_4$@Biochar Composite: Efficiency, Stability and Mechanism. *J. Environ. Chem. Eng.* **2021**, *9*, 105800. [CrossRef]
47. Tan, C.; Chen, C.; Danni, W.; Zhao, X.; Zhou, L.; Lu, X.; Liang, J.; Wang, W. A Novel $LaFeO_3$/g-C_3N_4/Ag_3PO_4 Dual Z-Scheme Heterojunction with Enhanced Light Absorption, Charge Carrier Separation and Transfer Capacity for Photodegradation of Antibiotics. *J. Alloys Compd.* **2024**, *1002*, 175211. [CrossRef]

materials

MDPI

Article

Preparation of TiO_2 Nanorods@Ni-Foam for Photocatalytic Decomposition of Acetaldehyde—In Situ FTIR Surface Investigation

Piotr Rychtowski *, Bartłomiej Prowans, Piotr Miądlicki , Maciej Trzeciak and Beata Tryba *

Department of Catalytic and Sorbent Materials Engineering, Faculty of Chemical Technology and Engineering, West Pomeranian University of Technology in Szczecin, Pułaskiego 10, 70-322 Szczecin, Poland; bartlomiej.prowans@zut.edu.pl (B.P.); piotr.miadlicki@zut.edu.pl (P.M.); maciej.trzeciak@zut.edu.pl (M.T.)
* Correspondence: piotr.rychtowski@zut.edu.pl (P.R.); beata.tryba@zut.edu.pl (B.T.)

Abstract: TNR@Ni-foam structures were prepared by an alkaline hydrothermal method in an autoclave in a strongly alkaline medium (10 M NaOH) at 150 °C with further acid washing (0.1 M HNO_3) and a second hydrothermal treatment in an autoclave at 180 °C. Two TiO_2 samples were used for preparation: anatase and P25 of mixed anatase and rutile phases. After the first step of hydrothermal treatment, a layered titanate structure was obtained ($Na_2Ti_3O_7$). Acid washing caused the substitution of Na^+ by H^+ and launched the formation of TNR. After the second hydrothermal treatment at 180 °C, for the optimal quantity of acid used for washing (10 mL per 0.75 g of TiO_2), titania was crystallized to an anatase phase with small quantities of brookite and rutile. The structures obtained from P25 exhibited more brookite and rutile than those based on the anatase precursor. The morphology of TNR@Ni-foam structures was observed by SEM. The obtained composites were tested for acetaldehyde photodegradation (240 ppm in air) during the continuous flow of gas (5 mL/min) through the reactor coupled with FTIR. The most active samples were those obtained from P25, which had a crystalline structure of TiO_2 and contained the lowest quantity of residue Na species.

Keywords: photocatalysis; TiO_2 nanostructures; acetaldehyde removal; alkalic-hydrothermal synthesis

1. Introduction

In our previous reports, it was demonstrated that titanium dioxide (TiO_2) loaded on nickel foam was an excellent material for photocatalytic air purification, especially for the acetaldehyde [1] and ethylene [2] decompositions. In the presence of nickel foam, an enhancement of TiO_2 photoactivity toward acetaldehyde and ethylene decompositions was observed under both ambient conditions and elevated temperatures [3]. The highest decomposition rate was achieved at 100 °C; under these conditions, the reaction yield on TiO_2-supported nickel foam was doubled in comparison with TiO_2 only. The performed photocatalytic tests of acetaldehyde and ethylene decompositions in the presence of some reactive scavengers indicated that superoxide anion radicals were the dominant species contributing to their decompositions. It was stated that these species were greatly generated at the interface of TiO_2 and nickel foam. Therefore, in the presence of nickel foam, the mineralization of both the acetaldehyde and ethylene was greatly enhanced. In such constructed composite, nickel foam acts as an electron supplier, but also improves the transfer of photogenerated charge carriers. What makes these types of composites unique is that they can be used in thermo-photocatalytic processes, where light and elevated temperature create a synergistic effect of thermocatalysis [4,5]. Moreover, the nickel foam structure is highly porous, which increases the available surface area for TiO_2 loading. The presence of free spaces in the foam structure allows the fluids to flow through the pores. This additional advantage allows the use of metal foam-based composites such as the photocatalytic filter, which purifies the fluid streams from the pollutants [6].

In the presented studies, TiO_2 nanorods@Ni-foam were synthesized and tested for acetaldehyde decomposition. TiO_2 nanorods (TNRs) have some advantages over TiO_2 nanoparticles, such as a higher available surface for molecule adsorption, increased harvesting of light, and better separation of free charges due to the optimal aspect ratio of reduction to oxidation sites [7]. TiO_2 of various exposed facets has different properties. Performed calculations of 5f-Ti atoms available for e^- trapping indicated that, for anatase crystal (101) exposed facets, they were the highest ($5.15191 \times 10^{18}/m^2$) and were medium for the (100) facet (nanorods) with $3.70300 \times 10^{18}/m^2$, and zero for the (001) facet. On the contrary, 2f-O atoms available for hole trapping were the highest for the (001) exposed facet of anatase ($6.96378 \times 1018/m^2$), medium for nanorods ($5.55450 \times 10^{18}/m^2$), and lowest for the (101) facet ($5.15191 \times 10^{18}/m^2$) [8]. Our previous studies [1] revealed that electron trapping and formation of $O_2^{-\bullet}$ radicals on TiO_2 was a key factor for acetaldehyde decomposition and mineralization. Additionally, hole-trapping sites in TiO_2 partly contributed to the acetaldehyde decomposition. Therefore, the TiO_2 nanorod structure should be a good candidate for acetaldehyde decomposition. Other researchers noticed improved mineralization of acetaldehyde on TiO_2 nanotubes versus TiO_2 nanoparticles [9]. Therefore, it was considered that elongation of TiO_2 toward the (100) facet should be beneficial for gas phase photocatalysis.

There are numerous reports on the preparation of TiO_2 nanorods (TNRs) via the hydrothermal method [10–12]. Most of them utilize substrates, such as expensive titanium isopropoxide (TTIP) [13] or volatile titanium tetrachloride ($TiCl_4$) [14], and they also require the use of concentrated acids [15]. Although the TNRs obtained that way are well structured in the context of their morphology and photocatalytic performance, this method is not suitable for the preparation of metal-based composites, since metal supports can not only be damaged but also completely dissolved. In order to successfully fabricate the nickel foam/TNRs composite, it is therefore necessary to use a synthesis without strong acids. Existing at the time of preparation, the methods of nickel foam/TNRs [16] are based on organic substrates. Organic-based syntheses are known to reduce the photocatalytic performance of obtained composites used for certain processes [17,18]. This is due to residual, unreacted substrates that form surface carbon deposits.

According to reported studies by other researchers [19], the alkaline synthesis of TiO_2 leads to the formation of an intermediate phase of the general formula $Na_{2\text{-}x}H_xTi_2O_5 \cdot 1.8\,H_2O$, which, upon the hydrothermal treatment and relevant pH, is transformed to TNRs. In this paper, we utilized a non-organic, two-step hydrothermal synthesis with the use of a strong base solution (10 M NaOH). This method was proved to be suitable for obtaining highly crystalline TNRs [20,21]. We utilized this method to prepare a thin titania layer on nickel foam, which resulted in the formation of highly porous, 3D Ni foam/TNR composites. These composites were then tested in the gas-phase continuous flow photocatalytic reactor coupled with FTIR in situ in the process of acetaldehyde photodegradation at 100 °C. This temperature was applied based on our previous experiments, in which we indicated that it was optimal.

In situ FTIR spectroscopy was used in our study as it is a significant technique in the field of gas-phase photocatalysis. Enabling real-time monitoring of transient intermediates provides essential mechanistic insights into the reactions occurring on photocatalytic materials [22,23]. As demonstrated in the work of Zhang et al. [24], using this technique allowed the monitoring of the mechanism of CO_2 photoreduction on treated TiO_2. In the work reported by Subbotina and Barsukov [25], this technique allowed some surface peroxides on the TiO_2 surface to be identified, which appeared to be the key intermediates during the photocatalytic oxidation of gaseous ethanol.

2. Materials and Methods

Two-step hydrothermal synthesis was utilized in the preparation of nickel foam/TiO_2 nanorods (TNRs). According to the morphological phase diagram of TiO_2-P25 transformation described by other researchers [21], the selected reaction conditions allow TiO_2 nanotubes/nanorods to be obtained. In order to determine when this approach is appropriate in the case of other types of TiO_2, in addition to P25, we used the other one, which exhibited only the anatase phase (KRONOClean®7050). The properties of both TiO_2-based materials are listed in Table 1.

Table 1. Selected properties of TiO_2-based materials used for synthesis [26,27].

Parameter \ Sample	KRONOClean®7050	P25
Bulk density (g/dm^3)	300	140
BET specific surface area (m^2/g)	225	35–65
Phase composition (anatase/rutile)	100/0	85/15
Average crystallite size (nm)	15	20

In the first step, 60 mL of 10 M NaOH (POCH Avantor; Gliwice, Poland) was mixed with 1.5 g of either P25 (Evonik; Essen, Germany) or KRONOClean®7050 (Kronos International; Leverkusen, Germany) in an ultrasonic bath for 15 min. The mixture was then transferred to a Teflon insert (100 mL) of a stainless-steel autoclave. Nickel foam (Nanografi Nano Technology; Ankara, Turkey) 2 × 2 × 0.1 cm was cleaned by rinsing with acetone (Chempur; Piekary Śląskie, Poland) and deionized water, then placed at an angle of 45 degrees to the autoclave wall, together with NaOH/TiO_2 suspension. The way it was performed is shown in Figure 1. The autoclave was heat-treated in an oven for 24 h at 150 °C. After cooling down to room temperature naturally, the whole mixture was vacuum filtered under 760 mbar and then rinsed with deionized water until a pH of 7 was reached. The pH was controlled via pH test strips (pH-Fix 7.0–14.0; MAcherey-Nagel; Dueren, Germany). Nickel foam was simply rinsed with deionized water in order to get rid of loosely adhered alkaline TiO_2 particles. After drying in an oven at 70 °C for 24 h, 0.75 g of obtained powder and nickel foam was mixed with a given amount of 0.1 M HNO_3 (Chempur; Piekary Śląskie, Poland) for 6 h in order to remove Na^+ cations. The following amounts of HNO_3 were applied: 0, 5, 10, and 15 mL. The mixture was then filtered under a vacuum and rinsed until a pH of 7 was reached. Both the obtained mixture and nickel foam were then put into the autoclave for the second hydrothermal treatment at 180 °C for 24 h. The mixture and nickel foam were then rinsed with deionized water and dried in an oven at 70 °C for 24 h. Both the powder and the nickel foam/TNRs obtained were further analyzed.

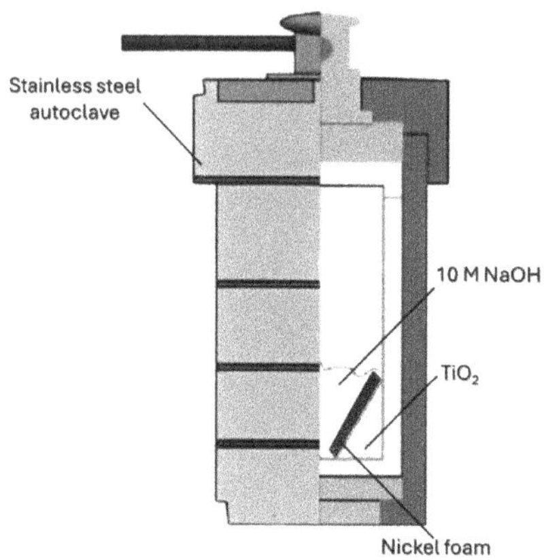

Figure 1. Preparation scheme of TNR-nickel foam composites.

X-ray diffraction (XRD) measurements of obtained TNRs were performed using a diffractometer (PANanalytical, Almelo, The Netherlands) with a Cu X-ray source, λ = 0.154439 nm. The data were collected in the 2θ range of 10–80°; step size of 0.013. The voltage and current parameters used were 35 kV and 30 mA, respectively. Phases were determined on the basis of the standard diffraction data of JCPDS: 01-071-1168 (anatase), 04-007-6488 (brookite), 01-088-1172 (rutile), and 00-059-0666 ($Na_2Ti_3O_7$) [28]. The phase composition was determined after previous background determination, K-alpha subtraction, and peak determination. Rietveld analysis was then performed. These operations were carried out using HighScore Plus software (version 3.0.5).

SEM (Scanning Electron Microscope) images were taken using model SU8020 (Hitachi, Tokio, Japan). Images were taken to study the morphology and topography of the TNRs/nickel foam composites. The samples were first attached to a circular platform using carbon adhesive tape. The images were taken at 15 kV and the vacuum level of each chamber was 2×10^{-7} Pa. The secondary electron detector SE—Top/Upper/Lower detectors and the SE/BSE or SE(TUL) signal mixing function were used.

X-ray fluorescent (XRF) measurements were performed via Epsilon3 spectrometer (PANanalytical, Almelo, The Netherlands) to determine the leftover percentage of sodium content in the powdered samples of TNRs.

Fourier Transform Infrared Spectroscopy in Attenuated Total Reflectance (FTIR-ATR) mode was used to analyze the prepared powder materials. The averages of 32 measurements of FTIR spectra were taken in a wavenumber range of 4000–400 cm^{-1}. The atmospheric air just before the measurement was used as the base spectrum.

The photocatalytic test system for acetaldehyde decomposition is shown in Figure 2. The tests were carried out in an in situ FTIR Nicolet iS50 (Thermo, Waltham, MA, USA) with a photocatalytic reactor (Harrick, Pleasantville, NY, USA) where 5 mL/min of ~240 ppm acetaldehyde in air was continuously fed. The gas mixture flowed through the set of 3 layers of circular TNR-Ni foams irradiated from above by the 365 nm UV LED light source with an optical power of 415 mW and optic fiber diameter of 5 mm (LABIS, Warsaw, Poland). The intensity of incident UV radiation was measured using a photo-radiometer, HD2102.1 (TEST-THERM, Kraków, Poland). The obtained UV intensity value measured on the surface of the reactor cover window equaled 20 W/m^2. The reactor temperature was set to 100 °C. The concentration of the outlet gas stream was measured by a gas chromatograph (SRI, Torrance, CA, USA) equipped with a flame ionization detector. Analyses were carried out under the following conditions: isothermal oven temperature of 130 °C; detector temperature of 250 °C; automatic sampling loop volume of 2 mL; and a metal capillary column (MXT-1301) of 15 m, with an ID of 0.53 mm and 3.00 µm. Both spectra and chromatograms were collected every 5 min for approximately 2 h.

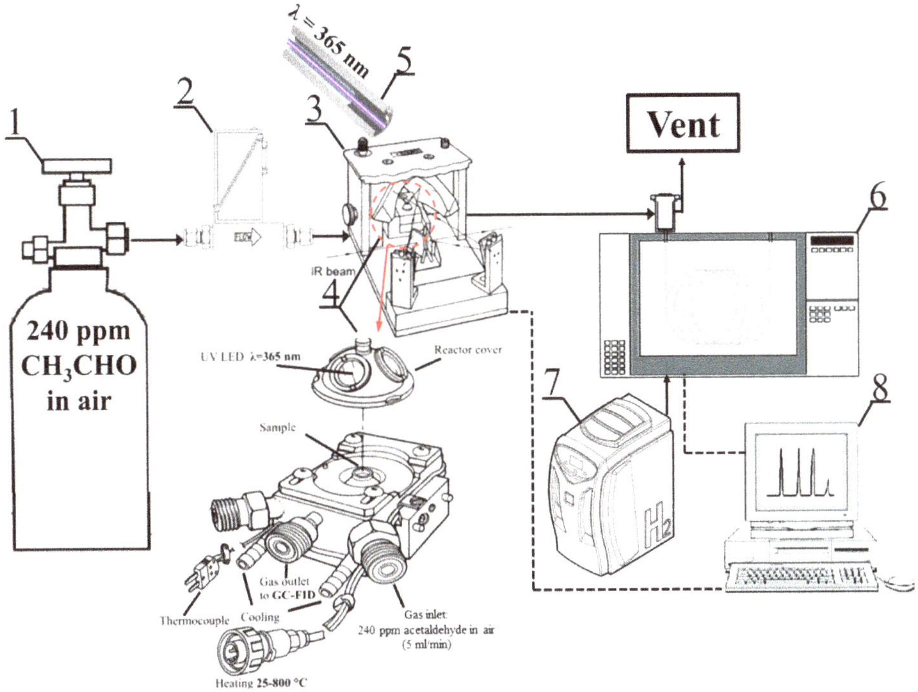

Figure 2. Schematic diagram of photocatalytic reaction system: 1—model gas bottle of 240 acetaldehyde in air, 2—mass flow meter, 3—in situ FTIR Praying Mantis™, 4—high-temperature reaction chamber, 5—optical fiber UV-LED light source, 6—gas chromatograph with FID, 7—hydrogen generator, 8—PC.

3. Results and Discussion

XRD (X-ray Diffraction) measurements were carried out to determine the phase composition of samples. It should be noted that these measurements were carried out using powdered titania samples. Therefore, the identified phase composition does not reflect exactly the composition of the thin titania layers coating nickel foams. Nevertheless, it provides knowledge about the approximated structure of the titania samples grown on a nickel foam surface. The presented diffractograms of titania samples in Figures 3 and 4 indicated that the flushing of titania samples by nitric acid strongly affected the formation of their crystal structure. Detected phases on diffractograms were anatase, brookite, rutile, and $Na_2Ti_3O_7$, which were determined on the basis of reference cards No.: 01-071-1168, 04-007-6488, 01-088-1172, and 00-059-0666, respectively. Table S1 provides a detailed summary of the obtained phase compositions, as determined by Rietveld refinement, and the analysis plots for each sample can be found in Figures S1–S10.

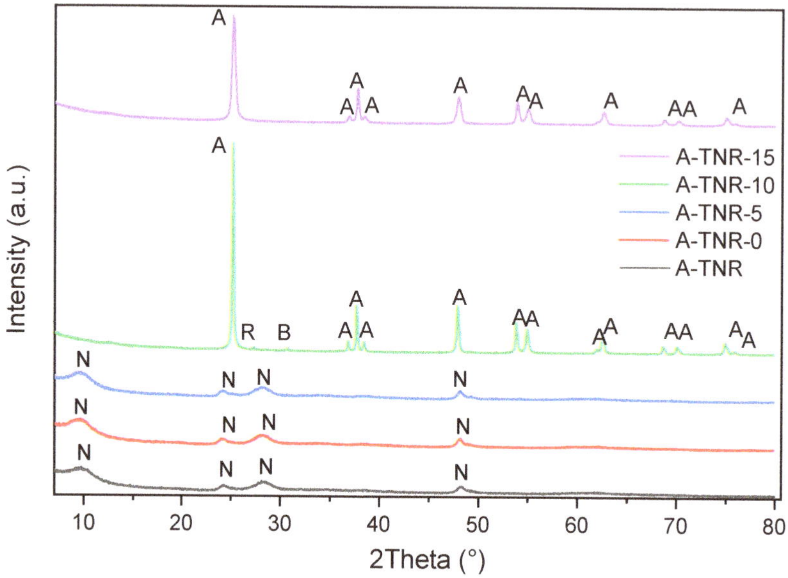

Figure 3. XRD diffractograms of anatase-originated TNRs (without nickel foam). A—anatase, B—brookite, R—rutile, N—$Na_2Ti_3O_7$.

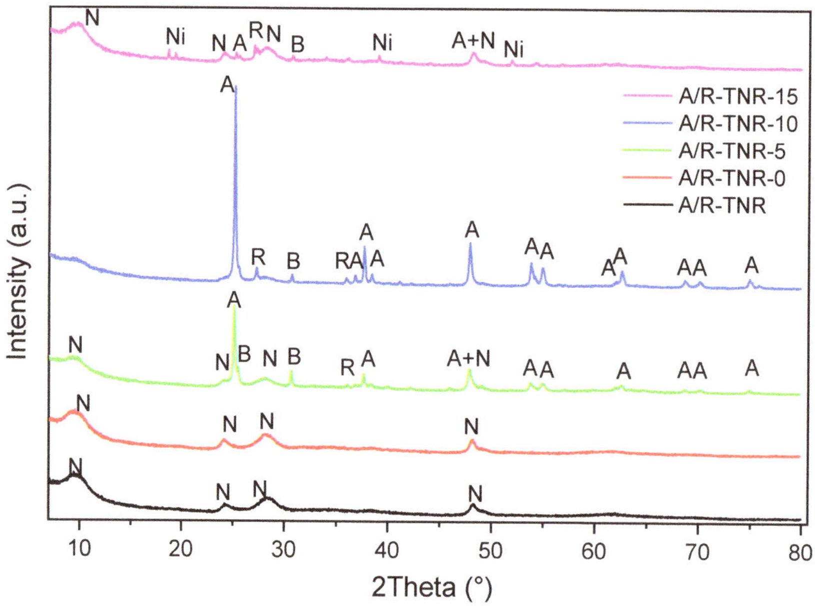

Figure 4. XRD diffractograms of anatase- and rutile-originated TNRs (without nickel foam). A—anatase, B—brookite, R—rutile, N—$Na_2Ti_3O_7$, Ni—$Ni(OH)_2$.

The main mechanism that takes place during alkaline syntheses was reported elsewhere [10,21,29,30] and can be summarized by the following reactions:

$$3TiO_2 + NaOH \rightarrow Na_2Ti_3O_7 + H_2O \quad (1)$$

$$Na_2Ti_3O_7 + 2HCl \rightarrow H_2Ti_3O_7 + NaCl \quad (2)$$

$$H_2Ti_3O_7 \rightarrow 3TiO_2 + H_2O \quad (3)$$

Reaction (1) takes place after the first step of the synthesis. Crystalline TiO_2 used for synthesis transforms into a layered compound containing Na^+ in the interplanar spaces. The ion can then be washed off by using acid (2), where H^+ ions substitute the sodium. Acid washing promotes the formation of NTs, which further curl up to give NRs. The intermediate compound crystallized into anatase, brookite, or rutile in the second hydrothermal synthesis step (3). It was reported that the phase and the morphology of the formed nanostructure are strongly dependent on the acid washing. Moreover, the protonated titanate nanostructures led to the formation of the anatase phase, whereas sodium-rich nanostructures led to the formation of a mixture of titania phases [10].

The obtained results showed that the crystal structure of the formed TiO_2 depended on the quantity of nitric acid used for synthesis. Titania samples originated from anatase (KRONOClean®7050) flushed by a small quantity of nitric acid such as 5 mL (A-TNR-5) or prepared without treatment with acid (A-TNR and A-TNR-0) showed the presence of a $Na_2Ti_3O_7$ phase. Contrary to that, the samples treated with 10 or 15 mL of HNO_3 (A-TNR-10 and A-TNR-15) showed the structure of anatase with a minority of brookite and rutile or anatase only (intensive reflex at a 2Θ angle of 25.2), respectively (Figure 3).

In the case of samples originating from the mixture of anatase and rutile (Figure 4), the highest content of the $Na_2Ti_3O_7$ phase was in the case of samples A/R-TNR and A/R-TNR-0. Flushing by the amount of 5 mL nitric acid (A/R-TNR-5) led to the formation of mostly anatase, but a significant amount of brookite and rutile phases were present too. Some leftover amounts of $Na_2Ti_3O_7$ were still present. The most crystalline was the sample treated with 10 mL of nitric acid (A/R-TNR-10) showing high-intensity reflex at a 2Θ angle of 25.2, and as previously stated, some trace amounts of brookite and rutile phases. Surprisingly, the highest amount of acid (15 mL) used (sample A/R-TNR-15) led to the return formation of the $Na_2Ti_3O_7$ phase by removing all the TiO_2 crystalline phases such as anatase, brookite, and rutile. It is worth noting that small intensity reflexes of $Ni(OH)_2$ were also detected, similarly as reported elsewhere [31]. The presence of the $Ni(OH)_2$ phase in titania powder can be explained by its splintering together with TiO_2 from the nickel foam. Most probably, $Ni(OH)_2$ was formed on nickel foam soaked in a strong alkali solution. The formation of Ni hydroxide on the surface of Ni foam in an alkaline solution (KOH) was described by Kai Wan et al. [32]. They reported that the formation of the NiOOH layer reduces interfacial electronic resistance. Therefore, it is stated that the formation of Ni hydroxide on the Ni foam surface can facilitate electron transfer between TiO_2 and Ni foam and contribute to the separation of free charges.

The first group of the samples tested were ones that originated from the anatase, untreated and treated with 5 mL of nitric acid (A-TNR-0 and A-TNR-5), presented in Figures 5 and 6, respectively. These composites exhibited similar and relatively even titania coating across the whole surface of regular nanostructures. These titania nanostructures are often referred to as nano-sheets. In addition, larger clusters of titania agglomerates in the form of lumps with a diameter of approximately a few to several micrometers can be observed for the A-TNR-0 and A-TNR-5 samples (Figure 5A vs. Figure 6A).

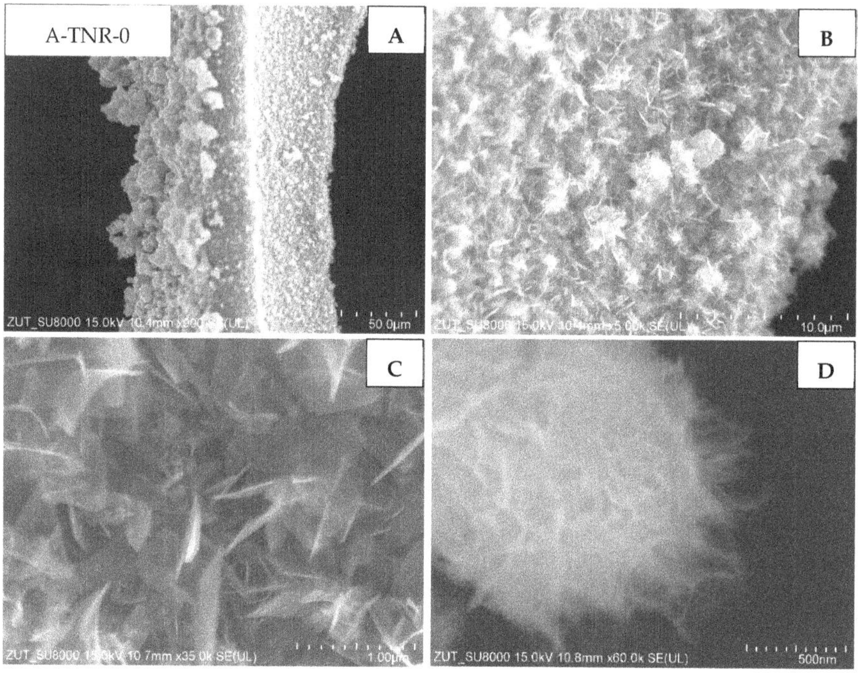

Figure 5. SEM images of A-TNR-0 in magnification: (**A**) 50 μm, (**B**) 10 μm, (**C**) 1 μm, (**D**) 500 nm.

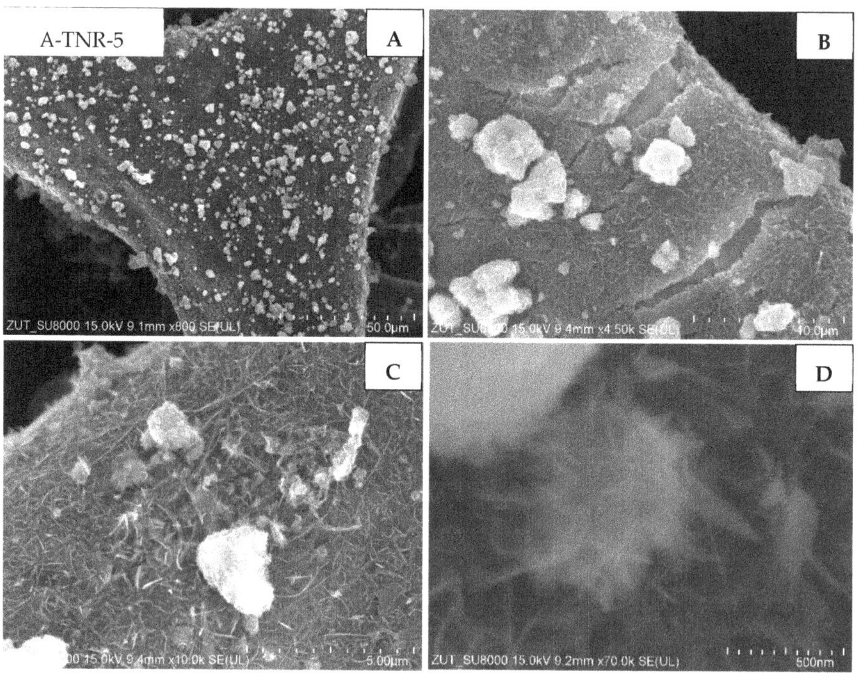

Figure 6. SEM images of A-TNR-5 in magnification: (**A**) 50 μm, (**B**) 10 μm, (**C**) 5 μm, (**D**) 500 nm..

Samples originated from anatase and treated with 10 and 15 mL of nitric acid (A-TNR-10 and A-TNR-15) are illustrated in Figures 7 and 8, respectively. The surface of both composites was uniformly coated with TiO_2 nanorods (Figures 7C and 8C). Interestingly, clusters of larger flower-like structures or bouquets with sizes much larger than the nanostructures on the underlying nickel foam itself were also visible (Figure 7A,B; Figure 8B,C). Similar structures were also observed elsewhere [33]. These clusters were significantly larger for sample A-TNR-10. The level of surface coating of A-TNR-15 (Figure 8A,B) allowed the observation of nickel grain borders, whereas, in the case of samples A-TNR-0, A-TNR-5, and A-TNR-10, this was not observed, indicating significantly thicker surface coating in those mentioned samples.

Figure 7. SEM images of A-TNR-10 in magnification: (**A**) 50 µm, (**B**) 10 µm, (**C**) 3 µm, (**D**) 1 µm.

The next composite studied (Figure 9) was the one that originated from the mixture of anatase and rutile and was untreated with nitric acid (A/R-TNR-0). This material was homogeneously covered with two-dimensional nanostructures, often referred to in the literature as nano-sheets or nano-ribbons [10]. Those structures are several nanometers thick, while their length is as large as 2 µm. In some areas, larger agglomerates with dimensions of several to tens of micrometers have been observed (Figure 9A,B). The observed nanostructures were irregularly arranged on the surface and bent or twisted.

The sample treated with 5 mL of nitric acid and originated from the mixture of anatase and rutile (A/R-TNR-5) is illustrated in Figure 10. In this case, TiO_2 was homogeneously coated on nickel foam and presented elongated particles narrowed at the end, similar to TiO_2 nanorods. Their average length was around 1.15 µm (Figure 10C). In Figure 10D, vertically oriented nanorods with rectangle shape and side dimensions of a few tens of nm are observed.

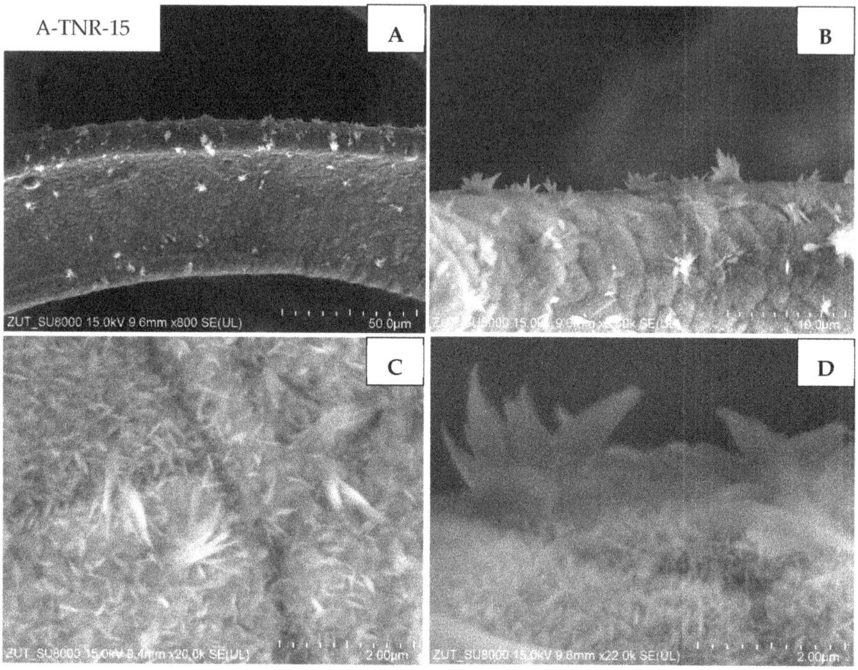

Figure 8. SEM images of A-TNR-15 in magnification: (**A**) 50 μm, (**B**) 10 μm, (**C**,**D**) 2 μm.

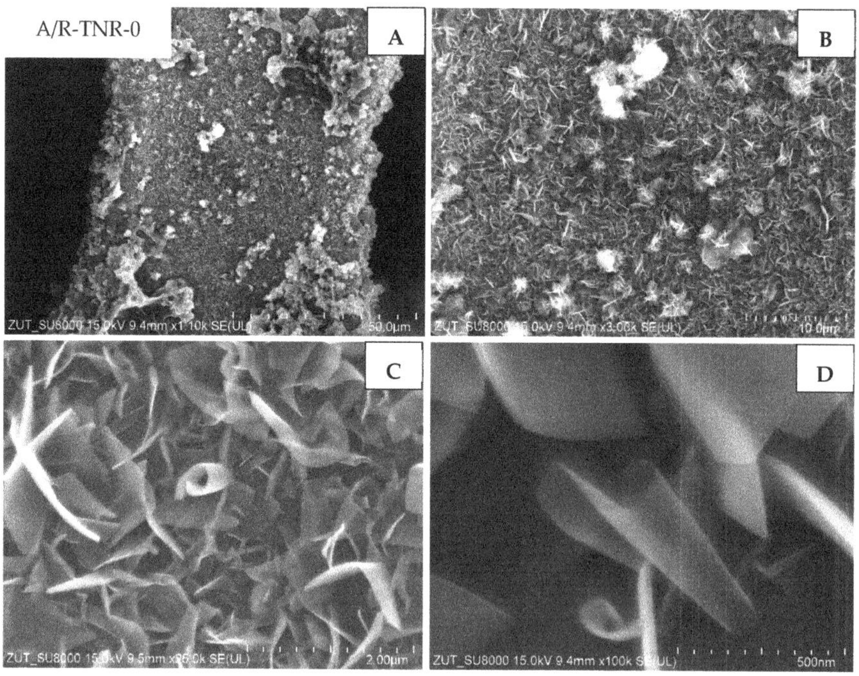

Figure 9. SEM images of A/R-TNR-0 in magnification: (**A**) 50 μm, (**B**) 10 μm, (**C**) 2 μm, (**D**) 500 nm.

Figure 10. SEM images of A/R-TNR-5 in magnification: (**A**) 50 μm, (**B**) 10 μm, (**C**) 4 μm, (**D**) 500 nm.

In Figure 11, the morphology of P25 treated with 10 mL of nitric acid (A/R-TNR-10) is presented. In this particular case, the nickel foam surface is also coated with TiO_2 nanorods. However, their dimensions are much larger than that previously observed: their average length was around 2.4 μm (Figure 11B,C). In Figure 11D, we can observe one of the nanorod planes vertically oriented toward the observer with a size reaching a couple of hundred nanometers (Figure 11D). It is worth noting that there are some areas of TiO_2 agglomerates (Figure 11A). TiO_2 nanorods, however, accounted for most of Ni foam coverage.

The last sample presented was the one that originated from P25 treated with 15 mL of nitric acid (A/R-TNR-15). This time, TiO_2 nanorods are in the mixture with some other nanostructures (Figure 12C,D). The obtained, additional structures can be described as star-like nanostructures. What is more, the nickel foam surface was also covered with macro-sized particles (Figure 12A) between which the aforementioned nanostructures are present.

The sodium content was determined via the XRF technique and is presented in Table 2. The leftover sodium was present in the studied TNR samples due to the use of a strong base during synthesis. Interestingly, only samples originating from the mixture of anatase and rutile (A/R-TNRs) indicated significant sodium content. This content was lower the greater the amount of nitric acid used for sample washing. The sodium content was highest when the sample was not washed, which occurred in the A/R-TNR-0 sample. On the other hand, samples originating from the anatase phase (A-TNRs) showed no sodium presence measured via XRF at all. This may be explained by the origin of the sample, which was obtained by the sulfate method, showing on its surface the presence of sulfate groups, which are base centers according to the Lewis theory. Consequently, they may already compete for surface active sites with the strong 10 M NaOH base acting on this type of TiO_2 at the preparation stage itself. It is therefore highly likely that a simple rinse with deionized water was sufficient in this case to flush out the sodium, which was weakly bound to the formed TNR surface. Samples derived from a mixture of anatase and rutile (A/R-TNR) apparently required harsher treatment (by nitric acid), as the sodium may have embedded itself in the structures of the resulting material, perhaps in the interplanar spaces. This was also observed by other groups [19,34].

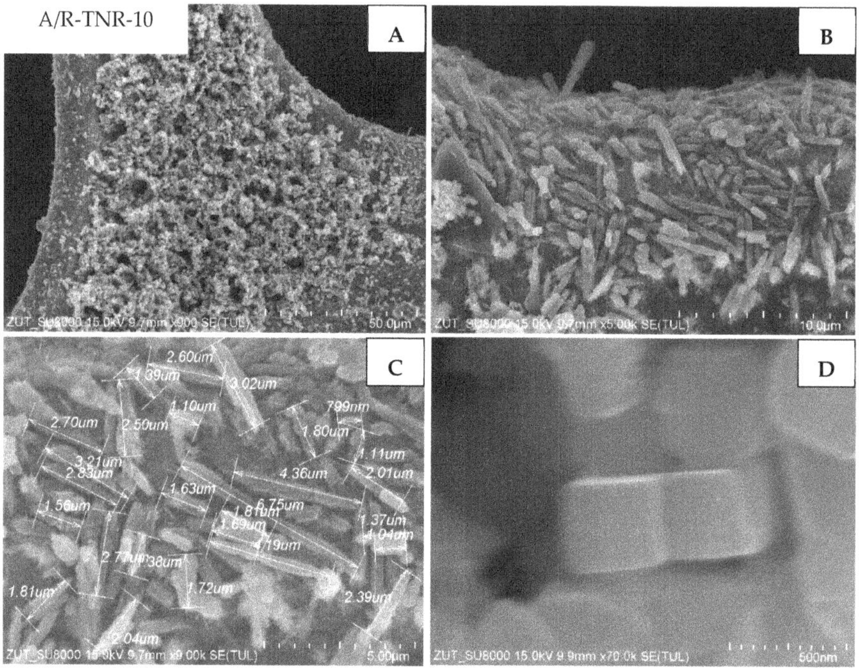

Figure 11. SEM images of A/R-TNR-10 in magnification: (**A**) 50 µm, (**B**) 10 µm, (**C**) 5 µm, (**D**) 500 nm.

Table 2. XRF sodium content.

Sample	Na Content (%)
A/R-TNR-0	4.662
A/R-TNR-5	2.34
A/R-TNR-10	1.032
A/R-TNR-15	0.595
A-TNR-0	0
A-TNR-5	0
A-TNR-10	0
A-TNR-15	0

FTIR-ATR analysis of the powder samples obtained is presented in Figure 13. Note that they were tested without the presence of nickel foam. One of the noticeable differences between the samples here is in the intensity of the band originating from the hydroxyl groups, located at a wavenumber of 1620 cm^{-1}. According to other studies, the transformation of the phases of TNRs during the heating is strongly related to the dehydration process of the titania layered structure [19]. This hypothesis strongly overlaps with our findings, as high hydroxylation of TiO_2 usually implies incomplete conversion of the titanate phase into anatase, rutile, or brookite. Accordingly, samples such as A-TNR, A/R-TNR-0, A/R-TNR-5, and A/R-TNR-15 exhibited the strongest intensity of the –OH band and were consequently not fully crystallized into TiO_2. At the same time, samples of the weakest

–OH band intensity, such as A-TNR-10, A-TNR-15, and A/R-TNR-10, were the ones that revealed a high-intensity anatase reflex on XRD patterns. The other band, which appeared at 1375 cm^{-1}, can be assigned to the vibrations of nitrate anions [35]. An additional band can be observed at 1332 cm^{-1}, which is most likely due to the presence of either carbonates or bicarbonates, and is more intensive, the more alkalic the sample surface is [36]. This band disappeared after acid washing with 10 mL and crystallization of TiO_2. However, under an excess of HNO_3 washing (15 mL), this band emerges again on the FTIR spectra of both titania samples. NaOH treatment caused disruption of the titania crystal structure and increased the affinity of the titania surface to CO_2 adsorption. The observed band at around 1550 cm^{-1} can be assigned to carbonate species and that at 1080 cm^{-1} to C-O vibrations. The intensity of these bands decreases with the rinsing of Na.

Figure 12. SEM images of A/R-TNR-15 in magnification: (**A**) 50 µm, (**B**) 10 µm, (**C**) 1 µm, (**D**) 500 nm.

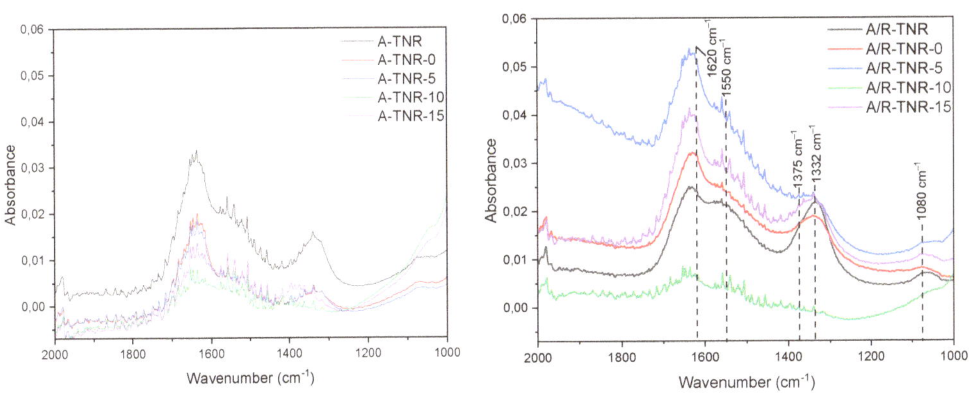

Figure 13. FTIR-ATR analyses of the powder materials obtained.

Photocatalytic tests for acetaldehyde removal are shown in Figure 14. These tests were performed at an elevated temperature (100 °C) to increase the overall process performance due to the catalytic activation of nickel foam. This phenomenon was reported in our previous paper [1]. Because of the relatively low flow rate of gas entering the reactor, which favors adsorption, the experiment could be divided into two stages: adsorption and photocatalytic decomposition. At the low flow rate used, the adsorption step took about 20 min. All the samples except one, originating from P25 untreated by the nitric acid (A/R-TNR-0), showed significant adsorption of acetaldehyde with a maximum of around 220 ppm for A-TNR-10. Although the acetaldehyde adsorption was high on the prepared samples, its photocatalytic decomposition was rather poor, only A/R-TNR-10 and A/R-TNR-15, originating from P25, showed noticeable activity. This was probably caused by the strong adsorption of formed carbonate species on the titania surface upon acetaldehyde oxidation and their resistance to further decomposition.

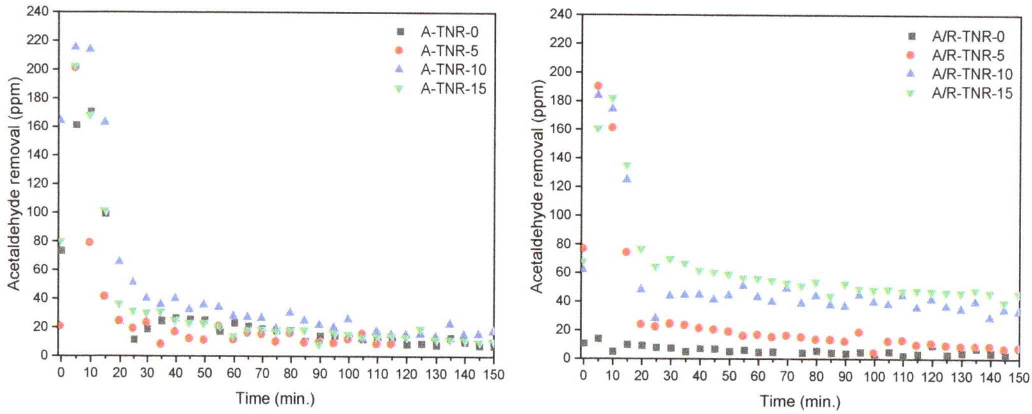

Figure 14. Photocatalytic decomposition of 240 ppm acetaldehyde in air using TNRs/nickel foam composites. Flow rate: 5 mL/min; temperature: 100 °C.

To investigate exactly what was happening on the surface of individual samples, in situ FTIR studies were performed.

In situ FTIR spectra were recorded directly during the adsorption/photocatalytic process of acetaldehyde removal. The most significant changes were observed in the wavenumber range of 2000–1200 cm^{-1}, so it was presented in detail for all spectra together with mappings. Both the appearance and disappearance of infrared bands can be observed here, indicating the loss of given surface groups while the growth of other bands was observed, which is evidence of the formation of intermediate products on the surface of the TNR/nickel foam composites studied. In Figure 15, an in situ FTIR analysis was presented during the aforementioned processes with the use of anatase-originated TNR, untreated by the nitric acid (A-TNR-0). Of all the samples, the largest number of new bands appeared here. As the photocatalytic process progresses, a significant increase in the band intensity at 1690 and 1653 cm^{-1} can be observed. Both can be assigned to ν(C=O) vibrations typical of acetic acid. Moreover, a significant decrease in the δ(OH) 1620 cm^{-1} band was observed, which can be assigned to the presence of surface water, and its reduction over time was due to the elevated temperature of the process. This is the case for all samples. Interestingly, this signal slightly elevates at the last process stage (120 min). Secondly, the increasing band at 1583 cm^{-1} can be assigned to $\nu_{as}(COO)$ and is due to the presence of acetate species. The next bands at 1428, 1417, 1390, and 1379 cm^{-1}, which are close to each other, can be assigned to $\nu_s(COO)$, $\delta_s(CH_3)$, $\nu_s(COO)$, and $\delta(CH_3)$, respectively. Their presence is evidence of the appearance of such compounds as acetaldehyde, acetic acid, and crotonaldehyde, with the former being the result of adsorption, and the others formed as intermediate products. Their intensity grows in time. A small intensity band at 1334 cm^{-1} indicates the $\delta_s(CH_3)$ mode and is due to the formation of acetate species. The quite intensely growing band at 1307 cm^{-1} can be assigned to ν(C-O), which results from the presence of either acetic acid or crotonaldehyde. The last band of relatively weak intensity at 1254 cm^{-1} of the ν(C-C) mode comes from the acetic acid [37–39].

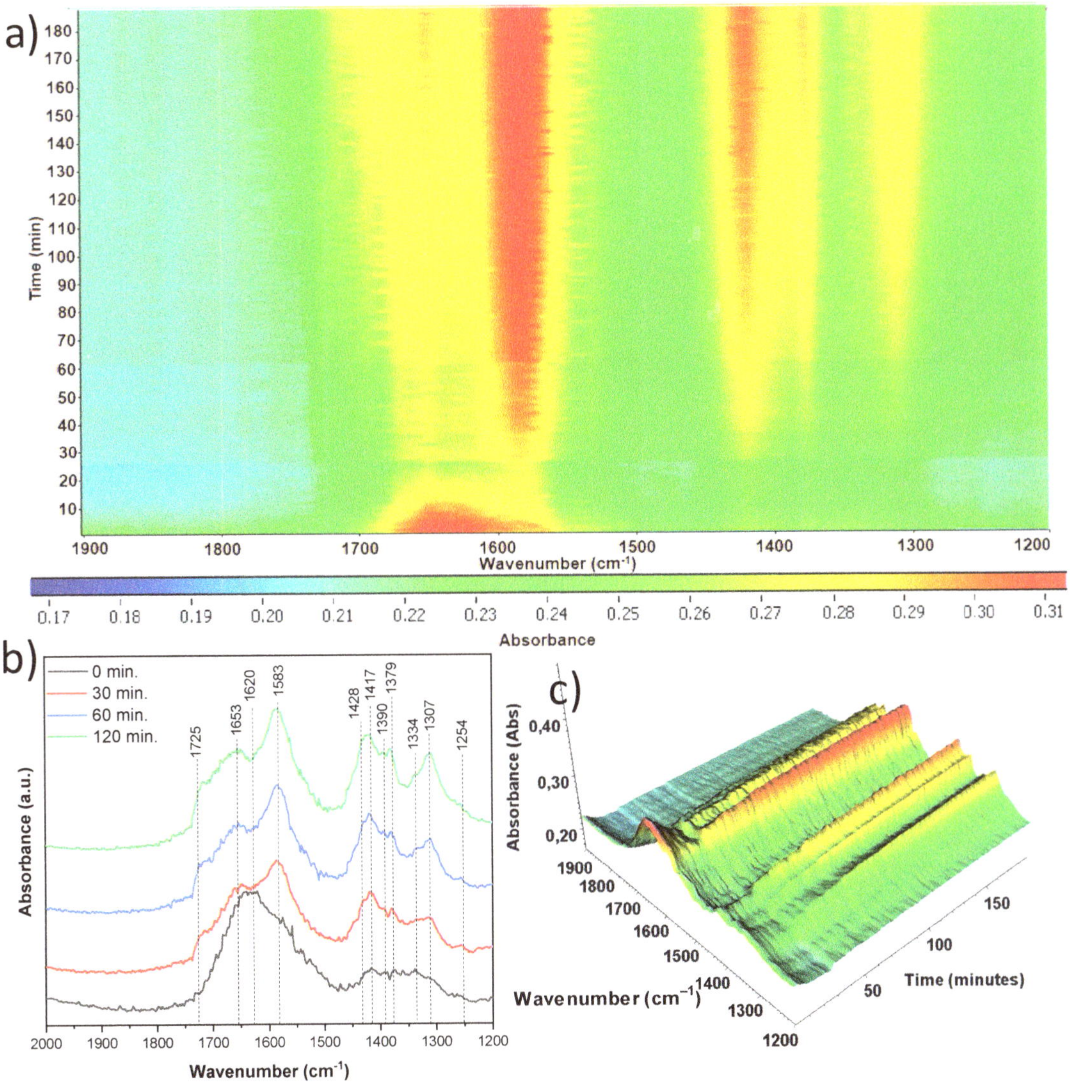

Figure 15. In situ FTIR of A-TNR-0 during the photocatalytic removal of acetaldehyde: (**a**) 2D mapping projection of the A = f(t, ν) relationship, (**b**) A = f(ν) graph after 0, 30, 60, and 120 min, (**c**) 3D mapping of the A = f(t, ν) relationship.

The next sample studied (Figure 16) originated from anatase and was treated with 5 mL of nitric acid (A-TNR-5), and indicated a much lower number of new bands grown than that observed previously. Surface hydroxyl group reduction is observed at 1620 cm^{-1}.

Simultaneously with the disappearing of hydroxyl groups, the new bands are observed at 1725 and 1653 cm^{-1}, which can be assigned to the stretching vibrations of the C=O group in the adsorbed formic acid and formaldehyde, respectively [40].

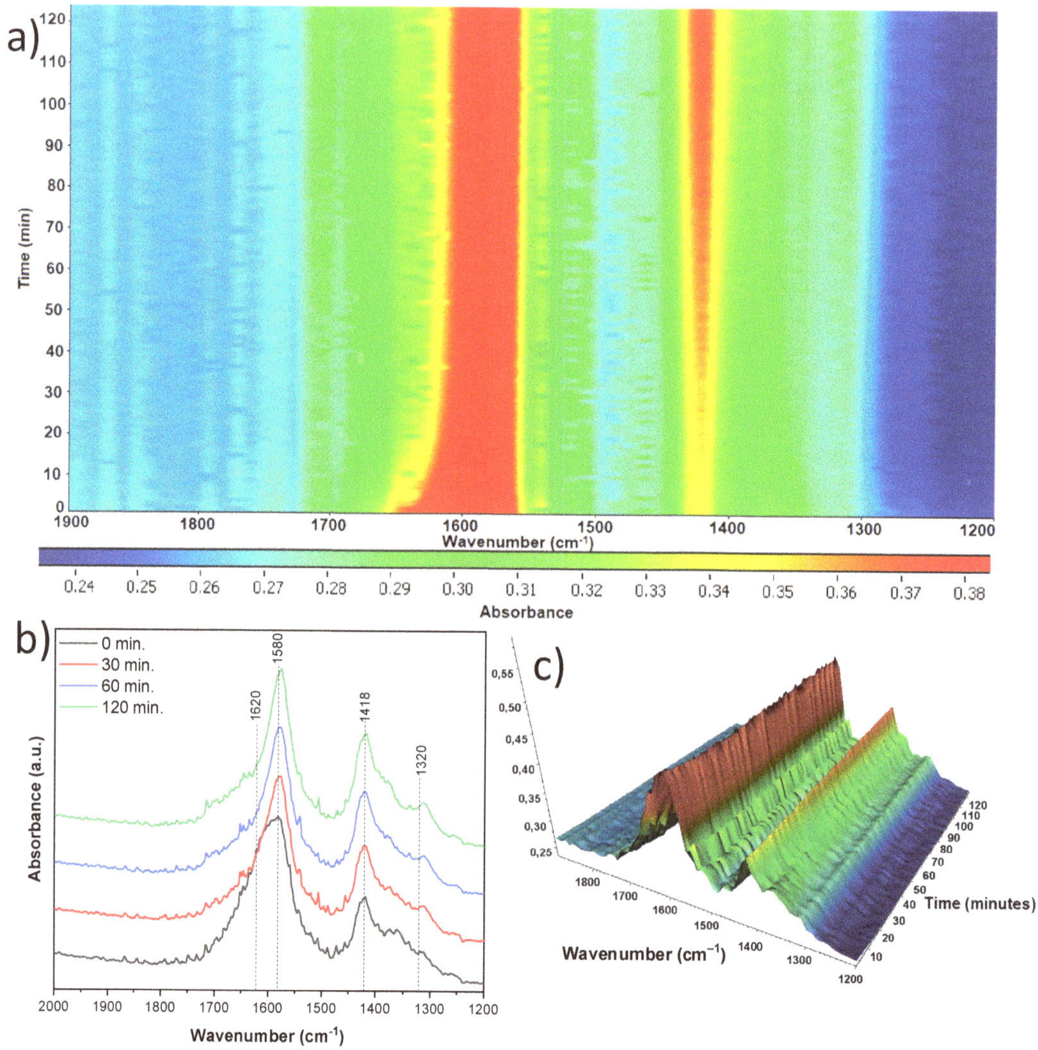

Figure 16. In situ FTIR of A-TNR-5 during the photocatalytic removal of acetaldehyde: (**a**) 2D mapping projection of the A = f(t, ν) relationship, (**b**) A = f(ν) graph after 0, 30, 60, and 120 min, (**c**) 3D mapping of the A = f(t, ν) relationship.

Very slight increase in the 1580 cm^{-1} band from the ν_{as}(COO) mode is observed, which indicates the formation of very low amounts of surface acetate species. The next band at 1418 cm^{-1} of the ν_s(COO) mode results from the significant increase in acetate species throughout the photocatalytic process. The last band of low intensity observed at 1320 cm^{-1} can be assigned to $\delta(CH_3)$, which comes from some of the organic species adsorbed to the surface [37–39].

The sample treated with 10 mL of nitric acid and originating from anatase (Figure 17) exhibited drastically lower band intensities in comparison to different samples. This was most likely due to the much thinner TiO_2 species layer on the surface of the nickel foam. Nickel is a strongly absorbing material in the infrared range, so the signals were significantly weakened. Once again, the 1620 cm^{-1} band of hydroxyl groups reduced with time. Interestingly, the 1592 cm^{-1} band of $\nu_{as}(CO)$ and the 1361 cm^{-1} band of $\nu_{ss}(CO)$ also reduced with time. Mino et al. [41] proved that these bands show CO_2 bound to the anatase surface via adsorption and as a form of monodentate carbonates. This may therefore be explained by the desorbing of carbon dioxide from the surface in our case. The band at 1430 cm^{-1} might result from the presence of acetate species and can be assigned to $\nu_s(COO)$. Additionally, the band at 1320 cm can be assigned to $\delta(CH_3)$, resulting from the adsorption of acetaldehyde or acetic acid [37–39].

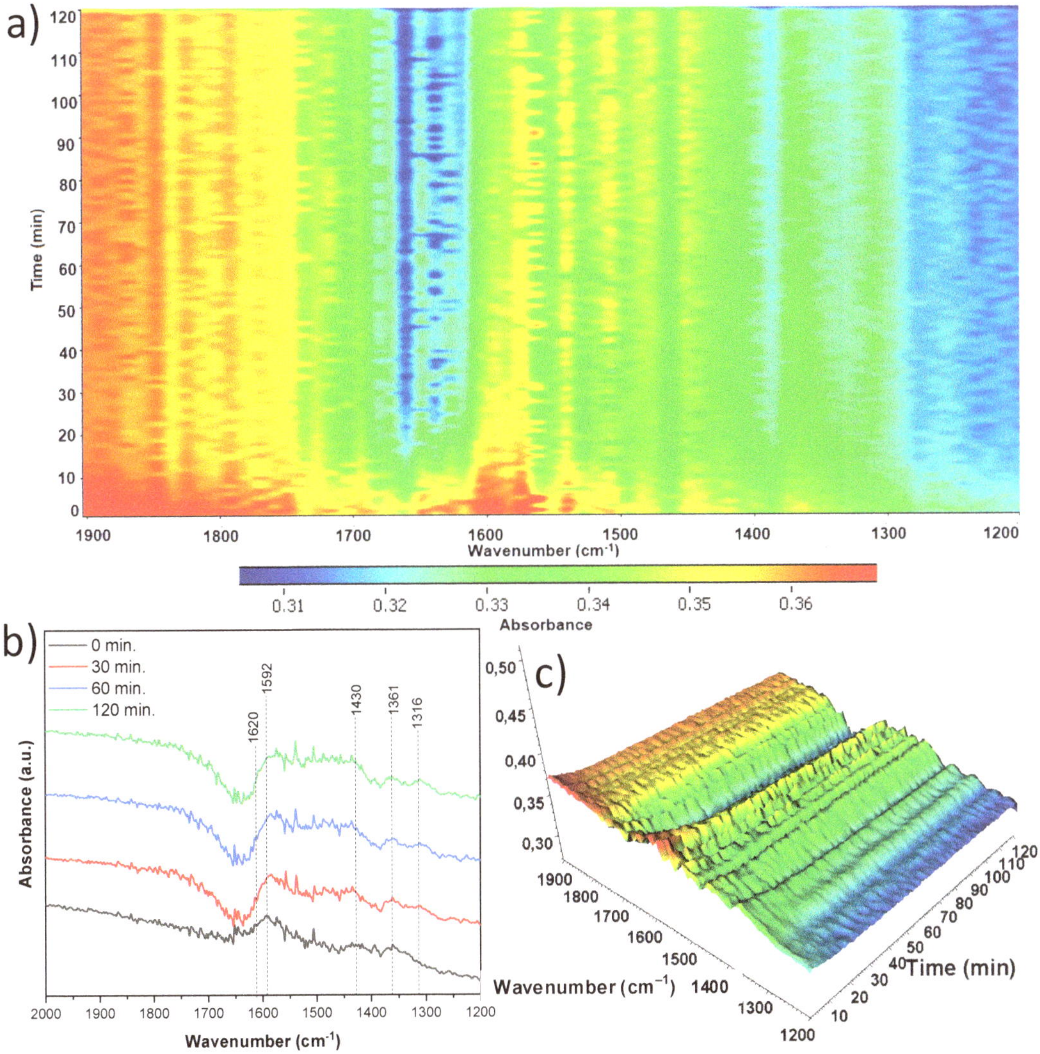

Figure 17. In situ FTIR of A-TNR-10 during the photocatalytic removal of acetaldehyde: (**a**) 2D mapping projection of the A = f(t, ν) relationship, (**b**) A = f(ν) graph after 0, 30, 60, and 120 min, (**c**) 3D mapping of the A = f(t, ν) relationship.

Although the sample treated with 15 mL of nitric acid and originating from anatase (Figure 18) was slightly different on the basis of its crystalline structure, its spectra were almost identical to the previously discussed sample. Again, both loss of hydroxyl groups (1620 cm^{-1}) and desorbing CO_2 are observed (1575 and 1353 cm^{-1}). A notable band intensity increase over time can be seen at 1313 cm^{-1}, which can again be assigned to the $\delta(CH_3)$ of adsorbed organic species [37–39].

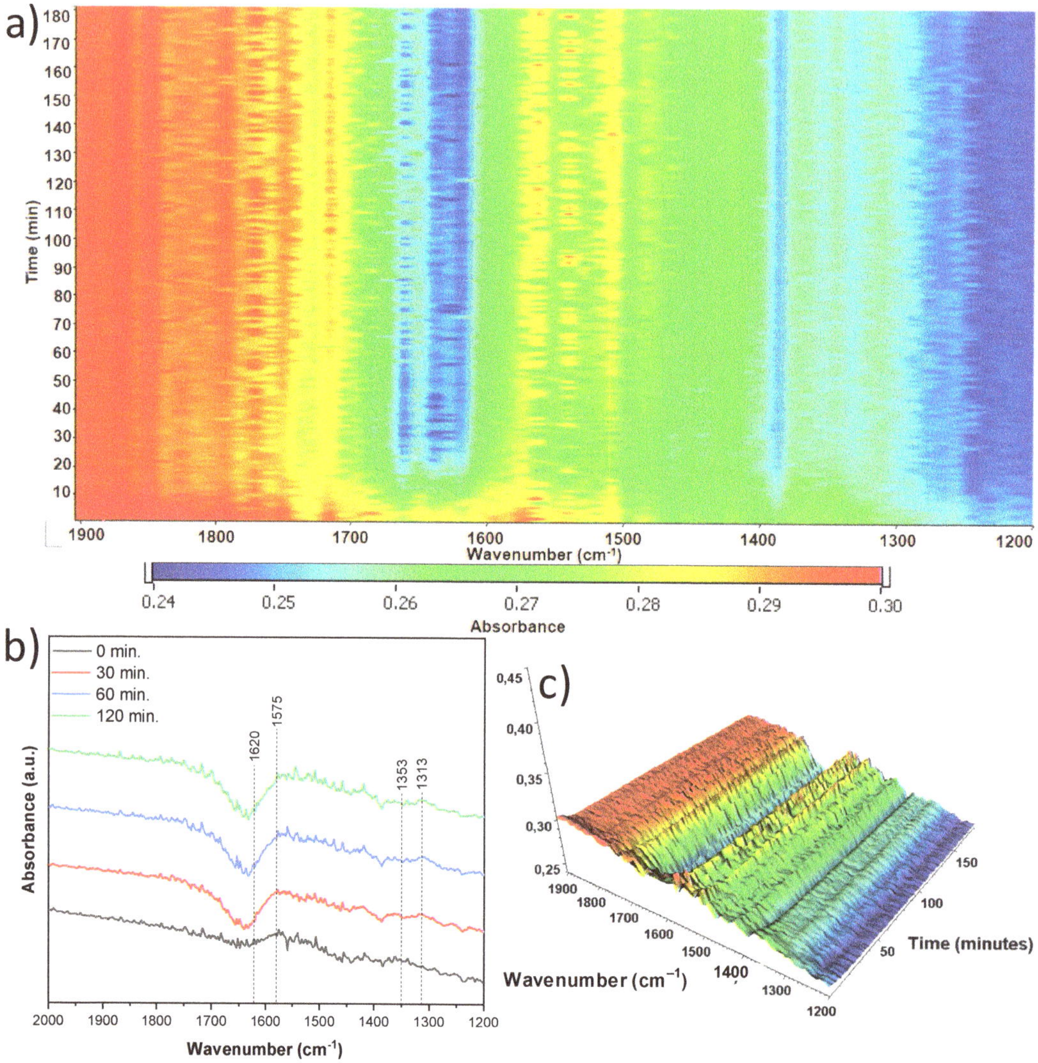

Figure 18. In situ FTIR of A-TNR-15 during the photocatalytic removal of acetaldehyde: (**a**) 2D mapping projection of the A = f(t, ν) relationship, (**b**) A = f(ν) graph after 0, 30, 60, and 120 min, (**c**) 3D mapping of the A = f(t, ν) relationship.

In the case of the sample originating from the mixture of anatase and rutile, untreated with nitric acid (Figure 19), the spectra changed the most at the very beginning of the process, but then remained the same. A notable increase in band intensity at 1770 cm^{-1} of $\nu_{as}(COO)$ can be observed, which cannot be assigned to only one molecule—its presence most likely resulted from the formation of acetic acid or acetate species. The band at 1652 cm^{-1} can be assigned to CO_2 bounded with hydroxyl groups [41]. The band at 1620 cm^{-1}, still most likely originating from hydroxyl groups attached to the surface, is stretched, and its intensity decreases dramatically. On the other hand, the intensity of the band at 1575 cm^{-1} decreases slightly at the beginning to return to the original level after a while. We therefore assume that this time they are acetate species thanks to the $\nu_{as}(COO)$ mode [37–39]. The band at 1369 cm^{-1} of $\nu_{ss}(CO)$ was present at the sample surface and decreased throughout the process, and this is most likely due to the desorption of CO_2 attached to the surface.

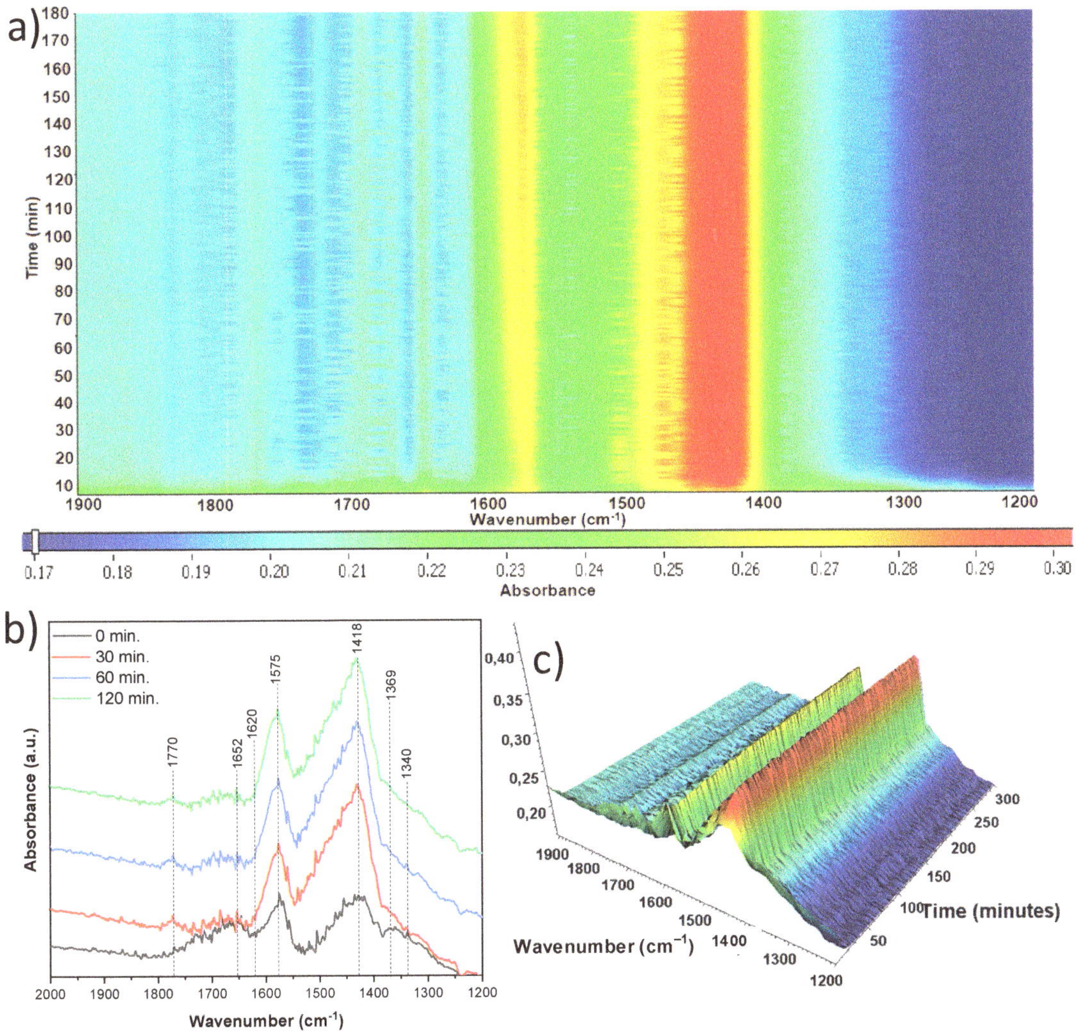

Figure 19. In situ FTIR of A/R-TNR-0 during the photocatalytic removal of acetaldehyde: (**a**) 2D mapping projection of the A = f(t, ν) relationship, (**b**) A = f(ν) graph after 0, 30, 60, and 120 min, (**c**) 3D mapping of the A = f(t, ν) relationship.

The next sample was the one that originated from the mixture of anatase and rutile and was treated with 5 mL of nitric acid (A/R-TNR-5). Two significant major bands (Figure 20) that decrease over the process duration can be noticed at 1620 and 1589 cm^{-1}, which can be assigned to the hydroxyl groups (δ(OH)) and carbon dioxide (ν_{as}(CO)), respectively. A noticeable ν_s(COO) band intensity increase was at 1410 cm^{-1} and was due to the formation of acetate species [37–39].

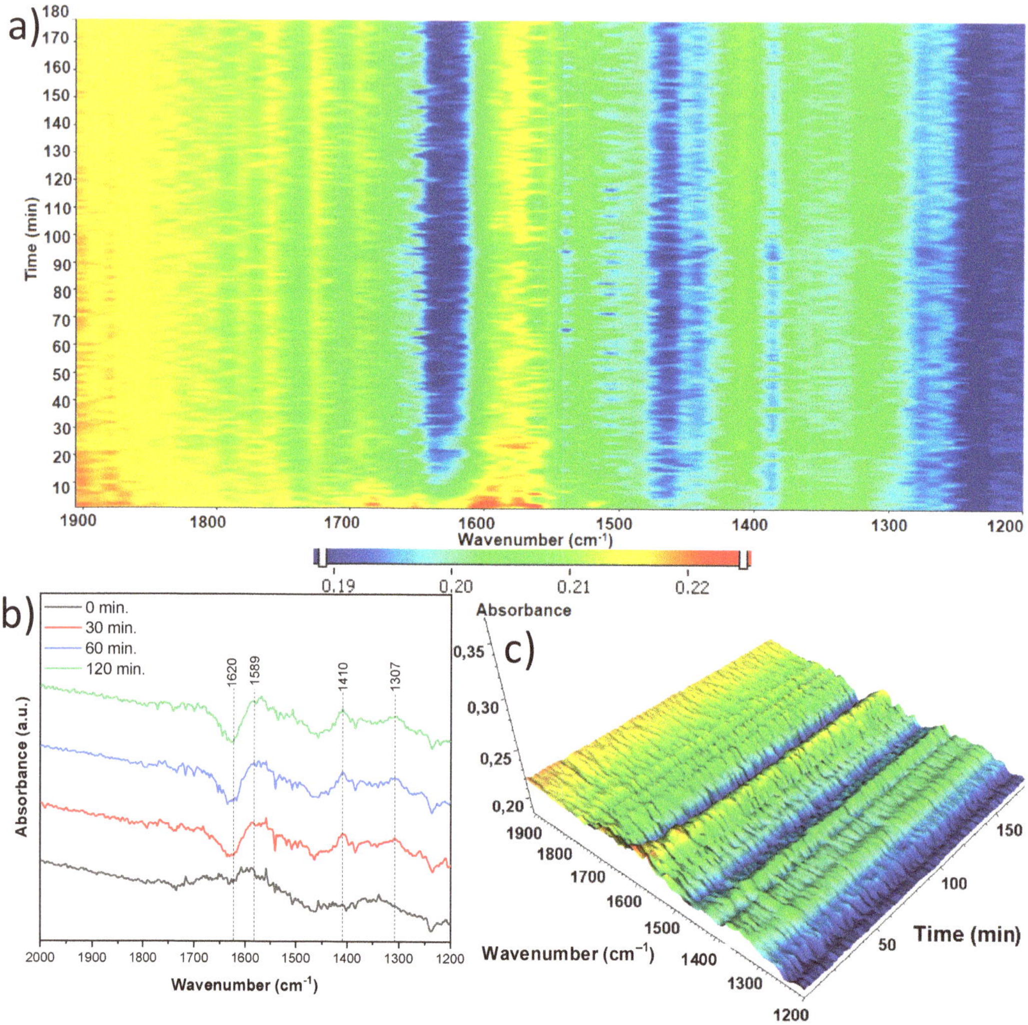

Figure 20. In situ FTIR of A/R-TNR-5 during the photocatalytic removal of acetaldehyde: (**a**) 2D mapping projection of the A = f(t, ν) relationship, (**b**) A = f(ν) graph after 0, 30, 60, and 120 min, (**c**) 3D mapping of the A = f(t, ν) relationship.

The next sample studied, which was treated with 10 mL of nitric acid and originated from the mixture of anatase and rutile (A/R-TNR-10), exhibited quite a stable surface (Figure 21). A slight decrease in intensity of bands at 1658 and 1620 cm^{-1} of ν_{as}(CO) and δ(OH), respectively, can be noticed. According to the literature [41], the former is derived from CO_2 bound to surface hydroxyl groups, and the latter is the hydroxyl groups themselves. These signals are therefore related to each other and also decrease together as the process proceeds. Some increasing bands at 1414 cm^{-1} of ν_s(COO) could be noticed, which were most likely due to the formation of surface acetate species [37–39]. The last two bands at 1371 and 1355 cm^{-1} were relatively stable over the process. Both bands are once again most likely linked to CO_2 attachment to the surface, with the former being $\nu_{ss}(CO)$, and the latter remaining unidentified.

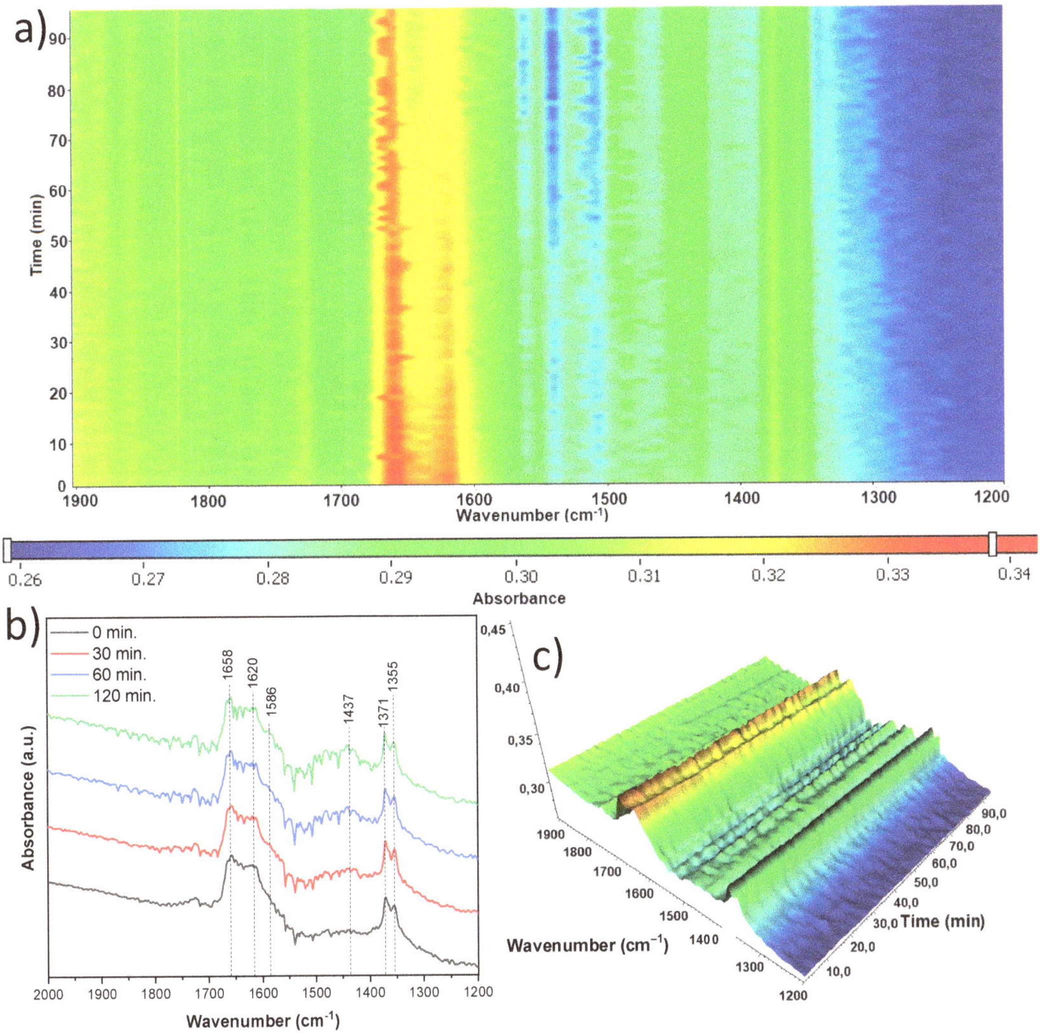

Figure 21. In situ FTIR of A/R-TNR-10 during the photocatalytic removal of acetaldehyde: (**a**) 2D mapping projection of the A = f(t, ν) relationship, (**b**) A = f(ν) graph after 0, 30, 60, and 120 min, (**c**) 3D mapping of the A = f(t, ν) relationship.

The last sample studied (Figure 22) was the one that originated from the mixture of anatase and rutile and was treated with 15 mL of nitric acid (A/R-TNR-15). Some small decrease in intensity of the 1620 cm^{-1} band of hydroxyl surface groups was observed. Close to it, another band (1588 cm^{-1}) can be observed, whose intensity increases over time and can be assigned to $\nu_{as}(COO)$ of either acetate or formate species. Additionally, three time bands increasing over at 1416, 1380, and 1308 cm^{-1} were observed, identified as the $\nu_s(COO)$ of acetate species, $\delta(CH_3)$ and $\delta_s(CH_3)$, respectively. The last two bands were most likely the cause of the crotonaldehyde formed on the surface [37–39].

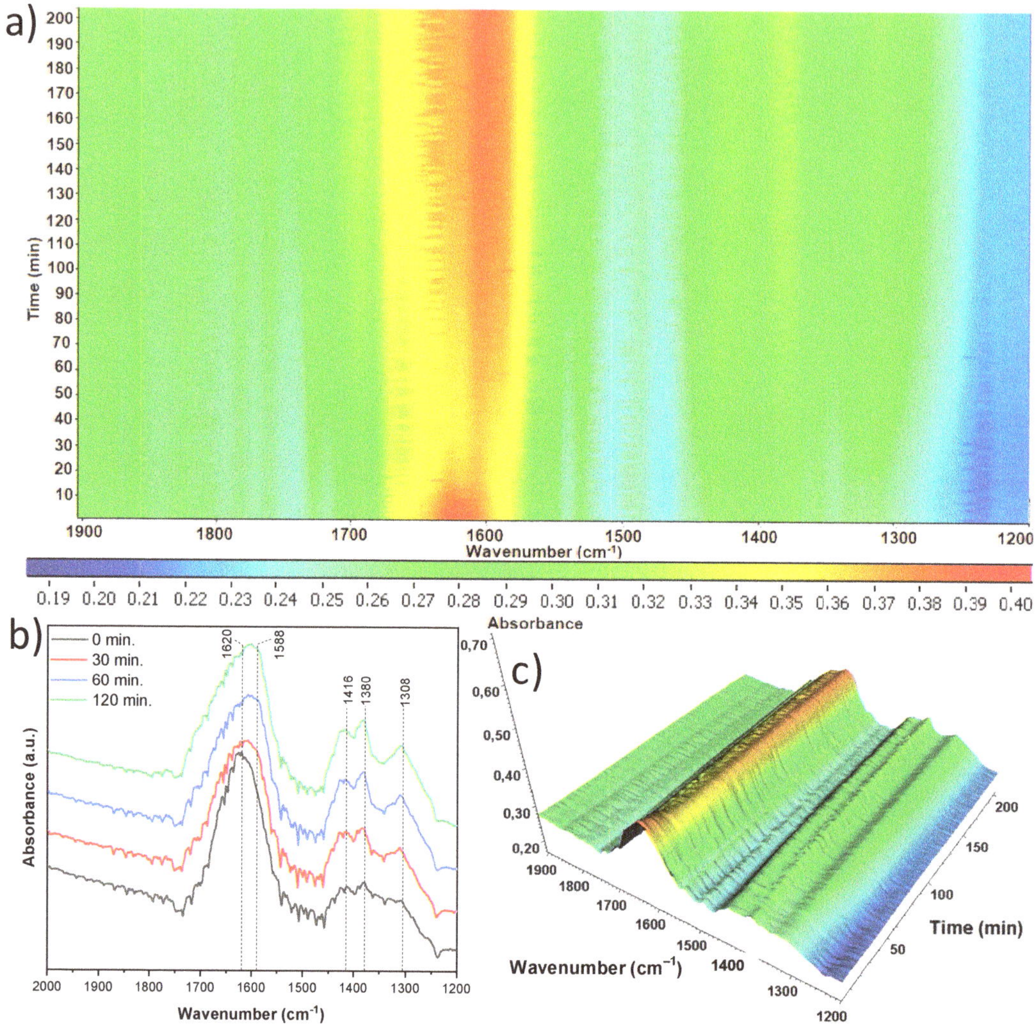

Figure 22. In situ FTIR of A/R-TNR-15 during the photocatalytic removal of acetaldehyde: (**a**) 2D mapping projection of the A = f(t, ν) relationship, (**b**) A = f(ν) graph after 0, 30, 60, and 120 min, (**c**) 3D mapping of the A = f(t, ν) relationship.

4. Conclusions

TNR@Ni-foam structures were prepared by the alkaline hydrothermal method, which consisted of two steps: (1) treatment of TiO_2 powder in an autoclave in 10 M NaOH at 150 °C for 24 h, then washing with 0.1 M HNO_3 and neutralization with deionized water; and (2) hydrothermal treatment in an autoclave at 180 °C for 24 h. In the first step of preparation, the layered titanate structure ($Na_2Ti_3O_7$) was formed. This obtained material was then washed with various quantities of HNO_3, which strongly affected the formation of TNR in the next step of hydrothermal treatment. It appeared that regardless of the titania precursor used (anatase or mixture of anatase and rutile), for 10 mL of HNO_3 solution used per 0.75 g of TiO_2, the highest crystalline structure of TNR was obtained. XRD measurements showed that these TNR structures consisted of anatase with small quantities of brookite and rutile. TNR@Ni-foam structures obtained from the P25 titania precursor contained more brookite and rutile than those prepared from anatase and had a structure of horizontally grown NR on Ni foam. In the case of the anatase precursor used, the hierarchical structures of titania nano-ribbons were observed on the Ni foam. These TNR structures were disrupted by an excess of HNO_3 used for titanate washing (15 mL per 0.75 g of TiO_2). XRD measurements showed that the crystallinity of TiO_2 decreased when an excess of HNO_3 was used during preparation. Additionally, in the case of the sample obtained from the P25 precursor and the HNO_3 washing of 15 mL, the $Ni(OH)_2$ phase was determined by the XRD pattern. In fact, the role of acid washing is the substitution of Na^+ to H^+ in layered $Na_2Ti_3O_7$ to initiate the process of TNR formation. After acid washing in the second step of hydrothermal treatment, there is a process of titanate dehydration and crystallization. However, when there is an excess of hydrogen ions in the solution, the protonation of TiO_2 takes place, and there is a possible increase in titania hydration. This effect was observed in the FTIR spectra observed for the prepared powdered titania nanostructures. Interestingly, the obtained TNR@Ni-foam structures from the anatase precursor did not contain any Na species (as indicated by XRF measurements), whereas those prepared from P25 had residue Na, which was lower for the higher quantity of HNO_3 used for titanate washing. The samples prepared from P25 with a small content of Na indicated higher photocatalytic activity toward acetaldehyde decomposition than those obtained from anatase-type TiO_2. One of the reasons is related to the phase composition; the presence of the brookite phase in combination with anatase is beneficial for acetaldehyde decomposition [42], and the samples obtained from P25 had a higher quantity of brookite. The other factor affecting the photocatalytic activity of these TNR@Ni-foam structures is related to the thickness of the titania layer and the acidity of the surface. P25 has a low density in comparison with KRONOClean®7050 and good dispersion properties, so can be abundantly attached to the Ni foam surface. The acidic surface of TiO_2 is detrimental to acetaldehyde decomposition [43]; moreover, strong adsorption of acetate species on certain titania centers upon acetaldehyde decomposition can deactivate the photocatalyst when the process is carried out under a continuous flow of reacted gases. It is assumed that TNR@Ni-foam structures prepared from P25 had a more alkaline surface than those obtained from the anatase precursor because of the presence of Na species and, for that reason, could be more active for acetaldehyde decomposition. TNR@Ni-foam structures obtained from the anatase precursor were more hydrophilic and revealed the desorption of water molecules from the surface upon photocatalytic process carried out at 100 °C, as was monitored by FTIR spectra (the band at 1620 cm^{-1}). Physically adsorbed water on the titania surface is detrimental to the photocatalytic process carried out in a gas phase because it limits the adsorption of organic pollutants on the titania surface and oxygen, which takes place during the formation of superoxide anion radicals. This was an additional reason for the weak activity of TNR@Ni-foam structures prepared from the anatase precursor. A thin layer of TiO_2 coating gives the possibility of acetaldehyde adsorption on the nickel foam surface. These adsorbed species do not proceed to photocatalytic decomposition. Generally, P25 appeared to be a good titania precursor for the preparation of TNR@Ni-foam composites, because of the stronger bound sodium species, which contributed to the formation of the layered $Na_2Ti_3O_7$ structure, which was an intermediate to TNR formulation. Moreover, mixed TiO_2 phases were crystallized with the dominant anatase phase, which was convenient due to its photocatalytic properties. It was shown that at the proposed conditions of TNR@Ni-foam preparation, $Ni(OH)_2$ species were formed, which can contribute to the enhancement of charge carrier separation due to their ability to reduce barriers for electron transfer. Therefore, the A/R-TNR-15 sample revealed high photocatalytic activity, although it had a smaller crystalline structure than A/R-TNR-10, but a thick titania coating.

Supplementary Materials: The following supporting information can be downloaded at: https://www.mdpi.com/article/10.3390/ma18050986/s1, Table S1. Phase composition summary of the studied samples determined from the Rietveld refinement. Figure S1. Rietveld refinement of XRD diffractogram of sample originated from anatase and before second hydrothermal step (A-TNR). Figure S2. Rietveld refinement of XRD diffractogram of sample originated from anatase and untreated by the nitric acid (A-TNR-0). Figure S3. Rietveld refinement of XRD diffractogram of sample originated from anatase and treated by 5 mL of nitric acid (A-TNR-5). Figure S4. Rietveld refinement of XRD diffractogram of sample originated from anatase and treated by 10 mL of nitric acid (A-TNR-10). Figure S5. Rietveld refinement of XRD diffractogram of sample originated from anatase and treated by 15 mL of nitric acid (A-TNR-15). Figure S6. Rietveld refinement of XRD diffractogram of sample originated from the mixture of anatase and rutile, before second hydrothermal step (A/R-TNR). Figure S7. Rietveld refinement of XRD diffractogram of sample originated from the mixture of anatase and rutile, untreated by the nitric acid (A/R-TNR-0). Figure S8. Rietveld refinement of XRD diffractogram of sample originated from the mixture of anatase and rutile, treated by 5 mL of nitric acid (A/R-TNR-5). Figure S9. Rietveld refinement of XRD diffractogram of sample originated from the mixture of anatase and rutile, treated by 10 mL of nitric acid (A/R-TNR-10). Figure S10. Rietveld refinement of XRD diffractogram of sample originated from the mixture of anatase and rutile, treated by 15 mL of nitric acid (A/R-TNR-15).

Author Contributions: Conceptualization, P.R. and B.T.; methodology, P.R., M.T. and B.T.; formal analysis, P.R., B.P., P.M. and M.T.; investigation, P.R., B.P., P.M., M.T. and B.T.; data curation, P.R. and P.M.; writing—original draft preparation, P.R. and B.T.; writing—review and editing, B.T.; visualization, P.R.; supervision, P.R. and B.T.; project administration, P.R.; funding acquisition, P.R. All authors have read and agreed to the published version of the manuscript.

Funding: This research was funded by the National Science Centre, grant number 2023/49/N/ST11/01737.

Institutional Review Board Statement: Not applicable.

Informed Consent Statement: Not applicable.

Data Availability Statement: All published data will be available in the open repository https://mostwiedzy.pl/ accessed on of 1 March 2025.

Acknowledgments: The authors would like to thank Marcin Sadłowski for XRD advice and Joanna Sreńscek-Nazzal for XRF analysis.

Conflicts of Interest: The authors declare no conflicts of interest.

References

1. Tryba, B.; Miądlicki, P.; Rychtowski, P.; Trzeciak, M.; Wróbel, R.J. The Superiority of TiO_2 Supported on Nickel Foam over Ni-Doped TiO_2 in the Photothermal Decomposition of Acetaldehyde. *Materials* **2023**, *16*, 5241. [CrossRef] [PubMed]
2. Trzeciak, M.; Miądlicki, P.; Tryba, B. Enhanced Degradation of Ethylene in Thermo-Photocatalytic Process Using TiO_2/Nickel Foam. *Materials* **2024**, *17*, 267. [CrossRef]
3. Honda, M.; Ochiai, T.; Listiani, P.; Yamaguchi, Y.; Ichikawa, Y. Low-Temperature Synthesis of Cu-Doped Anatase TiO_2 Nanostructures via Liquid Phase Deposition Method for Enhanced Photocatalysis. *Materials* **2023**, *16*, 639. [CrossRef] [PubMed]
4. Nair, V.; Muñoz-Batista, M.J.; Fernández-García, M.; Luque, R.; Colmenares, J.C. Thermo-Photocatalysis: Environmental and Energy Applications. *ChemSusChem* **2019**, *12*, 2098–2116. [CrossRef]
5. Mateo, D.; Cerrillo, J.L.; Durini, S.; Gascon, J. Fundamentals and Applications of Photo-Thermal Catalysis. *Chem. Soc. Rev.* **2021**, *50*, 2173–2210. [CrossRef] [PubMed]
6. Xing, C.; Yang, L.; Spadaro, M.C.; Zhang, Y.; Guardia, P.; Arbiol, J.; Liu, T.; Fan, X.; Fernández-García, M.; Llorca, J.; et al. Controllable Synthesis of Defective TiO_2 Nanorods for Efficient Hydrogen Production. *ACS Appl. Electron. Mater.* **2024**, *6*, 5833–5841. [CrossRef]
7. Ohno, T.; Higo, T.; Saito, H.; Yuajn, S.; Jin, Z.; Yang, Y.; Tsubota, T. Dependence of Photocatalytic Activity on Aspect Ratio of a Brookite TiO_2 Nanorod and Drastic Improvement in Visible Light Responsibility of a Brookite TiO_2 Nanorod by Site-Selective Modification of Fe^{3+} on Exposed Faces. *J. Mol. Catal. A Chem.* **2015**, *396*, 261–267. [CrossRef]

8. Dudziak, S.; Kowalkińska, M.; Karczewski, J.; Pisarek, M.; Gouveia, J.D.; Gomes, J.R.B.; Zielińska-Jurek, A. Surface and Trapping Energies as Predictors for the Photocatalytic Degradation of Aromatic Organic Pollutants. *J. Phys. Chem. C* **2022**, *126*, 14859–14877. [CrossRef]
9. Liu, Z.; Zhang, X.; Nishimoto, S.; Murakami, T.; Fujishima, A. Efficient Photocatalytic Degradation of Gaseous Acetaldehyde by Highly Ordered TiO_2 Nanotube Arrays. *Environ. Sci. Technol.* **2008**, *42*, 8547–8551. [CrossRef] [PubMed]
10. Gupta, T.; Samriti; Cho, J.; Prakash, J. Hydrothermal Synthesis of TiO_2 Nanorods: Formation Chemistry, Growth Mechanism, and Tailoring of Surface Properties for Photocatalytic Activities. *Mater. Today Chem.* **2021**, *20*, 100428. [CrossRef]
11. Ouda, A.A.; Mohamad Alosfur, F.K.; Ridha, N.J.; Abud, S.H.; Umran, N.M.; Al-aaraji, H.H.; Madlool, R.A. Facile Method to Synthesis of Anatase TiO_2 Nanorods. *J. Phys. Conf. Ser.* **2018**, *1032*, 012038. [CrossRef]
12. Singh, K.; Chen, C.-H.; Tai, L.-C.; Pang, S.-T.; Pan, T.-M. Enhancing pH Sensing Capabilities Through Hydroxylated Surface Groups on RuO *x* Flexible EGFET Sensor. *IEEE Sens. J.* **2024**, *24*, 13863–13869. [CrossRef]
13. Joo, J.; Kwon, S.G.; Yu, T.; Cho, M.; Lee, J.; Yoon, J.; Hyeon, T. Large-Scale Synthesis of TiO_2 Nanorods via Nonhydrolytic Sol−Gel Ester Elimination Reaction and Their Application to Photocatalytic Inactivation of *E. coli*. *J. Phys. Chem. B* **2005**, *109*, 15297–15302. [CrossRef] [PubMed]
14. Kolen'ko, Y.V.; Kovnir, K.A.; Gavrilov, A.I.; Garshev, A.V.; Frantti, J.; Lebedev, O.I.; Churagulov, B.R.; Van Tendeloo, G.; Yoshimura, M. Hydrothermal Synthesis and Characterization of Nanorods of Various Titanates and Titanium Dioxide. *J. Phys. Chem. B* **2006**, *110*, 4030–4038. [CrossRef]
15. Li, T.; Rui, Y.; Zhang, X.; Shi, J.; Wang, X.; Wang, Y.; Yang, J.; Zhang, Q. Anatase TiO_2 Nanorod Arrays as High-Performance Electron Transport Layers for Perovskite Solar Cells. *J. Alloys Compd.* **2020**, *849*, 156629. [CrossRef]
16. Li, M.; Wei, M.; Xie, X.; Dong, Q.; Gao, Q.; Cai, X.; Zhang, S.; Peng, F.; Fang, Y.; Yang, S. Ni Foam Supported TiO_2 Nanorod Arrays with CdS Branches: Type II and Z-Scheme Mechanisms Coexisted Monolithic Catalyst Film for Improved Photocatalytic H_2 Production. *Sol. RRL* **2022**, *6*, 2200187. [CrossRef]
17. Mohammed, A.M.; Becerikli, A.E.; Ristig, S.; Steinfeldt, N.; Strunk, J. Ordered Mesoporous TiO_2: The Effect of Structure, Residual Template and Metal Doping on Photocatalytic Activity. *Catalysts* **2023**, *13*, 895. [CrossRef]
18. Nateq, M.H.; Ceccato, R. Sol-Gel Synthesis of TiO_2 Nanocrystalline Particles with Enhanced Surface Area through the Reverse Micelle Approach. *Adv. Mater. Sci. Eng.* **2019**, *2019*, 1567824. [CrossRef]
19. Maxim, F.; Ferreira, P.; Vilarinho, P.M. Influence of the Neutralization Process on the Preparation of Titanate Nanotubes by Hydrothermal Synthesis. *J. Porous Mater.* **2011**, *18*, 37–45. [CrossRef]
20. Kasuga, T.; Hiramatsu, M.; Hoson, A.; Sekino, T.; Niihara, K. Formation of Titanium Oxide Nanotube. *Langmuir* **1998**, *14*, 3160–3163. [CrossRef]
21. Xiong, J.; He, L. Influence of Na^+ Content on the Structure and Morphology of TiO_2 Nanoparticles Prepared by Hydrothermal Transformation of Alkaline Titanate Nanotubes. *J. Exp. Nanosci.* **2017**, *12*, 384–393. [CrossRef]
22. Issa Hamoud, H.; Wolski, L.; Pankin, I.; Bañares, M.A.; Daturi, M.; El-Roz, M. In Situ and Operando Spectroscopies in Photocatalysis: Powerful Techniques for a Better Understanding of the Performance and the Reaction Mechanism. *Top. Curr. Chem.* **2022**, *380*, 37. [CrossRef]
23. Topsøe, N.-Y. In Situ FTIR: A Versatile Tool for the Study of Industrial Catalysts. *Catal. Today* **2006**, *113*, 58–64. [CrossRef]
24. Zhang, H.; Li, Y.; Wang, J.; Wu, N.; Sheng, H.; Chen, C.; Zhao, J. An Unprecedent Hydride Transfer Pathway for Selective Photocatalytic Reduction of CO_2 to Formic Acid on TiO_2. *Appl. Catal. B Environ.* **2021**, *284*, 119692. [CrossRef]
25. Subbotina, I.R.; Barsukov, D.V. Direct Evidence of the Key Role of UV-Formed Peroxide Species in Photocatalytic Gas–Solid Oxidation in Air on Anatase TiO_2 Particles. *Phys. Chem. Chem. Phys.* **2020**, *22*, 2200–2211. [CrossRef]
26. "KRONOClean 7050 Datasheet". DS2190EN/615EN KRONOS INTERNATIONAL Inc. 2015. Available online: https://www.ulprospector.com/documents/1368489.pdf?bs=28240&b=561007&st=20&r=eu&ind=inks (accessed on 15 January 2025).
27. "Product Information AEROXIDE TIO2 P 25" Evonic Industries 2016. Available online: https://mychem.ir/uploads/tds/25441.pdf (accessed on 24 January 2025).
28. Gates-Rector, S.; Blanton, T. The Powder Diffraction File: A Quality Materials Characterization Database. *Powder Diffr.* **2019**, *34*, 352–360. [CrossRef]
29. López Zavala, M.Á.; Lozano Morales, S.A.; Ávila-Santos, M. Synthesis of Stable TiO_2 Nanotubes: Effect of Hydrothermal Treatment, Acid Washing and Annealing Temperature. *Heliyon* **2017**, *3*, e00456. [CrossRef] [PubMed]
30. Zhao, J.-J.; Zhu, H.-R.; Huang, C.-J.; Yin, M.-H.; Li, G.-R. Recent Advances of Ethylene Glycol Oxidation Reaction: Catalytic Mechanism, Catalyst Design and Applications. *J. Mater. Chem. A* **2025**, *13*, 3236–3272. [CrossRef]
31. Wang, D.; Xu, R.; Wang, X.; Li, Y. NiO Nanorings and Their Unexpected Catalytic Property for CO Oxidation. *Nanotechnology* **2006**, *17*, 979–983. [CrossRef]
32. Wan, K.; Luo, J.; Zhang, X.; Subramanian, P.; Fransaer, J. In-Situ Formation of Ni (Oxy)Hydroxide on Ni Foam as an Efficient Electrocatalyst for Oxygen Evolution Reaction. *Int. J. Hydrogen Energy* **2020**, *45*, 8490–8496. [CrossRef]

33. Manjunatha, C.; Rastogi, C.K.; Manmadha Rao, B.; Girish Kumar, S.; Varun, S.; Raitani, K.; Maurya, G.; Karthik, B.; Swathi, C.; Sadrzadeh, M.; et al. Advances in Hierarchical Inorganic Nanostructures for Efficient Solar Energy Harvesting Systems. *ChemSusChem* **2024**, *17*, e202301755. [CrossRef] [PubMed]
34. Kim, G.-S.; Ansari, S.G.; Seo, H.-K.; Kim, Y.-S.; Shin, H.-S. Effect of Annealing Temperature on Structural and Bonded States of Titanate Nanotube Films. *J. Appl. Phys.* **2007**, *101*, 024314. [CrossRef]
35. Yi, X.; Sun, H.; Robertson, N.; Kirk, C. Nanoflower $Ni(OH)_2$ Grown in Situ on Ni Foam for High-Performance Supercapacitor Electrode Materials. *Sustain. Energy Fuels* **2021**, *5*, 5236–5246. [CrossRef]
36. Maluangnont, T.; Sangtawesin, T.; Pulphol, P.; Khamman, O.; Reunchan, P.; Gotoh, K.; Vittayakorn, N. Sodium-Poor, Hydroxyl-Rich, Defective $Na_2Ti_3O_7$ Prepared by γ-Irradiation and Its Enhanced Proton Conductivity. *Inorg. Chem.* **2024**, *63*, 18073–18082. [CrossRef] [PubMed]
37. Rotzinger, F.P.; Kesselman-Truttmann, J.M.; Hug, S.J.; Shklover, V.; Grätzel, M. Structure and Vibrational Spectrum of Formate and Acetate Adsorbed from Aqueous Solution onto the TiO_2 Rutile (110) Surface. *J. Phys. Chem. B* **2004**, *108*, 5004–5017. [CrossRef]
38. Mattsson, A.; Österlund, L. Adsorption and Photoinduced Decomposition of Acetone and Acetic Acid on Anatase, Brookite, and Rutile TiO_2 Nanoparticles. *J. Phys. Chem. C* **2010**, *114*, 14121–14132. [CrossRef]
39. Araña, J.; Garriga I Cabo, C.; Doña-Rodríguez, J.M.; González-Díaz, O.; Herrera-Melián, J.A.; Pérez-Peña, J. FTIR Study of Formic Acid Interaction with TiO_2 and TiO_2 Doped with Pd and Cu in Photocatalytic Processes. *Appl. Surf. Sci.* **2004**, *239*, 60–71. [CrossRef]
40. Rychtowski, P.; Tryba, B.; Skrzypska, A.; Felczak, P.; Sreńscek-Nazzal, J.; Wróbel, R.J.; Nishiguchi, H.; Toyoda, M. Role of the Hydroxyl Groups Coordinated to TiO_2 Surface on the Photocatalytic Decomposition of Ethylene at Different Ambient Conditions. *Catalysts* **2022**, *12*, 386. [CrossRef]
41. Mino, L.; Spoto, G.; Ferrari, A.M. CO_2 Capture by TiO_2 Anatase Surfaces: A Combined DFT and FTIR Study. *J. Phys. Chem. C* **2014**, *118*, 25016–25026. [CrossRef]
42. Tryba, B.; Jafari, S.; Sillanpää, M.; Nitta, A.; Ohtani, B.; Morawski, A.W. Influence of TiO_2 Structure on Its Photocatalytic Activity towards Acetaldehyde Decomposition. *Appl. Surf. Sci.* **2019**, *470*, 376–385. [CrossRef]
43. Tryba, B.; Wozniak, M.; Zolnierkiewicz, G.; Guskos, N.; Morawski, A.; Colbeau-Justin, C.; Wrobel, R.; Nitta, A.; Ohtani, B. Influence of an Electronic Structure of N-TiO_2 on Its Photocatalytic Activity towards Decomposition of Acetaldehyde under UV and Fluorescent Lamps Irradiation. *Catalysts* **2018**, *8*, 85. [CrossRef]

materials

MDPI

Article

WO_3/BiOBr S-Scheme Heterojunction Photocatalyst for Enhanced Photocatalytic CO_2 Reduction

Chen Li [1], Xingyu Lu [1], Liuyun Chen [1], Xinling Xie [1], Zuzeng Qin [1], Hongbing Ji [1,2] and Tongming Su [1,*]

[1] Guangxi Key Laboratory of Petrochemical Resource Processing and Process Intensification Technology, School of Chemistry and Chemical Engineering, Guangxi University, Nanning 530004, China; 15549366089@163.com (C.L.); 18736031369@163.com (X.L.); 18811313166@163.com (L.C.); xiexinling@126.com (X.X.); qinzuzeng@gxu.edu.cn (Z.Q.); jihb@mail.sysu.edu.cn (H.J.)

[2] State Key Laboratory Breeding Base of Green-Chemical Synthesis Technology, Institute of Green Petroleum Processing and Light Hydrocarbon Conversion, College of Chemical Engineering, Zhejiang University of Technology, Hangzhou 310014, China

* Correspondence: sutm@gxu.edu.cn

Abstract: The photocatalytic CO_2 reduction strategy driven by visible light is a practical way to solve the energy crisis. However, limited by the fast recombination of photogenerated electrons and holes in photocatalysts, photocatalytic efficiency is still low. Herein, a WO_3/BiOBr S-scheme heterojunction was formed by combining WO_3 with BiOBr, which facilitated the transfer and separation of photoinduced electrons and holes and enhanced the photocatalytic CO_2 reaction. The optimized WO_3/BiOBr heterostructures exhibited best activity for photocatalytic CO_2 reduction without any sacrificial reagents, and the CO yield reached 17.14 μmol g^{-1} after reaction for 4 h, which was 1.56 times greater than that of BiOBr. The photocatalytic stability of WO_3/BiOBr was also improved.

Keywords: WO_3; BiOBr; heterojunction; photocatalytic; CO_2

1. Introduction

The massive use of fossil fuel energy increases the amount of greenhouse gases, such as CO_2, in the air, causing greenhouse effects, melting glaciers, and energy shortages [1–3]. Converting CO_2 into chemical raw materials is a practical way to solve the environmental and energy crisis [4]. Among many technologies, photocatalytic reduction of CO_2 to a chemical feedstock is a feasible way to achieve a green sustainable carbon cycle [5]. However, the low visible light absorption efficiency and recombination of the photoinduced electron-hole pairs in photocatalysts severely limit the efficiency of photocatalytic CO_2 conversion [6,7]. Therefore, it is very significant to improve the photogenerated charge carriers separation in photocatalysts.

The construction of heterojunctions is considered to be beneficial for improving photocatalytic efficiency. Among various heterojunctions, S-scheme heterojunctions are widely studied due to their stronger redox ability [8]. In general, the S-scheme heterojunction includes an oxidation catalyst (OP) and a reduction catalyst (RP). When the RP is combined with the OP, the electrons will flow from the RP to the OP due to the higher Fermi level of the RP, resulting in the formation of an internal electric field at the interface. Under light irradiation, the electrons in the conduction band (CB) of OP combine with the holes in the valence band (VB) of RP due to the presence of an internal electric field, and the holes and electrons in the OP and RP can be used for oxidation or reduction reactions, respectively, thus promoting the separation of photogenerated electron-holes [9–11]. Therefore, the design of S-scheme heterojunctions with high charge carriers separation efficiency and strong redox capacity is a promising approach for improving photocatalytic efficiency.

Among many photocatalysts, bismuth-based photocatalysts have good photoelectric properties and are widely used in the fields of organic matter photodegradation and CO_2 photoreduction [12,13]. Layered bismuth halides (BiOX, X = Cl, Br, and I) have been extensively studied due to their high photocatalytic performance [14]. BiOX consists of a bismuth–oxygen layer $(Bi_2O_2)^{2+}$ and two alternatingly arranged Br layers. Such a layered structure can form an internal electric field to shorten the transmission distance of the photogenerated electrons, which is very favorable for photocatalytic reactions [15–17]. Notably, BiOBr has aroused wide concern due to its narrow band gap and suitable band structure. However, the rapid recombination of photoinduced charge carriers in BiOBr limits its application in photocatalytic CO_2 reduction [18]. To improve the photocatalytic efficiency of BiOBr, numerous ways have been proposed, such as regulating oxygen vacancies [19], metal doping [20], and hybridization with various semiconductors [21–23]. The formation of an S-scheme heterojunction can be a feasible way to improve the photocatalytic efficiency of BiOBr [24–26]. The use of a suitable oxidized semiconductor combined with BiOBr to construct an S-scheme heterojunction can shorten the transfer distance of charge carriers, prolong the carrier lifetime, and enhance the redox ability. For instance, WO_3 is widely used as the oxidizing photocatalyst due to its advantages of low cost, excellent photoelectric performance, and strong oxidation ability [27,28]. Moreover, WO_3 has a suitable band structure and can form S-scheme heterojunctions with BiOBr, which is expected to enhance the performance of photocatalytic CO_2 reduction.

In this work, WO_3/BiOBr heterojunctions were prepared and used for photocatalytic CO_2 reduction. Under visible light ($\lambda \geq 400$ nm), the optimized WO_3/BiOBr exhibited the best CO yield of 17.14 μmol g^{-1} irradiation for 4 h without the addition of sacrificial agents, which was 1.56 times higher than that of BiOBr. Combining the activity data and characterization results, the S-scheme charge transfer mechanism was demonstrated for WO_3/BiOBr, which enhances the rapid transfer and separation of photoinduced charge carriers.

2. Experimental

2.1. Synthesis of the Photocatalyst

2.1.1. Materials

$Bi(NO_3)_3 \cdot 5H_2O$ was obtained from Guangdong Guanghua Technology Co., Ltd. (Shantou, China). Polyvinylpyrrolidone (PVP, K-30) was purchased from Shanghai McLean Biochemical Technology Co., Ltd. (Shanghai, China). KBr was purchased from Tianjin Guangfu Technology Development Co., Ltd. (Tianjin, China). WO_3 was purchased from Tianjin Bodi Chemical Co., Ltd. (Tianjin, China). All the chemical agents involved in this work were analytically pure and were used without further purification.

2.1.2. Preparation of BiOBr

First, 5 mmol of $Bi(NO_3)_3 \cdot 5H_2O$ and 0.2 g of PVP were mixed in 40 mL deionized (DI) water and stirred for 60 min. At the same time, 5 mmol of KBr was dissolved in 40 mL of DI water and stirred for 60 min. After that, the above solution was mixed and stirred together, and the pH was adjusted to 6. The mixture was then stirred for 1 h and heated at 160 °C for 12 h. After cooling down, the mixture was washed with DI water several times and dried at 60 °C for 12 h in a vacuum drying oven.

2.1.3. Preparation of WO_3/BiOBr

First, 0.2 g of BiOBr was dispersed in 40 mL DI water for 30 min by ultrasonication, and then a certain amount of WO_3 (3, 5, 10 wt%) was added to the above suspension and stirred at 60 °C for 10 h by water bath. The sample was then washed with DI water and dried by vacuum drying at 60 °C for 12 h. The samples with different mass ratios were named $3WO_3$/BiOBr, $5WO_3$/BiOBr, and $10WO_3$/BiOBr.

2.2. Photocatalytic CO_2 Reduction

First, 30 mg photocatalyst was dispersed evenly in 1.5 mL DI water by ultrasonication, and then the photocatalyst was evenly spread on a quartz sheet and dried at 60 °C in a vacuum drying oven. For the photocatalytic CO_2 reduction, the quartz sheet with the photocatalyst was placed at the bottom of the reactor. After that, the wet CO_2 was injected into the reactor for 30 min, with a flow quantity of 40 mL min^{-1} to ensure that the air was completely removed. A 300 W xenon lamp (CEL-HXF300, Beijing China Education Au-light Co., Ltd., Beijing, China) was applied as the light source for the reaction and a 400 nm cutoff filter was used to filter out light below 400 nm. The light intensity of the light source was determined to be 178 mW cm^{-2} by using an optical power meter (CENP2000, Beijing China Education Au-light Co., Ltd., Beijing, China). The reaction temperature was maintained at 25 °C by using circulating cooling water. Every hour, 0.5 mL of gas product was detected on a GC-2030 gas chromatograph (Shimadzu, Kyoto, Japan).

2.3. Characterization

The XRD patterns were obtained on a Bruker D8 X-ray diffractometer (BRUKER AXS GMBH, Karlsruhe, Germany). Fourier transform infrared spectra and in situ infrared spectra were recorded on a Bruker Tensor II infrared spectrometer (Bruker, Karlsruhe, Germany). SEM images were obtained on a ZEISS GeminiSEM 300 instrument (ZEISS, Oberkochen, Germany). TEM images and elemental mapping images were recorded on a JEOL JEM-F200 transmission electron microscope (JEOL, Tokyo, Japan). XPS spectra were performed on a Thermo Scientific K-Alpha instrument (Thermo Fisher Scientific, Waltham, MA, USA). Ultraviolet-visible diffuse reflectance spectra were analyzed by a TU-19 spectrophotometer (PERSEE ANALYTICS, Beijing, China). N_2 adsorption and desorption curves were obtained on a Tristar II physical adsorption instrument (Micromeritics, Norcross, GA, USA). Time-resolved fluorescence spectra (TRPL) were analyzed on an Edinburgh FLS 1000 spectrometer (Edinburgh Instruments, Livingston, UK).

2.4. Photoelectrochemical Measurements

A CHI 760E electrochemical workstation was used for the photoelectrochemical measurements (Shanghai Chenhua Instrument Co., Ltd., Shanghai, China). The test was carried out in a three-electrode system, with an Ag/AgCl electrode, a Pt electrode, and a photocatalyst electrode used as the reference electrode, the counter electrode, and the working electrode, respectively. In addition, 0.5 M Na_2SO_4 was used as the electrolyte. To prepare the working electrode, 20 mg of photocatalyst was added to 400 μL absolute ethanol and 20 μL of Nafion® solution (Shanghai McLean Biochemical Technology Co., Ltd, Shanghai, China), and the mixture was ultrasonicated at 25 °C for 2 h. Then, 20 μL of the mixed solution was uniformly dropped on FTO conductive glass with a central area of 1 cm^2 (2 cm × 2 cm) and the working electrode was dried naturally. The electrochemical impedance spectrum (EIS) was tested at an alternating amplitude of 5 mV and with a frequency from 0.01 to 1,000,000 Hz. The transient photocurrent response was tested with a 300 W xenon lamp (equipped with a 400 nm cutoff filter) as the light source. Mott–Schottky curves were obtained at frequencies of 1000, 1500, and 2000 Hz.

3. Results and Discussion

3.1. Structure and Morphology

The crystalline phase and composition of BiOBr, WO_3, and xWO_3/BiOBr were measured by XRD. From Figure 1a, the peaks of WO_3 are consistent with the triclinic phase WO_3 (PDF#20−1323) [29]. In the XRD patterns of BiOBr, the diffraction peaks at 10.9°, 25.1°, 31.7°, 32.2°, 46.2°, and 57.1° are ascribed to the (001), (101), (102), (110), (200), and (212) planes of the tetragonal phase BiOBr, respectively (PDF#09−0393) [30]. Notably, from Figures 1a and S1, enlarged regions from 20° to 30° in XRD patterns of xWO_3/BiOBr are shown in Figure S1, with the peaks at 23.1°, 23.6°, 24.4°, and 26.8° ascribed to the (002), (020), (200), and (120) planes of WO_3 (PDF#20−1323), and the peak at 25.1° is attributed to the (101) plane of BiOBr (PDF#09−0393). Both the peaks of WO_3 and BiOBr can be found in the XRD patterns of xWO_3/BiOBr, indicating that the xWO_3/BiOBr composites were successfully prepared.

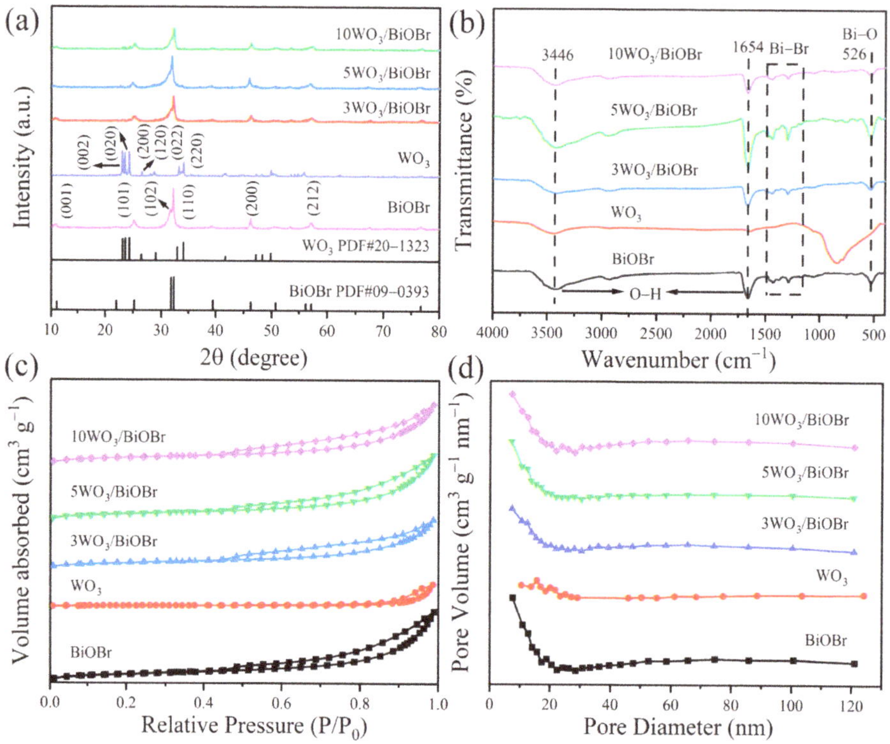

Figure 1. XRD patterns (**a**), FT–IR spectra (**b**), N_2 adsorption–desorption isotherms (**c**), and pore size distributions (**d**) of BiOBr, WO_3, and xWO_3/BiOBr.

FT–IR spectra were used to further study the structure of the obtained samples. Figure 1b shows that the peaks at 526 cm^{-1} are assigned to the tensile vibration of the Bi–O bond [31,32], and the peaks located in the range of 1000~1500 cm^{-1} region are ascribed to the asymmetric and symmetric vibrations of the Bi–Br bond [33,34]. The peaks of WO_3 at 826 cm^{-1} are attributed to the W–O bond [35]. In addition, the peaks at 1654 cm^{-1} are attributed to the bending vibrations of the adsorbed H_2O on the surface of BiOBr, and the peaks at 3436 cm^{-1} correspond to the stretching vibrations of the adsorbed OH groups [36]. However, due to the low content of WO_3, the peaks of WO_3 cannot be seen from the FT–IR spectra of xWO_3/BiOBr.

The specific surface area and average pore diameter of the samples were studied by N_2 adsorption and desorption. Figure 1c shows that all the samples have the type IV isotherms and type H3 hysteresis loops, indicating that the samples are mesoporous material [37]. The specific surface areas of BiOBr, WO_3, $3WO_3$/BiOBr, $5WO_3$/BiOBr, and $10WO_3$/BiOBr are 19.64, 3.54, 14.81, 17.96, and 17.15 $m^2\ g^{-1}$ (Table S1), respectively, indicating that the specific surface area of xWO_3/BiOBr composites is slightly lower than that of BiOBr but much higher than that of WO_3. Among the xWO_3/BiOBr composites, $5WO_3$/BiOBr exhibits the largest specific surface area and can provide more reaction sites for photocatalytic reactions [38].

Scanning electron microscopy (SEM) and transmission electron microscopy (TEM) were used to investigate the morphology and microstructure of the catalyst (Figure 2).

BiOBr shows the nanoflower structure assembled from many nanosheets, while WO_3 has a polyhedral structure. From the SEM images of $5WO_3$/BiOBr, we can see that the BiOBr nanoflower was dispersed and covered on the surface of WO_3, which is beneficial for increasing the contact interface between WO_3 and BiOBr. In addition, the SEM images of $3WO_3$/BiOBr and $10WO_3$/BiOBr are also shown in Figure S2. The microstructure of $5WO_3$/BiOBr was further investigated by HRTEM. Figure 2e and 2f are the enlarged regions of the white boxes in Figure 2d and 2e, respectively. As shown in Figure 2f, the lattice fringe spacing of 0.33 and 0.28 nm is attributed to the (120) planes of WO_3 [39] and (102) planes of BiOBr [40], respectively. Notably, a tight contact interface between WO_3 and BiOBr is clearly observed, indicating the formation of the WO_3/BiOBr heterojunction. Moreover, the lattice fringe of WO_3 can also be observed close to the lattice fringe of BiOBr, indicating the successful construction of the heterointerface between WO_3 and BiOBr. In addition, according to the EDS elemental mapping of $5WO_3$/BiOBr (Figure 2g–k), the Bi, O, Br, and W elements are evenly distributed in the $5WO_3$/BiOBr composite. According to the EDX spectrum (Figure S3), the Bi, Br, O, and W elements exist in $5WO_3$/BiOBr and the mass fraction and atomic fraction of W are 2.23% and 1.31%. These results indicate the successful synthesis of $5WO_3$/BiOBr composites.

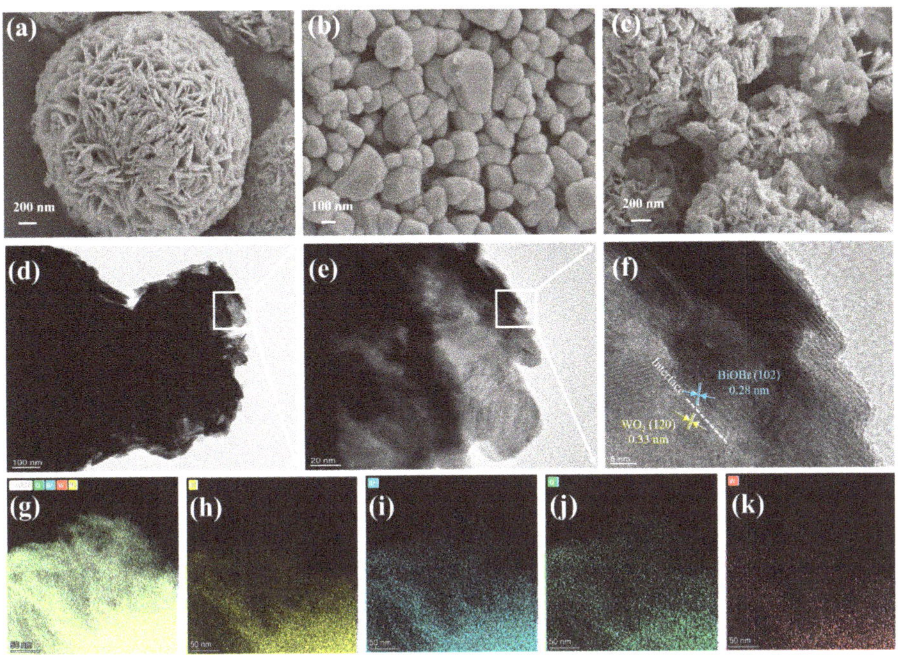

Figure 2. SEM images of BiOBr (**a**), WO_3 (**b**), and $5WO_3$/BiOBr (**c**). TEM and HRTEM images (**d–f**), HAADF image (**g**), and corresponding EDX mapping profiles of Bi (**h**), Br (**i**), O (**j**), and W (**k**) of $5WO_3$/BiOBr.

3.2. Surface Chemical State

The surface chemical state of the photocatalysts was investigated by XPS spectra. The XPS survey spectra of BiOBr, WO_3, and $5WO_3$/BiOBr display that Bi, O, Br, and W elements are present in $5WO_3$/BiOBr, which demonstrates the successful synthesis of $5WO_3$/BiOBr (Figure S4). The enlarged spectral regions from 0 eV to 100 eV in the XPS survey spectra of BiOBr, WO_3, and $5WO_3$/BiOBr are shown in Figure S4b. The peaks of Bi 5d, O 2s, W $4f_{7/2}$, and W $4f_{5/2}$ are found at 25.9, 25.9, 35.9, and 37.9 eV, respectively. Overtly, no WO_3 signal was found in enlarged spectral regions from 0 eV to 100 eV of BiOBr.

Figure 3 shows the XPS spectra of Bi 4f, Br 3d, O 1s, and W 4f. As can be seen from the XPS spectra of BiOBr (Figure 3a), the two peaks at 158.6 and 163.9 eV are ascribed to Bi $4f_{7/2}$ and Bi $4f_{5/2}$ for Bi^{3+}, respectively. Compared to BiOBr, the binding energies of Bi $4f_{7/2}$ and Bi $4f_{5/2}$ in $5WO_3$/BiOBr shifted to 158.8 and 164.1 eV, respectively [41,42]. From the Br 3d spectra of BiOBr and $5WO_3$/BiOBr (Figure 3b), two peaks at 68.8 and 67.7 eV are attributed to Br $3d_{3/2}$ and Br $3d_{5/2}$, respectively, indicating the presence of Br^- [43], while the binding energies of Br $3d_{3/2}$ and Br $3d_{5/2}$ in $5WO_3$/BiOBr shift to 69.0 and 67.9 eV, respectively. The shift of the Bi $4f_{5/2}$, Bi $4f_{7/2}$, Br $3d_{3/2}$, and Br $3d_{5/2}$ in $5WO_3$/BiOBr indicates that the electrons were migrated from BiOBr to WO_3 after BiOBr was contacted with WO_3.

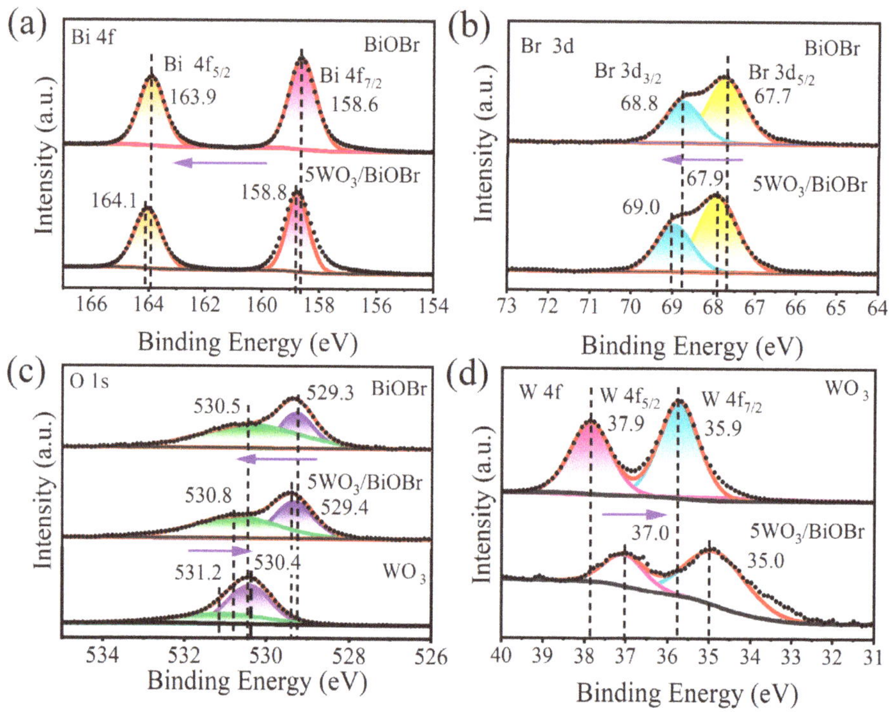

Figure 3. XPS spectra of Bi 4f (**a**), Br 3d (**b**), O 1 s (**c**), and W 4f (**d**) in BiOBr, WO_3, and $5WO_3$/BiOBr.

As shown in the O 1s spectra of BiOBr, WO_3, and $5WO_3$/BiOBr (Figure 3c), the binding energy of W–O is located at 530.4 and 531.2 eV for WO_3 [44], the binding energy of Bi–O is located at 529.3 and 530.5 eV for BiOBr [45], and the peaks shift to 529.4 and 530.8 eV, respectively, in the O 1s XPS spectrum of $5WO_3$/BiOBr, indicating that the chemical environment was changed after the formation of the WO_3/BiOBr heterojunction. In addition, two peaks located at 37.9 and 35.9 eV were found in the W 4f XPS spectrum of WO_3, which correspond to W $4f_{5/2}$ and W $4f_{7/2}$ for W^{6+} [46], respectively (Figure 3d). In the XPS spectra of $5WO_3$/BiOBr, W $4f_{5/2}$ and W $4f_{7/2}$ shifted to 37.0 and 35.0 eV, respectively, indicating that the electron cloud density of the W in WO_3 was increased. The above result indicates that the electrons in BiOBr can migrate to WO_3 through the interface after WO_3 was coupled with BiOBr, and a built-in electric field was constructed at the WO_3/BiOBr interface [47,48].

3.3. Light Absorption Capacity and Band Structure

The light absorption capacity and band structures of WO_3, BiOBr, and $5WO_3$/BiOBr were analyzed by UV–Vis DRS and UPS spectroscopy (Figure 4). As shown in Figure 4a,

WO_3 shows obvious visible light absorption. However, no significant changes were observed for the light absorption capacity of BiOBr after the introduction of WO_3. The band gaps of BiOBr and WO_3 were determined to be 2.92 and 2.59 eV by the Kubelka-Munk method, respectively (Figure 4b), which is in accordance with the results in the literature [49,50].

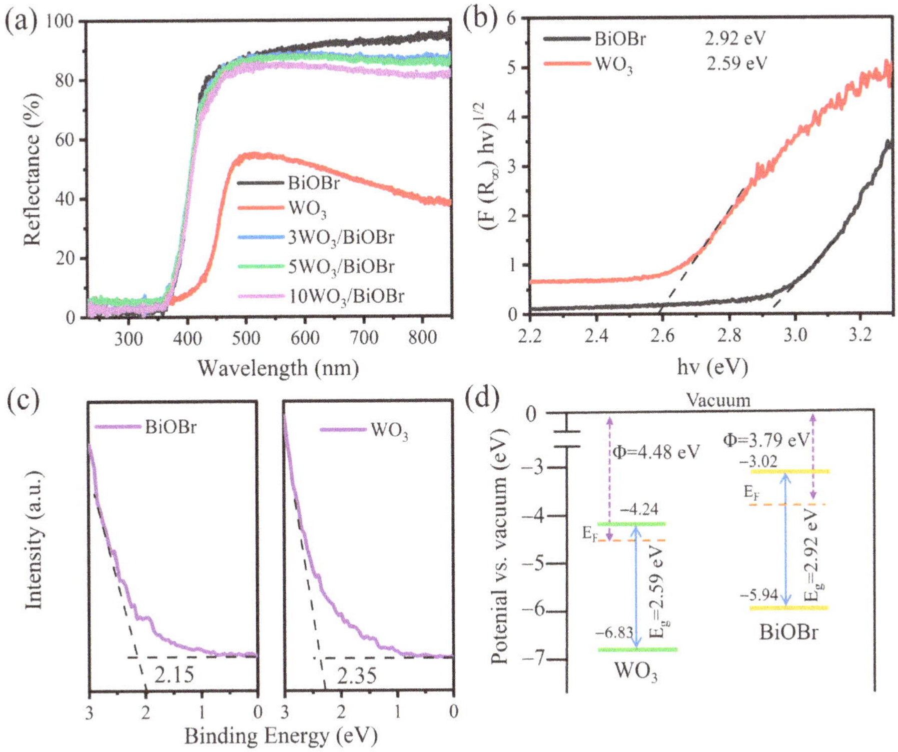

Figure 4. UV–Vis DRS of BiOBr, WO_3, and xWO_3/BiOBr (**a**), band gap of BiOBr and WO_3 (**b**), and UPS spectra (**c**) and band structure (**d**) of WO_3 and BiOBr.

UPS spectra were carried out to analyze the work function (Φ) and valence band (VB) of the catalyst (Figures 4c and S5). As shown in the UPS spectra of BiOBr and WO_3 (Figure S5), the secondary electron cutoff edges (E_{cutoff}) of BiOBr and WO_3 are 17.43 and 16.74 eV, respectively. As a result, the work functions of BiOBr and WO_3 were calculated to be 3.79 and 4.48 eV, respectively, according to Equation S1 [51]. In addition, the value of the work function is equal to the difference value between the vacuum level (E_v) and the Fermi level (E_f). Herein, the value of the vacuum level is identified as 0 eV [52], and the Fermi levels of BiOBr and WO_3 were calculated to be −3.79 and −4.48 eV relative to the vacuum level. In addition, as shown in Figure 4c, the valence band (VB) values of BiOBr and WO_3 are 2.15 and 2.35 eV relative to the Fermi level, respectively. Therefore, the VB values of BiOBr and WO_3 were calculated to be −5.94 and −6.83 eV (vs. vacuum), respectively [53]. Therefore, the conduction band (CB) values of BiOBr and WO_3 were determined to be −3.02 and −4.24 eV (vs. vacuum), respectively [54]. As shown in the band structures of BiOBr and WO_3 (Figure 4d), the electrons in BiOBr will transfer to WO_3 due to the higher work function of WO_3, and a built-in electric field can be formed at the WO_3/BiOBr interface, which is in favor of enhancing the separation of photoinduced electron-hole pairs.

In addition, Mott–Schottky plots were obtained to analyze the flat-band potentials (E_{fb}) of the samples. From Figure S6, the positive slopes of the Mott–Schottky plots indicate that both BiOBr and WO_3 are n-type semiconductors [55]. The E_{fb} values of BiOBr and WO_3 were determined to be −0.45 and −0.26 V (vs. Ag/AgCl, pH = 7), respectively. According to Equation (S2) [56], the E_{fb} values of BiOBr and WO_3 were calculated to be 0.16 and 0.35 V (vs. NHE, pH = 0), respectively. In general, for n-type semiconduction materials, the flat band potential is approximately equal to the Fermi level [57]. Therefore, the E_f values of BiOBr and WO_3 were 0.16 and 0.35 V (vs. NHE, pH = 0), indicating that the Fermi level of WO_3 is lower than that of BiOBr. This confirms that the electrons in BiOBr can migrate to WO_3 due to the higher Fermi level of BiOBr, which is in accordance with the UPS results.

3.4. Charge Transfer and Separation

The charge transfer kinetics of BiOBr and $5WO_3/BiOBr$ were analyzed by time-resolved photoluminescence (TRPL) spectra (Figure 5a). In general, the carrier lifetime is estimated by the photoluminescent decay time, and the average fluorescence lifetime can be calculated by Equation (1) [58,59], where A_1 and A_2 are pre-exponential factors and τ_1 and τ_2 represent the lifetimes of radiative and nonradiative transitions, respectively. Figure 5a shows that the average fluorescence lifetimes of BiOBr and $5WO_3/BiOBr$ are 0.86 ns and 1.98 ns, respectively. The prolongation of the fluorescence lifetime in $5WO_3/BiOBr$ indicates that the separation efficiency of photoinduced electrons and holes was enhanced after the $WO_3/BiOBr$ interface was formed [60].

$$\tau_a = \frac{A_1\tau_1^2 + A_2\tau_2^2}{A_1\tau_1 + A_2\tau_2} \quad (1)$$

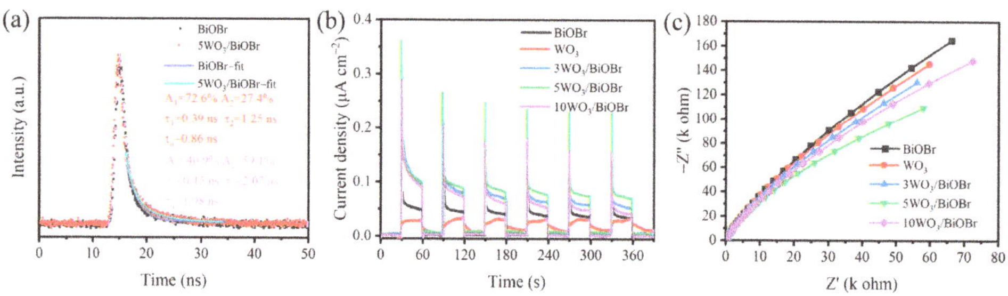

Figure 5. TRPL spectra of BiOBr and $5WO_3/BiOBr$ (**a**), transient photocurrent density (**b**), and EIS Nyqui st plots (**c**) of BiOBr, WO_3, and $xWO_3/BiOBr$.

The transient photocurrent densities were also investigated under visible light ($\lambda \geq 400$ nm) to analyze the transfer and separation of the photogenerated charge carriers. As displayed in Figure 5b, the transient photocurrent density of $5WO_3/BiOBr$ is higher than that of other samples, indicating that the charge separation efficiency in $5WO_3/BiOBr$ is higher. In addition, the EIS Nyquist plots of BiOBr, WO_3, and $xWO_3/BiOBr$ were obtained to explore the transfer resistance of the electrons. In general, the arc radius of the Nyquist curve is proportional to the impedance of the photocatalyst [61]. As shown in Figure 5c, the radius of $5WO_3/BiOBr$ is smaller than that of WO_3, BiOBr, $3WO_3/BiOBr$, and $10WO_3/BiOBr$, indicating that the charge transfer resistance of $5WO_3/BiOBr$ was reduced after BiOBr coupling with WO_3. Among them, $5WO_3/BiOBr$ exhibited the highest photocurrent density and the smallest impedance arc radius, indicating the most efficient separation of photogenerated electron-hole pairs and the fastest photogenerated charge transfer on $5WO_3/BiOBr$. Based on the above discussion, the formation of the $WO_3/BiOBr$ heterojunction is beneficial for improving the charge separation efficiency and improving the photocatalytic performance.

3.5. Photocatalytic CO_2 Reduction

The photocatalytic CO_2 reduction activity test over BiOBr and xWO_3/BiOBr was carried out under visible light (Figure 6). Figure 6a,b show that the photocatalytic CO_2 reduction activity of xWO_3/BiOBr was greater than BiOBr. Moreover, the CO production over xWO_3/BiOBr first increased and then decreased with the increase of the WO_3 amount. Among them, 5WO_3/BiOBr exhibited the best CO production of 17.14 μmol g^{-1} after 4 h reaction, which is 1.56 times greater than that of BiOBr (10.96 μmol g^{-1}). The higher photocatalytic performance of 5WO_3/BiOBr can be ascribed to the enhanced separation of photogenerated charge carriers at the WO_3/BiOBr interface. To demonstrate the advantages of 5WO_3/BiOBr, the photocatalytic CO_2 reduction performance of the BiOBr-based photocatalysts are listed in Table S2. Compared with the other photocatalysts shown in Table S2, 5WO_3/BiOBr presented satisfactory photocatalytic CO_2 reduction activity. Among them, 5WO_3/BiOBr exhibited the highest photocurrent density and the smallest impedance arc radius, indicating the most efficient separation of photogenerated electron-hole pairs and the fastest photogenerated charge transfer on 5WO_3/BiOBr. Therefore, the CO production rate of 5WO_3/BiOBr is higher than that of 3WO_3/BiOBr and 10WO_3/BiOBr. In addition, when the amount of WO_3 is 3 wt%, a small number of heterojunctions cannot effectively promote the separation of photogenerated electrons and holes. However, when the amount of WO_3 is 5 wt%, excessive WO_3 will limit the light absorption capacity of BiOBr and cover the active site of CO_2 reduction of BiOBr.

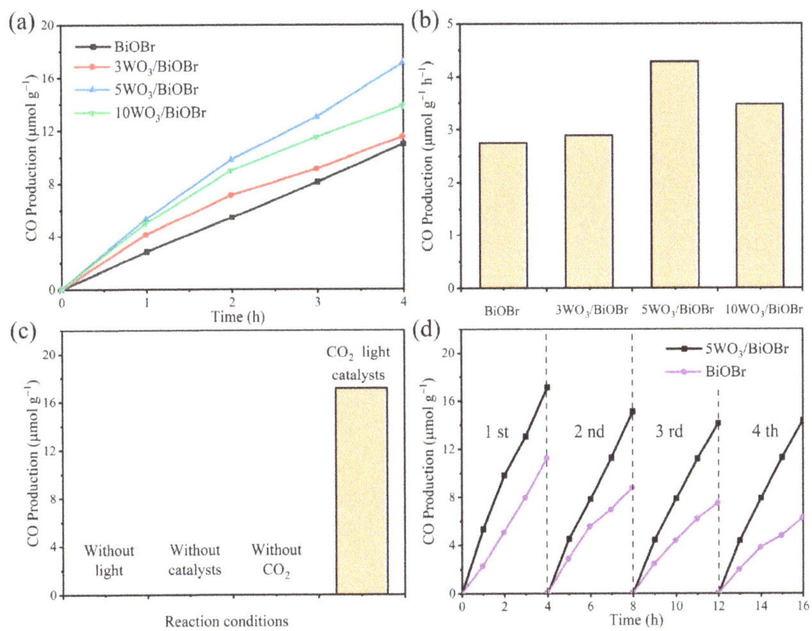

Figure 6. Time course of photocatalytic CO_2 reduction over BiOBr and xWO_3/BiOBr (**a**,**b**), photocatalytic CO_2 reduction over 5WO_3/BiOBr under different conditions (**c**), and cycle test of BiOBr and 5WO_3/BiOBr photocatalytic CO_2 reduction to CO (**d**).

Controlled experiments were conducted to confirm the influence factors and C source of the product. From Figure 6c, when the photocatalytic reaction was carried out without light, a catalyst, or CO_2, no CO product was detected, indicating that light, catalyst, and CO_2 are the necessary conditions for photocatalytic CO_2 reduction reaction. The stability of BiOBr and 5WO_3/BiOBr was also studied. As shown in Figure 6d, after four cycles, the CO production over BiOBr and 5WO_3/BiOBr decreased by 44.1% and 16.2%, respectively, indicating that the combination of WO_3 with BiOBr can improve the stability of BiOBr during the photocatalytic reaction process. The SEM, XRD, and FT–IR of $5WO_3$/BiOBr were carried out to further analyze the stability of $5WO_3$/BiOBr (Figures S7 and S8). Figure S7 shows that no obvious change can be found in the XRD patterns and FT–IR spectra of $5WO_3$/BiOBr before and after the reaction, indicating that the structure of the catalyst was stable after the photocatalytic reaction. In addition, the morphology of $5WO_3$/BiOBr did not change before or after the reaction (Figure S8). The above results show that $5WO_3$/BiOBr is stable for photocatalytic CO_2 reduction.

3.6. In Situ FTIR Spectra

In situ DRIFTS spectra were used to investigate the photocatalytic CO_2 reduction reaction over $5WO_3$/BiOBr (Figure 7). Figure 7a,b show the in situ DRIFTS spectra of CO_2 and H_2O adsorption on BiOBr and $5WO_3$/BiOBr in the dark. The peaks at 1663 cm^{-1} and 1654 cm^{-1} are the signals of the water adsorbed on the catalyst surface [62]. The peaks at 1267, 1445, and 1466 cm^{-1} are ascribed to bicarbonate (HCO_3^-) [63], the peaks at 1296 and 1312 cm^{-1} are attributed to the monolithic carbonate group (m-CO_3^{2-}) [64], and the peaks at 1267 and 1363 cm^{-1} are assigned to bidentate carbonate (b-CO_3^{2-}) [65]. In addition, the peak at 1701 cm^{-1} was ascribed to $COOH^-$. In general, $COOH^-$ is the core intermediate for the generation of CO and CH_4, and its formation time is a critical step for photocatalytic CO_2 reduction to CO [66].

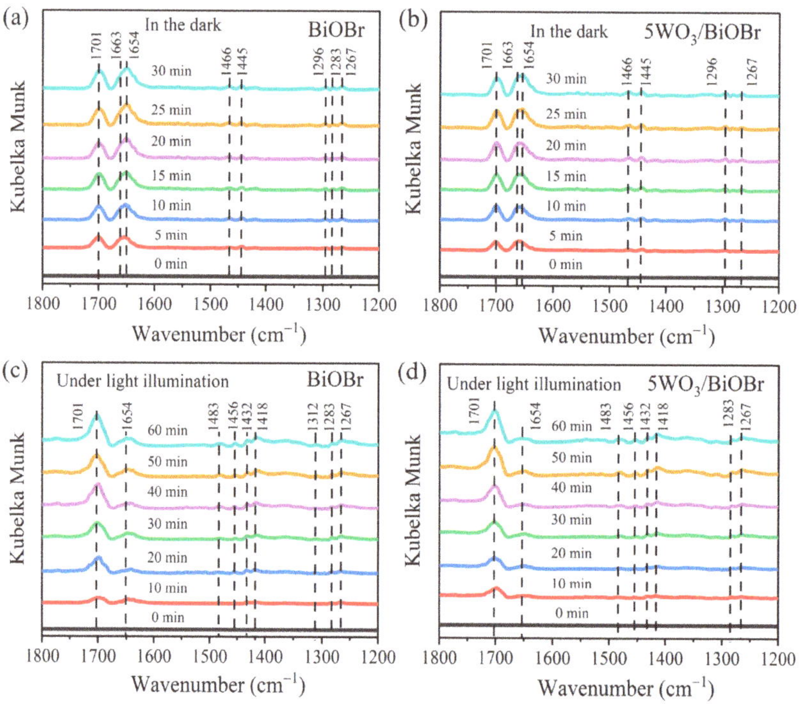

Figure 7. In situ DRIFTS spectra of CO_2 and H_2O adsorption on BiOBr (**a**) and $5WO_3$/BiOBr (**b**) in the dark. In situ DRIFTS spectra of CO_2 and H_2O adsorption on BiOBr (**c**) and $5WO_3$/BiOBr (**d**) under light irradiation.

Figure 7c,d show the in situ DRIFTS spectra of CO_2 and H_2O adsorption on BiOBr and $5WO_3$/BiOBr under visible light irradiation. As shown in Figure 7c,d, the characteristic peaks at 1418, 1432, and 1456 cm^{-1} are attributed to HCO_3^-, and the peaks in 1483 cm^{-1} are attributed to m-CO_3^{2-}, indicating that new carbon species can be formed under light irradiation. Moreover, the concentration of the CO_2 intermediates were improved with the increase of the irradiation time. Notably, the concentration of $COOH^-$ was significantly improved under light irradiation, which is conducive to the photocatalytic CO_2 reduction reaction.

3.7. Possible Photocatalytic Mechanism

Based on the above discussion, the reaction mechanism of photocatalytic CO_2 reduction over xWO_3/BiOBr composites was proposed. As displayed in Figure 8a, BiOBr acts as the reducing photocatalyst, while WO_3 is an oxidizing photocatalyst, and the Fermi level of BiOBr is higher than that of WO_3. When BiOBr and WO_3 are in contact with each other, electrons spontaneously transfer from BiOBr to WO_3 until the Fermi level reaches equilibrium. In addition, at the interface, BiOBr and WO_3 are positively and negatively charged, respectively. An electron depletion region is formed at the BiOBr interface and the energy band bends upward. To the contrary, an electron accumulation zone is formed at the WO_3 interface and the energy band bends downward. In this case, an internal electric field is formed at the WO_3/BiOBr interfaces (Figure 8b). Under light irradiation, the electrons in the BiOBr and WO_3 valence bands are excited and then jump to their conduction band. Subsequently, the electrons accumulated in the CB of WO_3 combine with the holes in the VB of BiOBr (Figure 8c), which follow the S-scheme charge transfer mechanism. In addition, the electrons accumulated in the CB of BiOBr can participate in the photocatalytic CO_2 reduction, while the holes in the VB of WO_3 can trigger the H_2O oxidation. That is, the S-scheme charge transfer mechanism not only separates photogenerated electron-holes efficiently and quickly, but also maintains the strong redox ability of WO_3/BiOBr composites, which enhances the photocatalytic CO_2 performance.

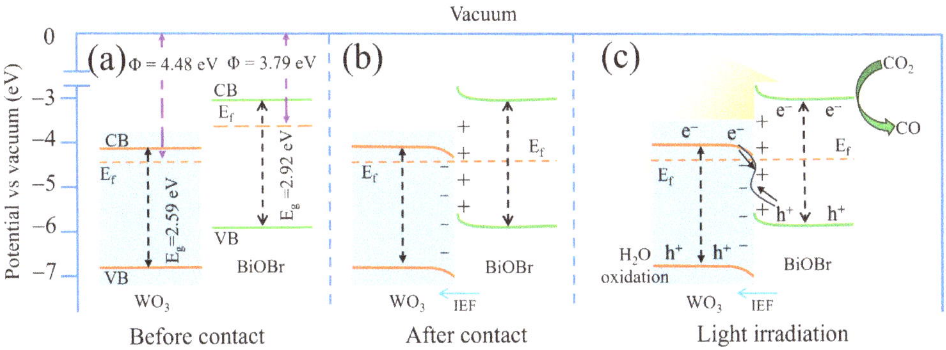

Figure 8. Band energy positions of WO_3 and BiOBr before (**a**) and after (**b**) contact, S-scheme charge transfer mechanism in WO_3/BiOBr composites under light irradiation (**c**).

4. Conclusions

In summary, WO_3/BiOBr S-scheme heterojunctions were synthesized for photocatalytic CO_2 reduction. The optimized WO_3/BiOBr heterostructures exhibited enhanced photocatalytic CO_2 reduction performance without any sacrificial reagents, and the CO yield reached 17.14 μmol g^{-1} after reaction for 4 h, which was 1.56 times greater than that of BiOBr. The photocatalytic stability of WO_3/BiOBr was also improved. The enhanced photocatalytic performance can be attributed to the S-scheme charge transfer mechanism, which effectively improves the separation efficiency of photogenerated charge carriers, thus promoting the photocatalytic CO_2 reduction. This study provides new insights into the construction of efficient and stable S-scheme heterojunction photocatalysts for photocatalytic CO_2 reduction.

Supplementary Materials: The following supporting information can be downloaded at: https://www.mdpi.com/article/10.3390/ma17133199/s1, Figure S1: Enlarged regions from 20° to 30° in XRD patterns of WO_3, BiOBr (a), and xWO_3/BiOBr (b). Figure S2: SEM images of $3WO_3$/BiOBr (a) and $10WO_3$/BiOBr (b). Figure S3: EDX spectrum of $5WO_3$/BiOBr. Figure S4: XPS survey spectra (a) and enlarged spectral regions from 0 eV to 100 eV (b) of BiOBr, WO_3, and $5WO_3$/BiOBr. Figure S5: UPS spectra of WO_3 (a) and BiOBr (b). Figure S6: Mott-Schottky plots of BiOBr (a) and WO_3 (b) at the frequencies of 1000 Hz, 1500 Hz, and 2000 Hz. Figure S7: XRD patterns (a) and FT-IR spectra (b) of $5WO_3$/BiOBr before and after reaction. Figure S8: SEM images of $5WO_3$/BiOBr before (a) and after (b) reaction. Table S1: Specific surface area and average pore size of the samples. Table S2: Comparison of photocatalytic CO_2 reduction performance over the BiOBr-based photocatalysts. References [45,55,66–70] are cited in the Supplementary Materials.

Author Contributions: Conceptualization, T.S.; Methodology, L.C. and X.X.; Formal analysis, L.C. and T.S.; Investigation, C.L., X.L. and L.C.; Resources, T.S. and H.J.; Data curation, C.L.; Writing—original draft, C.L.; Writing—review & editing, X.X., Z.Q., H.J. and T.S.; Supervision, T.S.; Funding acquisition, Z.Q. and T.S. All authors have read and agreed to the published version of the manuscript.

Funding: This work was supported by the National Natural Science Foundation of China (22208065), Guangxi Natural Science Foundation (2022GXNSFBA035483), National Natural Science Foundation of China (22078074), Opening Project of Guangxi Key Laboratory of Petrochemical Resource Processing and Process Intensification Technology (2023K012), and Special Funding for "Guangxi Bagui Scholars".

Institutional Review Board Statement: Not applicable.

Informed Consent Statement: Not applicable.

Data Availability Statement: The original contributions presented in the study are included in the article/Supplementary Material, further inquiries can be directed to the corresponding author.

Conflicts of Interest: The authors declare no conflicts of interest.

References

1. Yang, H.; Dang, Y.; Cui, X.; Bu, X.; Li, J.; Li, S.; Sun, Y.; Gao, P. Selective synthesis of olefins via CO_2 hydrogenation over transition-metal-doped iron-based catalysts. *Appl. Catal. B* **2023**, *321*, 122050. [CrossRef]
2. He, F.; Zhu, B.; Cheng, B.; Yu, J.; Ho, W.; Macyk, W. 2D/2D/0D $TiO_2/C_3N_4/Ti_3C_2$ MXene composite S-scheme photocatalyst with enhanced CO_2 reduction activity. *Appl. Catal. B* **2020**, *272*, 119006. [CrossRef]
3. Wang, S.; Han, X.; Zhang, Y.; Tian, N.; Ma, T.; Huang, H. Inside-and-Out Semiconductor Engineering for CO_2 Photoreduction: From Recent Advances to New Trends. *Small Struct.* **2020**, *2*, 2000061. [CrossRef]
4. Liang, J.; Yu, H.; Shi, J.; Li, B.; Wu, L.; Wang, M. Dislocated Bilayer MOF Enables High-Selectivity Photocatalytic Reduction of CO_2 to CO. *Adv. Mater.* **2023**, *35*, 2209814. [CrossRef] [PubMed]
5. Yang, M.; Wang, P.; Li, Y.; Tang, S.; Lin, X.; Zhang, H.; Zhu, Z.; Chen, F. Graphene aerogel-based NiAl-LDH/g-C_3N_4 with ultratight sheet-sheet heterojunction for excellent visible-light photocatalytic activity of CO_2 reduction. *Appl. Catal. B* **2022**, *306*, 121065. [CrossRef]
6. Chen, L.; Li, H.; Li, H.; Li, H.; Qi, W.; Zhang, Q.; Zhu, J.; Zhao, P.; Yang, S. Accelerating photogenerated charge kinetics via the g-C_3N_4 Schottky junction for enhanced visible-light-driven CO_2 reduction. *Appl. Catal. B* **2022**, *318*, 121863. [CrossRef]
7. Jo, Y.; Kim, C.; Moon, G.-H.; Lee, J.; An, T.; Choi, W. Activation of peroxymonosulfate on visible light irradiated TiO_2 via a charge transfer complex path. *Chem. Eng. J.* **2018**, *346*, 249–257. [CrossRef]
8. Xu, F.; Meng, K.; Cheng, B.; Wang, S.; Xu, J.; Yu, J. Unique S-scheme heterojunctions in self-assembled $TiO_2/CsPbBr_3$ hybrids for CO_2 photoreduction. *Nat. Commun.* **2020**, *11*, 4613. [CrossRef]
9. Li, L.; Ma, D.; Xu, Q.; Huang, S. Constructing hierarchical $ZnIn_2S_4$/g-C_3N_4 S-scheme heterojunction for boosted CO_2 photoreduction performance. *Chem. Eng. J.* **2022**, *437*, 135153. [CrossRef]
10. Liapun, V.; Hanif, M.B.; Sihor, M.; Vislocka, X.; Pandiaraj, S.; Unnikrishnan, V.K.; Thirunavukkarasu, G.K.; Edelmannová, M.F.; Reli, M.; Monfort, O.; et al. Versatile application of $BiVO_4/TiO_2$ S-scheme photocatalyst: Photocatalytic CO_2 and Cr(VI) reduction. *Chemosphere* **2023**, *337*, 139397. [CrossRef]
11. Yue, X.; Cheng, L.; Fan, J.; Xiang, Q. 2D/2D $BiVO_4/CsPbBr_3$ S-scheme heterojunction for photocatalytic CO_2 reduction: Insights into structure regulation and Fermi level modulation. *Appl. Catal. B* **2022**, *304*, 120979. [CrossRef]
12. Ye, L.; Deng, Y.; Wang, L.; Xie, H.; Su, F. Bismuth-Based Photocatalysts for Solar Photocatalytic Carbon Dioxide Conversion. *ChemSusChem* **2019**, *12*, 3671–3701. [CrossRef] [PubMed]
13. Zhao, X.; Xia, Y.; Wang, X.; Wen, N.; Li, H.; Jiao, X.; Chen, D. Promoted photocarriers separation in atomically thin $BiOCl/Bi_2WO_6$ heterostructure for solar-driven photocatalytic CO_2 reduction. *Chem. Eng. J.* **2022**, *449*, 137874. [CrossRef]

14. Zhao, J.; Miao, Z.; Zhang, Y.; Wen, G.; Liu, L.; Wang, X.; Cao, X.; Wang, B. Oxygen vacancy-rich hierarchical BiOBr hollow microspheres with dramatic CO_2 photoreduction activity. *J. Colloid Interface Sci.* **2021**, *593*, 231–243. [CrossRef]
15. Jiang, L.; Li, J.; Li, Y.; Wu, X.; Zhang, G. Promoted charge separation from nickel intervening in $[Bi_2O_2]^{2+}$ layers of Bi_2O_2S crystals for enhanced photocatalytic CO_2 conversion. *Appl. Catal. B* **2021**, *294*, 120249. [CrossRef]
16. Meng, L.; Qu, Y.; Jing, L. Recent advances in BiOBr-based photocatalysts for environmental remediation. *Chinese Chem. Lett.* **2021**, *32*, 3265–3276. [CrossRef]
17. Wu, J.; Li, X.; Shi, W.; Ling, P.; Sun, Y.; Jiao, X.; Gao, S.; Liang, L.; Xu, J.; Yan, W.; et al. Efficient Visible-Light-Driven CO_2 Reduction Mediated by Defect-Engineered BiOBr Atomic Layers. *Angew Chem. Int.* **2018**, *57*, 8719–8723. [CrossRef] [PubMed]
18. Zhang, B.; Zhang, M.; Zhang, L.; Bingham, P.A.; Tanaka, M.; Li, W.; Kubuki, S. BiOBr/MoS_2 catalyst as heterogenous peroxymonosulfate activator toward organic pollutant removal: Energy band alignment and mechanism insight. *J. Colloid Interface Sci.* **2021**, *594*, 635–649. [CrossRef]
19. Li, X.; Li, K.; Ding, D.; Yan, J.; Wang, C.; Carabineiro, S.A.C.; Liu, Y.; Lv, K. Effect of oxygen vacancies on the photocatalytic activity of flower-like BiOBr microspheres towards NO oxidation and CO_2 reduction. *Sep. Purif. Technol.* **2023**, *309*, 123054. [CrossRef]
20. Wang, K.; Cheng, M.; Xia, F.; Cao, N.; Zhang, F.; Ni, W.; Yue, X.; Yan, K.; He, Y.; Shi, Y.; et al. Atomically Dispersed Electron Traps in Cu Doped BiOBr Boosting CO_2 Reduction to Methanol by Pure H_2O. *Small* **2023**, *19*, 2207581. [CrossRef]
21. Wang, Z.; Cheng, B.; Zhang, L.; Yu, J.; Tan, H. BiOBr/NiO S-Scheme Heterojunction Photocatalyst for CO_2 Photoreduction. *Sol. Rrl* **2021**, *6*, 2100587. [CrossRef]
22. Yan, C.; Xu, M.; Cao, W.; Chen, Q.; Song, X.; Huo, P.; Zhou, W.; Wang, H. Fabricated S-scheme BiOBr/Cu_2O heterojunction photocatalyst for adjusting conversion of CO_2 to CH_4. *J. Environ. Chem. Eng.* **2023**, *11*, 111479. [CrossRef]
23. Zhang, T.; Maihemllti, M.; Okitsu, K.; Talifur, D.; Tursun, Y.; Abulizi, A. In situ self-assembled S-scheme BiOBr/pCN hybrid with enhanced photocatalytic activity for organic pollutant degradation and CO_2 reduction. *Appl. Surf. Sci.* **2021**, *556*, 149828. [CrossRef]
24. Meng, A.; Cheng, B.; Tan, H.; Fan, J.; Su, C.; Yu, J. TiO_2/polydopamine S-scheme heterojunction photocatalyst with enhanced CO_2-reduction selectivity. *Appl. Catal. B* **2021**, *289*, 120039. [CrossRef]
25. Rana, S.; Kumar, A.; Sharma, G.; Dhiman, P.; Garcia-Penas, A.; Stadler, F.J. Recent advances in perovskite-based Z-scheme and S-scheme heterojunctions for photocatalytic CO_2 reduction. *Chemosphere* **2023**, *339*, 139765. [CrossRef] [PubMed]
26. Wang, L.; Zhu, B.; Zhang, J.; Ghasemi, J.B.; Mousavi, M.; Yu, J. S-scheme heterojunction photocatalysts for CO_2 reduction. *Matter* **2022**, *5*, 4187–4211. [CrossRef]
27. Li, X.; Song, X.; Ma, C.; Cheng, Y.; Shen, D.; Zhang, S.; Liu, W.; Huo, P.; Wang, H. Direct Z-Scheme WO_3/Graphitic Carbon Nitride Nanocomposites for the Photoreduction of CO_2. *ACS Appl. Nano Mater.* **2020**, *3*, 1298–1306. [CrossRef]
28. Sun, Y.; Han, Y.; Song, X.; Huang, B.; Ma, X.; Xing, R. CdS/WO_3 S-scheme heterojunction with improved photocatalytic CO_2 reduction activity. *J. Photochem. Photobiol. B* **2022**, *233*, 112480. [CrossRef] [PubMed]
29. Jiang, X.; Ding, Y.; Zheng, S.; Ye, Y.; Li, Z.; Xu, L. In-Situ Generated $CsPbBr_3$ Nanocrystalson O-Defective WO_3 for Photocatalytic CO_2 Reduction. *ChemSusChem* **2022**, *15*, e202102295. [CrossRef]
30. Jiang, R.; Zhu, H.Y.; Li, J.B.; Fu, F.Q.; Yao, J.; Jiang, S.T.; Zeng, G.M. Fabrication of novel magnetically separable BiOBr/$CoFe_2O_4$ microspheres and its application in the efficient removal of dye from aqueous phase by an environment-friendly and economical approach. *Appl. Surf. Sci.* **2016**, *364*, 604–612. [CrossRef]
31. Shan, W.; Hu, Y.; Bai, Z.; Zheng, M.; Wei, C. In situ preparation of g-C_3N_4/bismuth-based oxide nanocomposites with enhanced photocatalytic activity. *Appl. Catal. B Environ.* **2016**, *188*, 1–12. [CrossRef]
32. Zhang, M.; Lai, C.; Li, B.; Huang, D.; Zeng, G.; Xu, P.; Qin, L.; Liu, S.; Liu, X.; Yi, H.; et al. Rational design 2D/2D BiOBr/CDs/g-C_3N_4 Z-scheme heterojunction photocatalyst with carbon dots as solid-state electron mediators for enhanced visible and NIR photocatalytic activity: Kinetics, intermediates, and mechanism insight. *J. Cata* **2019**, *369*, 469–481. [CrossRef]
33. Gao, Z.; Yao, B.; Yang, F.; Xu, T.; He, Y. Preparation of BiOBr-Bi heterojunction composites with enhanced photocatalytic properties on BiOBr surface by in-situ reduction. *Mat. Sci. Semiconr. Proc.* **2020**, *108*, 104882. [CrossRef]
34. Li, W.; Zou, Y.; Geng, X.; Xiao, F.; An, G.; Wang, D. Constructing highly catalytic oxidation over BiOBr-based hierarchical microspheres: Importance of redox potential of doped cations. *Mol. Cat.* **2017**, *438*, 19–29. [CrossRef]
35. Thiyagarajan, K.; Muralidharan, M.; Sivakumar, K. Defects Induced Magnetism in WO_3 and Reduced Graphene Oxide-WO_3 Nanocomposites. *J. Supercond. Nov. Magn.* **2017**, *31*, 117–125. [CrossRef]
36. Du, C.; Nie, S.; Feng, W.; Zhang, J.; Qi, M.; Liang, Y.; Wu, Y.; Feng, J.; Dong, S.; Liu, H.; et al. Hydroxyl regulating effect on surface structure of BiOBr photocatalyst toward high-efficiency degradation performance. *Chemosphere* **2022**, *287*, 132246. [CrossRef] [PubMed]
37. Tang, Q.Y.; Yang, M.J.; Yang, S.Y.; Xu, Y.H. Enhanced photocatalytic degradation of glyphosate over 2D CoS/BiOBr heterojunctions under visible light irradiation. *J. Hazard. Mater.* **2021**, *407*, 124798. [CrossRef] [PubMed]
38. Tao, W.; Tang, Q.; Hu, J.; Wang, Z.; Jiang, B.; Xiao, Y.; Song, R.; Guo, S. Construction of a hierarchical BiOBr/C_3N_4 S-scheme heterojunction for selective photocatalytic CO_2 reduction towards CO. *J. Mater. Chem. A* **2023**, *11*, 24999–25007. [CrossRef]
39. Zhang, K.; Zhang, Y.; Zhang, D.; Liu, C.; Zhou, X.; Yang, H.; Qu, J.; He, D. Efficient photocatalytic water disinfection by a novel BP/BiOBr S-scheme heterojunction photocatalyst. *Chem. Eng. J.* **2023**, *468*, 143581. [CrossRef]
40. Li, S.; Hu, S.; Jiang, W.; Zhang, J.; Xu, K.; Wang, Z. In situ construction of WO_3 nanoparticles decorated Bi_2MoO_6 microspheres for boosting photocatalytic degradation of refractory pollutants. *J. Colloid. Interf. Sci.* **2019**, *556*, 335–344. [CrossRef]

41. Xie, J.; Lu, Z.; Feng, Y.; Huang, J.; Hu, J.; Hao, A.; Ca, Y. A small organic molecule strategy for remedying oxygen vacancies by bismuth defects in BiOBr nanosheet with excellent photocatalytic CO_2 reduction. *Nano Res.* **2024**, *17*, 297–306. [CrossRef]
42. Jin, Y.; Li, F.; Li, T.; Xing, X.; Fan, W.; Zhang, L.; Hu, C. Enhanced internal electric field in S-doped BiOBr for intercalation, adsorption and degradation of ciprofloxacin by photoinitiation. *Appl. Catal. B* **2022**, *302*, 120824. [CrossRef]
43. Hao, J.; Zhang, Y.; Zhang, L.; Shen, J.; Meng, L.; Wang, X. Restructuring surface frustrated Lewis acid-base pairs of BiOBr through isomorphous Sn doping for boosting photocatalytic CO_2 reduction. *Chem. Eng. J.* **2023**, *464*, 142536. [CrossRef]
44. Albukhari, S.M.; Al-Hajji, L.A.; Ismail, A.A. Minimizing CO_2 emissions by photocatalytic CO_2 reduction to CH_3OH over Li_2MnO_3/WO_3 heterostructures under visible illumination. *Environ. Res.* **2024**, *241*, 117573. [CrossRef] [PubMed]
45. Qin, W.; Yang, Q.; Ye, H.; Xie, Y.; Shen, Z.; Guo, Y.; Deng, Y.; Ling, Y.; Yu, J.; Luo, G.; et al. Novel 2D/2D BiOBr/$Zn(OH)_2$ photocatalysts for efficient photoreduction CO_2. *Sep. Purif. Technol.* **2023**, *306*, 122721. [CrossRef]
46. Yaseen, M.; Jiang, H.; Li, J.; Yu, X.; Ashfaq Ahmad, M.; Nauman Ali, R.; Wang, L.; Yang, J.; Liu, Q. Synergistic effect of Z-scheme and oxygen vacancy of CeO_2/WO_3 heterojunction for enhanced CO_2 reduction activity. *Appl. Surf. Sci.* **2023**, *631*, 157360. [CrossRef]
47. Liu, Y.; Guo, J.-g.; Wang, Y.; Hao, Y.-j.; Liu, R.-h.; Li, F.-t. One-step synthesis of defected $Bi_2Al_4O_9/\beta$-Bi_2O_3 heterojunctions for photocatalytic reduction of CO_2 to CO. *Green Energy Environ.* **2021**, *6*, 244–252. [CrossRef]
48. Xiao, Y.; Maimaitizi, H.; Okitsu, K.; Tursun, Y.; Abulizi, A. Sonochemical Fabrication of S-scheme Hierarchical CdS/BiOBr Heterojunction Photocatalyst with High Performance for Carbon Dioxide Reduction. *Part. Part. Syst. Charact.* **2022**, *39*, 2200019. [CrossRef]
49. Chico-Vecino, M.; Murillo-Sierra, J.C.; Pino-Sandoval, D.A.; Hinojosa-Reyes, L.; Maya-Treviño, M.L.; Contreras, D.; Hernández-Ramírez, A. Preparation of WO_3/In_2O_3 heterojunctions and their performance on the CO_2 photocatalytic conversion in a continuous flow reactor. *J. Environ. Chem. Eng.* **2023**, *11*, 110372. [CrossRef]
50. Huang, W.; Hua, X.; Zhao, Y.; Li, K.; Tang, L.; Zhou, M.; Cai, Z. Enhancement of visible-light-driven photocatalytic performance of BiOBr nanosheets by Co^{2+} doping. *J. Mater. Sci. Mater. Electron.* **2019**, *30*, 14967–14976. [CrossRef]
51. Ma, Y.; Li, J.; Cai, J.; Zhong, L.; Lang, Y.; Ma, Q. Z-scheme g-C_3N_4/ZnS heterojunction photocatalyst: One-pot synthesis, interfacial structure regulation, and improved photocatalysis activity for bisphenol A. *Colloids Surf. A* **2022**, *653*, 130027. [CrossRef]
52. Kong, W.; Zhu, D.; Zhang, Y.; Luo, R.; Ma, J.; Lei, J.; Ju, H. Electron Donor Coordinated Metal-Organic Framework to Enhance Photoelectrochemical Performance. *Angew. Chem. Int.* **2023**, *62*, 202308514. [CrossRef] [PubMed]
53. Radha, R.; Kulangara, R.V.; Elaiyappillai, E.; Sridevi, J.; Balakumar, S. Modulation in the Band Dispersion of Bi_2WO_6 Nanocrsytals Using the Electronegativity of Transition Elements for Enhanced Visible Light Photocatalysis. *Cryst. Growth Des.* **2019**, *19*, 6224–6238. [CrossRef]
54. Yu, T.; Liu, Q.; Zhu, Z.; Wu, W.; Liu, L.; Zhang, J.; Gao, C.; Yang, T. Construction of a photocatalytic fuel cell using a novel Z-scheme MoS_2/rGO/Bi_2S_3 as electrode degraded antibiotic wastewater. *Sep. Purif. Technol.* **2021**, *277*, 119276. [CrossRef]
55. Gaopeng, L.; Lin, W.; Bin, W.; Xingwang, Z.; Jinman, Y.; Pengjun, L.; Wenshuai, Z.; Ziran, C.; Jiexiang, X. Synchronous activation of Ag nanoparticles and BiOBr for boosting solar-driven CO_2 reduction. *Chin. Chem. Lett.* **2023**, *34*, 107962.
56. Sun, M.; Zhu, C.; Wei, S.; Chen, L.; Ji, H.; Su, T.; Qin, Z. Phosphorus-Doped Hollow Tubular g-C_3N_4 for Enhanced Photocatalytic CO_2 Reduction. *Materials* **2023**, *16*, 6665. [CrossRef] [PubMed]
57. Juntrapirom, S.; Anuchai, S.; Thongsook, O.; Pornsuwan, S.; Meepowpan, P.; Thavornyutikarn, P.; Phanichphant, S.; Tantraviwat, D.; Inceesungvorn, B. Photocatalytic activity enhancement of g-C_3N_4/BiOBr in selective transformation of primary amines to imines and its reaction mechanism. *Chem. Eng. J.* **2020**, *394*, 124934. [CrossRef]
58. Chen, K.; Wang, X.; Li, Q.; Feng, Y.-N.; Chen, F.-F.; Yu, Y. Spatial distribution of $ZnIn_2S_4$ nanosheets on g-C_3N_4 microtubes promotes photocatalytic CO_2 reduction. *Chem. Eng. J.* **2021**, *418*, 129476. [CrossRef]
59. Guo, H.; Wan, S.; Wang, Y.; Ma, W.; Zhong, Q.; Ding, J. Enhanced photocatalytic CO_2 reduction over direct Z-scheme $NiTiO_3$/g-C_3N_4 nanocomposite promoted by efficient interfacial charge transfer. *Chem. Eng. J.* **2021**, *412*, 128646. [CrossRef]
60. Zhao, M.; Qin, J.; Wang, N.; Zhang, Y.; Cui, H. Reconstruction of surface oxygen vacancy for boosting CO_2 photoreduction mediated by BiOBr/CdS heterojunction. *Sep. Purif. Technol.* **2024**, *329*, 125179. [CrossRef]
61. Wu, J.; Xie, Y.; Ling, Y.; Si, J.; Li, X.; Wang, J.; Ye, H.; Zhao, J.; Li, S.; Zhao, Q.; et al. One-step synthesis and Gd^{3+} decoration of BiOBr microspheres consisting of nanosheets toward improving photocatalytic reduction of CO_2 into hydrocarbon fuel. *Chem. Eng. J.* **2020**, *400*, 125944. [CrossRef]
62. Yin, S.; Zhao, X.; Jiang, E.; Yan, Y.; Zhou, P.; Huo, P. Boosting water decomposition by sulfur vacancies for efficient CO_2 photoreduction. *Energy Environ. Sci.* **2022**, *15*, 1556–1562. [CrossRef]
63. Meng, J.; Duan, Y.; Jing, S.; Ma, J.; Wang, K.; Zhou, K.; Ban, C.; Wang, Y.; Hu, B.; Yu, D.; et al. Facet junction of BiOBr nanosheets boosting spatial charge separation for CO_2 photoreduction. *Nano Energy* **2022**, *92*, 106671. [CrossRef]
64. Wang, B.; Zhao, J.; Chen, H.; Weng, Y.-X.; Tang, H.; Chen, Z.; Zhu, W.; She, Y.; Xia, J.; Li, H. Unique Z-scheme carbonized polymer dots/$Bi_4O_5Br_2$ hybrids for efficiently boosting photocatalytic CO_2 reduction. *Appl. Catal. B Environ.* **2021**, *293*, 120182. [CrossRef]
65. Bai, P.; Zhao, Y.; Li, Y. Efficient photocatalytic CO_2 reduction coupled with selective styrene oxidation over a modified g-C_3N_4/BiOBr composite with high atom economy. *Green Chem.* **2024**, *26*, 2290–2299. [CrossRef]
66. Lin, W.; Gaopeng, L.; Bin, W.; Xin, C.; Chongtai, W.; Zixia, L.; Jiexiang, X.; Huaming, L. Oxygen Vacancies Engineering–Mediated BiOBr Atomic Layers for Boosting Visible Light-Driven Photocatalytic CO_2 Reduction. *Solar Rrl* **2020**, *5*, 2000480.

67. Guan, C.; Hou, T.; Nie, W.; Zhang, Q.; Duan, L.; Zhao, X. Enhanced photocatalytic reduction of CO_2 on BiOBr under synergistic effect of Zn doping and induced oxygen vacancy generation. *J. Colloid Interface Sci.* **2023**, *633*, 177–188. [CrossRef] [PubMed]
68. Jiayu, Z.; Yupei, L.; Xiaojing, W.; Jun, Z.; Yinsu, W.; Fatang, L. Simultaneous Phosphorylation and Bi Modification of BiOBr for Promoting Photocatalytic CO_2 Reduction. *ACS Sustain. Chem. Eng.* **2019**, *7*, 14953–14961.
69. Xin, F.; ChuYa, W.; Longpeng, Z.; Qi, Z.; Hengdeng, Z.; Qi, W.; Guangcan, Z. Efficient visible-light-driven CO_2 reduction mediated by novel Au-doped BiOBr nanosheets. *J. Environ. Chem. Eng.* **2023**, *11*, 109986.
70. Ying, K.X.; Cathie, L.W.P.; WeeJun, O.; SiangPiao, C.; Rahman, M.A. Oxygen-Deficient BiOBr as a Highly Stable Photocatalyst for Efficient CO_2 Reduction into Renewable Carbon-Neutral Fuels. *ChemCatChem* **2016**, *8*, 3074–3081.

materials

MDPI

Article

Effects on Metallization of n+-Poly-Si Layer for N-Type Tunnel Oxide Passivated Contact Solar Cells

Qinqin Wang [1,*], Beibei Gao [1,2], Wangping Wu [3], Kaiyuan Guo [1], Wei Huang [1] and Jianning Ding [1,*]

1 Institute of Technology for Carbon Neutralization, Yangzhou University, Yangzhou 225009, China; wqqwh517@126.com (B.G.); mz120220597@stu.yzu.edu.cn (K.G.); mx120220364@stu.yzu.edu.cn (W.H.)
2 Jinko Solar Co., Ltd., Haining 314400, China
3 Electrochemistry and Corrosion Laboratory, School of Mechanical Engineering, Changzhou University, Changzhou 213164, China; wwp3.14@163.com
* Correspondence: wangqinqin@yzu.edu.cn (Q.W.); dingjn@yzu.edu.cn (J.D.)

Abstract: Thin polysilicon (poly-Si)-based passivating contacts can reduce parasitic absorption and the cost of n-TOPCon solar cells. Herein, n^+-poly-Si layers with thicknesses of 30~100 nm were fabricated by low-pressure chemical vapor deposition (LPCVD) to create passivating contacts. We investigated the effect of n^+-poly-Si layer thickness on the microstructure of the metallization contact formation, passivation, and electronic performance of n-TOPCon solar cells. The thickness of the poly-Si layer significantly affected the passivation of metallization-induced recombination under the metal contact ($J_{0,metal}$) and the contact resistivity (ρ_c) of the cells. However, it had a minimal impact on the short-circuit current density (J_{sc}), which was primarily associated with corroded silver (Ag) at depths of the n^+-poly-Si layer exceeding 40 nm. We introduced a thin n^+-poly-Si layer with a thickness of 70 nm and a surface concentration of 5×10^{20} atoms/cm^3. This layer can meet the requirements for low $J_{0,metal}$ and ρ_c values, leading to an increase in conversion efficiency of 25.65%. This optimized process of depositing a phosphorus-doped poly-Si layer can be commercially applied in photovoltaics to reduce processing times and lower costs.

Keywords: n^+-poly-Si layer; n-TOPCon solar cell; metallization; parasitic absorption; Ag pastel

1. Introduction

A tunnel oxide passivated contact (TOPCon) solar cell with an ultrathin silicon oxide (SiO_x) film and a phosphorus-doped polysilicon (n^+-poly-Si) layer has the potential for a high theoretical efficiency limit of 28.7% [1]. The market share of TOPCon cells is significant according to the ITRPV [2]. An n-TOPCon cell (size: 330.15 cm^2) at JinkoSolar has achieved the highest conversion efficiency recorded thus far, 26.89% [3]. The n^+-poly-Si/SiO_x layer with high carrier selectivity exhibits superior passivation quality at the metal/silicon contact, distinguishing it from Passivated Emitter and Rear Cell (PERC) solar cells [4,5]. The ultrathin SiO_x layer has a large tunneling barrier for holes [6]. The doped poly-Si layer separates the metal/silicon contact from minority charge carriers, resulting in low recombination and band bending. The doped poly-Si is typically crystallized from an amorphous silicon layer (a-Si) after a high-temperature heat treatment process. Normally, the desired a-Si layer can be prepared using methods such as plasma-enhanced chemical vapor deposition (PECVD) [7,8], low-pressure CVD (LPCVD) [9,10], atmospheric pressure CVD [11], hot wire CVD [12], sputtering [13,14], and electron beam evaporation [15]. Published studies focused on poly-Si passivating contacts fabricated using the standard industrial approach based on LPCVD and ex situ phosphorus (P) doping. This process has high throughput and is simple [4,16–20]. Studies in the literature investigated the parameters of passivating contacts, including a tunneling SiO_x film [21], the structural properties and thickness of the poly-Si film [22–24], and the metallization of the passivating contact [16]. These studies have provided valuable insights into the impacts of LPCVD process parameters on the structure and electrical properties of poly-Si films [17,25–29].

The doped poly-Si layer could be helpful in achieving good passivation performance and low contact resistivity by controlling interfacial defects in the SiO_x and the doping concentration of the poly-Si layer [30,31]. After maximizing contact passivation using a SiO_x/poly-Si layer, the lowest dark saturation current densities (J_0) of emitters in cells with passivation layers comprising an n^+-poly-Si/SiO_x contact and a boron-doped poly-Si/SiO_x contact were 0.66 fA/cm^2 and 4.4 fA/cm^2, respectively [32]. At present, researchers and engineers are making further efforts to improve cell efficiency. One important method is minimizing the parasitic absorption losses of the poly-Si layer for solar cells. However, a poly-Si layer with a thickness of 100 nm is necessary under current metallization conditions [4,33]. At the same time, various studies have been conducted to prevent metal finger penetration by using a barrier interlayer, such as titanium nitride (TiN) [34,35] or SiO_x [36,37], created through a low-temperature method like Laser-Enhanced Contact Optimization (LECO) [38]. Although there are many positive outcomes regarding the formation of planar (n) poly-Si passivating contacts, one of the challenging and critical tasks associated with poly-Si layers is determining how to balance parasitic absorption and contact in poly-Si layers [24,39–41].

In this work, n-TOPCon cells containing P-doped poly-Si with passivating contacts were fabricated. The thickness of the poly-Si layer was changed using the LPCVD method, and the influence of the poly-Si layer thickness on J_0, metallization recombination ($J_{0,metal}$), and contact resistivity (ρ_c) were studied; a microstructure analysis of the contact formation was conducted; and further, the *I–V* parameters of the n-TOPCon solar cells were investigated, including their efficiency (E_{ff}), open-circuit voltage (V_{oc}), fill factor (FF), series resistance (R_{ser}), and short-circuit current density (J_{sc}).

2. Experimental

2.1. Fabrication

n-TOPCon cells were obtained from commercially available n-type Czochralski silicon (Cz-Si) wafers. Texture was generated in an alkaline (KOH)/textured additive solution. The textured wafers were placed into a quartz tube furnace containing BCl_3 gas using a LYDOP™ system. Subsequently, the B-doping side was treated with a Hymson laser with a beam width of 90–100 μm to obtain front B-selective emitters. After the laser treatment, the rear side was cleansed with a KOH/polished additive solution. One 1.6 ± 0.2 nm thick layer of SiO_x and an intrinsic a-Si 30–100 nm thick layer on the rear side were prepared by LPCVD. Subsequently, the a-Si layer was crystallized to form poly-Si in a high-temperature tube furnace at 850 °C for 20–30 min using a gas mixture of $POCl_3$, O_2, and N_2 which could be simultaneously doped with P. Phosphorus silicate glass (PSG) was removed using a 5% HF solution for 5 min, and the P-doping concentration profiles of the SiO_x/P-doped poly-Si films were determined by electrochemical capacitance–voltage (ECV) profiling. Wafers with different P-doped poly-Si film thicknesses using two-sided passivation of the SiN_x layer were used as J_0 samples. Similar samples with polished surfaces were screen-printed with Ag paste lines on one side only and nine different pitches were used as $J_{0,metal}$ samples; these which were placed with the fingers facing up and then sintered in a sintering furnace at a peak temperature of 730 °C [42–44]. P-doped poly-Si contacts were fabricated with four different thicknesses (see Table 1). Figure 1 displays the structure of an n-TOPCon solar cell.

Table 1. Process conditions and performance of wafers.

<table>
<tr><th rowspan="2">Conditions (Poly Thickness)</th><th>SiO$_x$ Layer Formation Process</th><th colspan="4">a-Si Layer Formation Process</th><th colspan="3">Phosphorus Doped Poly-Si Layer Formation Process</th><th colspan="5">Main Result Data</th></tr>
<tr><th>$t_{oxidation}$ (min)</th><th>T (°C)</th><th>t (min)</th><th>G_{SiH4} (sccm)</th><th>ECV (nm)</th><th>$T_{deposition}$ (°C)</th><th>G_{POCl3} (sccm)</th><th>$T_{drive\text{-}in}$ (°C)</th><th>$R_\square$ (Ω/sq)</th><th>ρ_c (mΩ·cm^2)</th><th>J_0 (fA/cm^2)</th><th>PL (a.u)</th><th>$J_{0,metal}$ (fA/cm^2)</th></tr>
<tr><td>100 nm</td><td rowspan="4">8</td><td rowspan="4">600</td><td>24</td><td rowspan="4">850</td><td>100</td><td rowspan="4">850</td><td rowspan="4">1100–1200</td><td rowspan="4">850–860</td><td>45</td><td>2.5</td><td>4.6</td><td>31,090</td><td>26</td></tr>
<tr><td>70 nm</td><td>19</td><td>70</td><td>52</td><td>2.9</td><td>4.1</td><td>32,258</td><td>304</td></tr>
<tr><td>50 nm</td><td>16</td><td>50</td><td>54</td><td>3.5</td><td>4.7</td><td>33,262</td><td>413</td></tr>
<tr><td>30 nm</td><td>13</td><td>30</td><td>57</td><td>3.9</td><td>4.7</td><td>33,216</td><td>545</td></tr>
</table>

$t_{oxidation}$: post-oxidation duration; T: temperature; t: time; $T_{drive\text{-}in}$: drive-in temperature; ρ_c: contact resistivity; J_0: emitter dark saturation current density; G: gas flow.

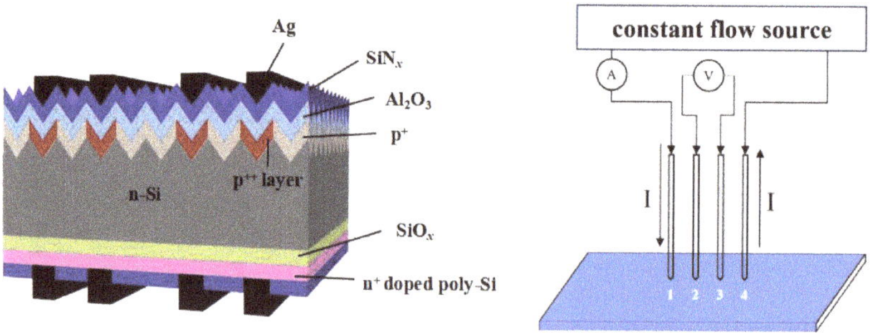

(a) (b)

Figure 1. (**a**) Structure of n-TOPCon solar cell and (**b**) schematic diagram of four-probe sheet resistance tester.

After cleaning the wafers and removing the BSG and PSG, a clean P-doped poly-Si/SiO$_x$ layer was obtained on one side only by etching the poly-Si wraparound side in a mixed solution of KOH and polishing additives. The front and rear of the cells were passivated with 4 nm thick Al$_2$O$_3$ through ALD/78 nm thick SiN$_x$ and 75 nm thick SiN$_x$ through PECVD, respectively. The fingers and busbars of the wafers were metallized using a commercial Ag paste, using an H-patterned grid which was screen-printed on both sides with a 16-busbar configuration. Finally, the cells with Ag fingers were heat-treated at a peak temperature of 730 °C.

2.2. Characterization

The current–voltage (I–V) parameters of the in-house standard cells were measured using a Wavelabs tester. The J_0 and $J_{0,metal}$ values of the samples were measured using a WCT-120 Sinton and extracted at an excess carrier density of 3×10^{15} cm^{-3} (Boulder, CO, USA) [45]. The SiO$_x$/P-doped poly-Si profiles of the monitor wafers were measured by an ECV device (WEP CVP21), using a 0.1 M NH$_4$F solution as an etchant. We used a four-probe test rig to determine the sheet resistance (Figure 1b). The ρ_c values were measured using the transfer-length method (TLM; GP-4 TEST) from the H-patterns printed on one side of the J_0 samples. The microstructures and chemical composition of the contact interfaces were observed by energy-filtered scanning electron microscopy (FESEM-EDS, Regulus8230, HITCHI, Tokyo, Japan) and transmission electron microscopy (Thermo Scientific Talos™ STEM, 200 kV, FEI Talos F200X, Waltham, MA, USA). The optical reflectance values and losses of the cells were analyzed using a PVE300-IVT (Bentham Instruments, Reading, UK) instrument and the Current Loss Analysis Calculator V1.4 (Series, from the Solar Energy Research Institute of Singapore, based on the Yablonovitch limit of 46.43 mA/cm^2 [46]).

3. Results and Discussion

3.1. Microstructure and Performance

The n^+-poly-Si layers of different thicknesses influenced the n^+-poly-Si/SiO_x profiles, sheet resistance ($R_\square$), J_0, ρ_c, and $J_{0,metal}$ of the n-TOPCon cells, and the results are shown in Table 1. Figure 2 displays the ECV profiles and sheet resistances for different thicknesses of n^+-poly-Si/SiO_x layers. In Figure 2a, n^+-poly-Si layers with different thicknesses have almost the same surface concentration of >5 × 10^{20} atoms/cm^3 and nearly the same "knee-shaped" tail ECV profile; these were doped at the same high temperature. On an n-type silicon wafer, when the thickness of the n^+-poly-Si layer was reduced from 100 to 30 nm, the sheet resistance ($R_\square$) of the cell increased from 45 to 57 Ω/sq. This increase occurred because the thickness of the poly-Si layer determined the total doping amount (Figure 2b). Although the surface doping concentration is the same, ECV curves of poly-Si layers with different thickness are integrated into the total doping amount. It can be found that as the thickness of the poly-Si layer increases, the total doping amount also increases, leading to a decrease in sheet resistance.

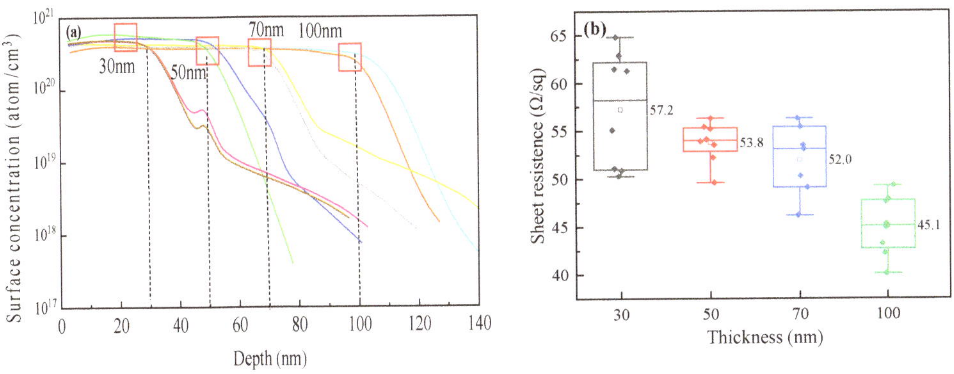

Figure 2. ECV profiles (**a**) and sheet resistances (**b**) of cells with different n^+-poly-Si/SiO_x layer thicknesses. (The solid pattern represents the test data, and the hollow pattern represents the average value).

The J_0 values of n^+-poly-Si/SiO_x layers with different poly-Si thicknesses exhibit a slight change of approximately 0.5 fA/cm^2 when the thickness increases from 30 to 100 nm. The penetration depth is determined from the inflection point of the SiO_x layer to a doping concentration of N = 1 × 10^{19} atoms/cm^3 (Figure 3). It can be seen from Figure 3a that as the poly-Si thickness increases, the penetration depth of p-doping increases, and the inflection point concentration decreases. When the thickness of the poly-Si layer is 30 nm, the inflection point concentration is higher, resulting in a large Auger recombination so the J_0 value becomes large. When the thickness of the poly-Si layer is 100 nm, the inflection point concentration is low, the passivation effect is poor, and J_0 increases. In order to study the passivation properties of different poly-Si thicknesses, PL tests were performed. The results showed that PL intensity decreased slightly with an increase in thickness. The PL values also showed a minor change of approximately 1000 a.u. This indicates that compared to a 100 nm thick layer of n^+-poly-Si/SiO_x, a 30 nm thick layer can also provide effective field-effect passivation. This is attributed to the good passivation of the SiO_x layer and a 15~20 nm penetration depth of P doped into the Si substrate.

$J_{0,\text{measured}}$ plots of n^+-poly-Si layers of various thicknesses with different metallization fractions are depicted in Figure 4. The values of $J_{0,metal}$ are determined through a straightforward linear interpolation of the measured data points [42–44]. The results indicate that the n^+-poly-Si layers with a thickness of 100 nm have the lowest $J_{0,\text{measured}}$ values, approximately 26 fA/cm^2. This could be attributed to the thick layer's ability to withstand the depth of the corrosion of the poly-Si layer by Ag paste. The thinner the poly layer, the more it is destroyed by the paste, which greatly reduces the passivation performance. However, when the thickness of the layer was decreased from 70 to 30 nm, $J_{0,metal}$ increased from 304 to 545 fA/cm^2. It is well known that metal contact recombination is related to doping concentration and corrosion depth. When the poly-Si layer thickness is more than 70 nm, the doping concentration increases (Figure 2a), and a thick poly-Si layer provides good resistance against the corrosion of the slurry. When the poly-Si layer is thin, it cannot meet the demand for metallization. The thinner the poly layer, the more the paste destroys the poly layer, which greatly reduces the passivation performance.

Cross-sectional SEM images of the interface between the screen-printed Ag bulk and the n$^+$-poly-Si/SiO$_x$ contacts are shown in Figure 5. The polished cross-section of the cell shows a 100 nm thick n$^+$-poly-Si layer uniformly covered by the glass layer beneath the Ag bulk, with some voids. The interface between a Ag finger and the n$^+$-poly Si/SiO$_x$ layer is obvious in Figure 5a; it can be observed that the Ag bulks have a block structure and that the interlayer n$^+$-poly Si/SiO$_x$ layer seems to be non-uniform (Figure 5b). The thickness of the interlayer is thin in some regions; however, there are some thick interlayers in the dotted-line region (Figure 5c). It is hypothesized that the numerous small white particles on the local area of the poly-Si layer's surface are the result of Ag particle precipitation. Meanwhile, we further show the chemical composition of the thick interlayer in Figure 5c, the EDX elemental mapping images are displayed in Figure 6.

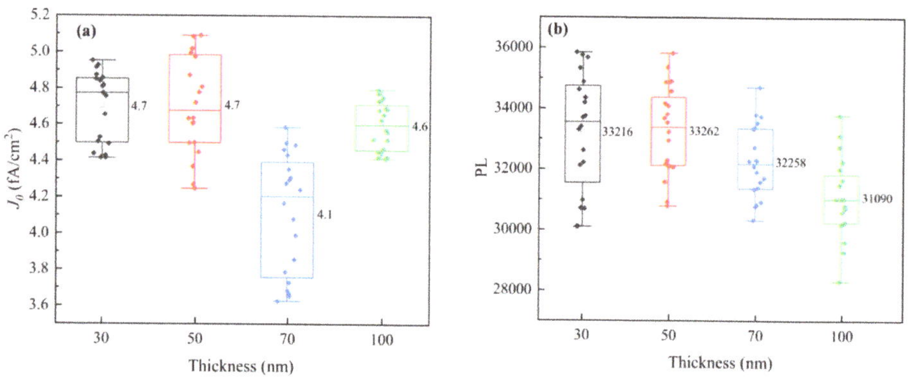

Figure 3. J_0 (**a**) and PL (**b**) plots of cells. (The solid pattern represents the test data, and the hollow pattern represents the average value).

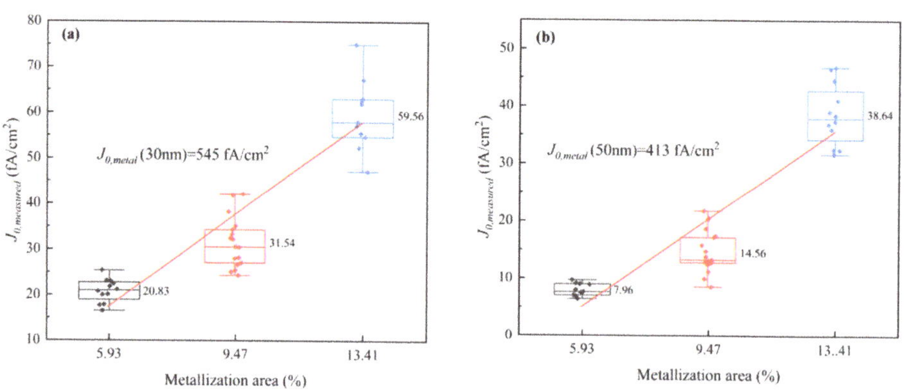

Figure 4. *Cont.*

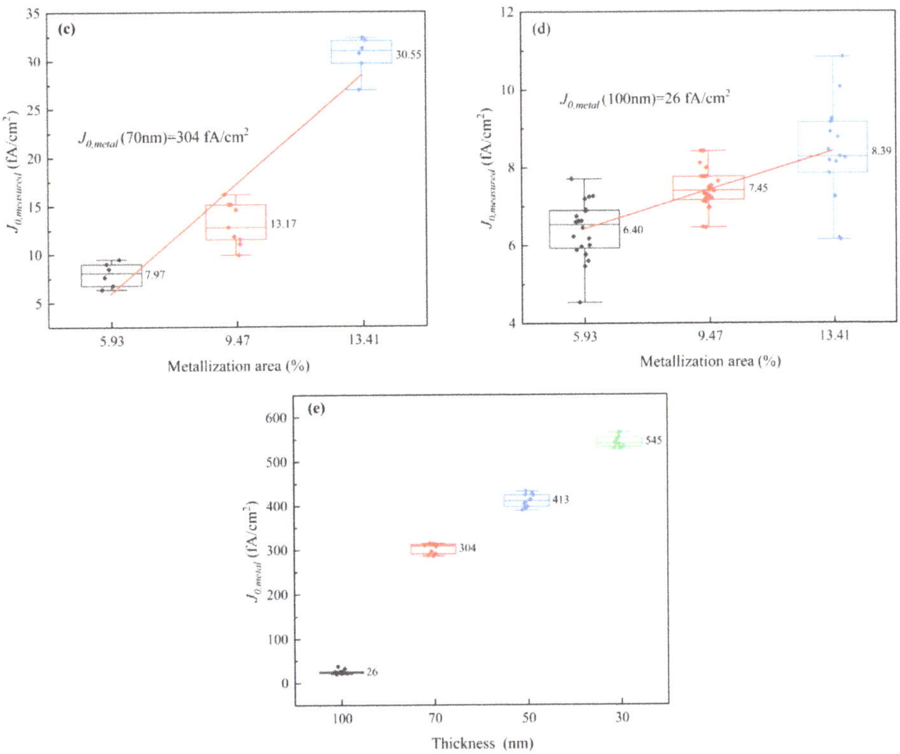

Figure 4. $J_{0,\text{measured}}$ plots for n$^+$-poly-Si layers (**a**) 30 nm, (**b**) 50 nm, (**c**) 70 nm, and (**d**) 100 nm thick with different metallization fractions ranging from 5.9% to 17.3% and (**e**) $J_{0,metal}$ of different thickness of n$^+$-poly-Si layers. (The solid pattern represents the test data, and the hollow pattern represents the average value).

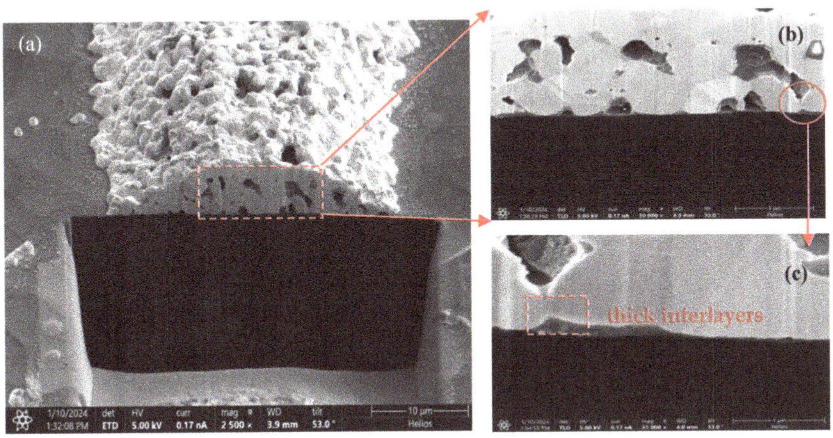

Figure 5. Cross-sectional SEM images of the metal contact on (**a**) an n$^+$-poly-Si layer, the interface of Ag–Si fingers (**b**); and numerous little white particles on the local area of the surface of the poly-Si layer (**c**) shown in (**b**).

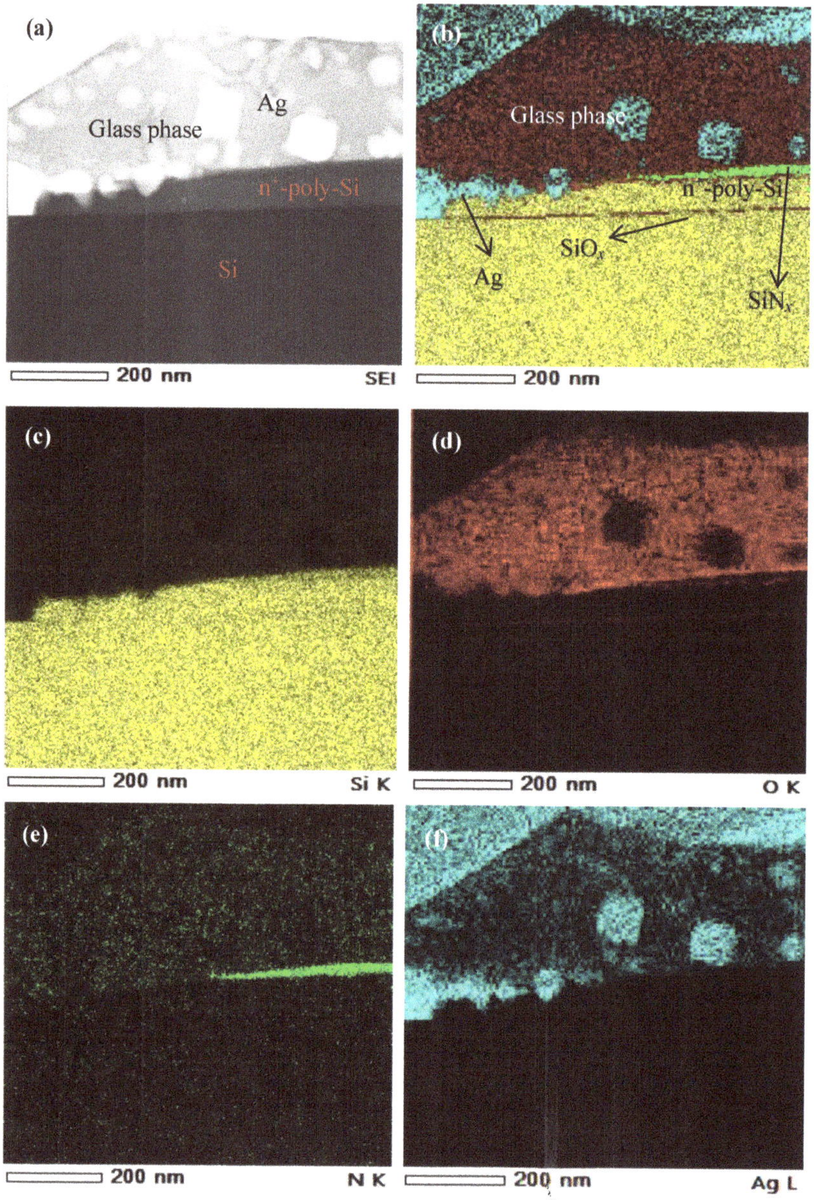

Figure 6. (**a**) HAADF-STEM image of the formed contact of an n+ poly-Si layer from Figure 5c; (**b**) corresponding EDX elemental mapping and EDX elemental mapping of Si (**c**), O (**d**), N (**e**), and Ag (**f**).

A large number of white schistose Ag particles embedded in the SiO*x*-based glass phase are visible in a HAADF-STEM (High-Angle Annular Dark-Field Scanning Transmission Electron Microscopy, HITCHI, Japan) image of the contact interface between the Ag grid finger and the polished surface of the n^+-poly-Si layer (Figure 6a). The corresponding elemental distribution was measured using EDX spectroscopy and is shown in Figure 6b as an overlay of Ag, Si, N, and O; the matrix elements are shown separately in Figure 6b–f.

The partial SiN_x layer can still be observed at the interface between the Ag finger and the n^+-poly-Si/SiO_x layer, which could influence the contact performance (Figure 6e). Some Ag particles have corroded the poly-Si layer, which is crucial for metallization recombination (Figure 6f). However, the SiO_x tunneling layer is clearly present in Figure 6b,d.

As shown in the EDX line scan (Figure 7), the corrosion depth of the Ag particles is approximately 0.06 μm, which is consistent with the $J_{0,metal}$ analysis mentioned above. This imposes a constraint on the minimum thickness of the n^+-poly-Si layer. Therefore, when the thickness of the poly-Si layer is thin, it cannot resist the corrosion caused by the Ag paste. This leads to the formation of metal–semiconductor contact to some extent, resulting in a higher $J_{0,metal}$.

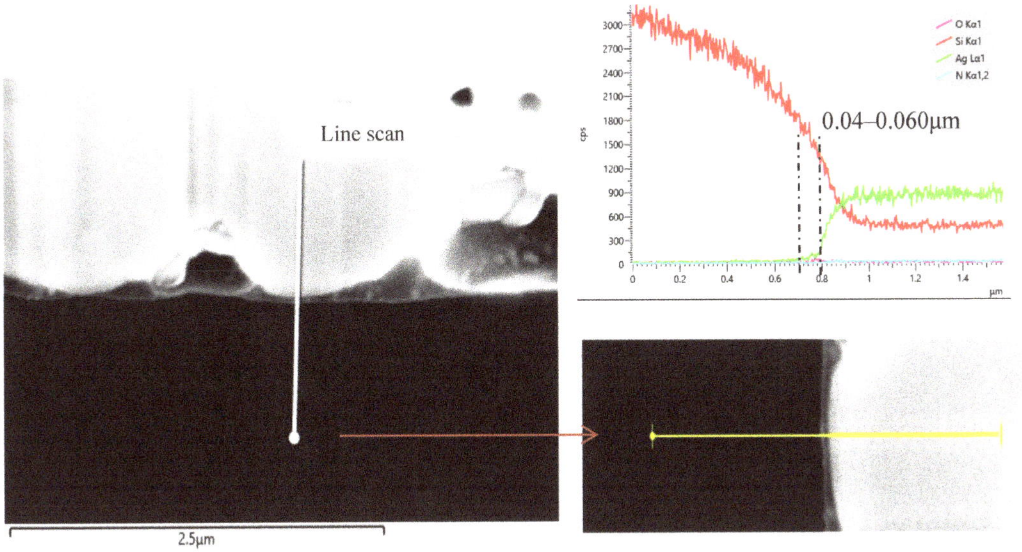

Figure 7. STEM/EDX energy element line scan images of the Ag–Si interface.

TEM was conducted and revealed the composition of the contact cross-section formed between the screen-printed Ag bulk and the n^+-poly-Si/SiO_x layers. As shown in Figure 8, the thickness of the n^+-poly-Si/SiO_x layer is 50 nm. The Ag–Si alloy can be observed in specific regions of the n^+-poly-Si layer which are only 10 nm away from the SiO_x layer. This means that the depth of the Ag–Si alloy can reach up to 40 nm, which is consistent with Figure 7. There are also some Ag embryos in certain areas of the n^+-poly-Si layer (Figure 8b). This can explain why the contact resistance is low but the metallization recombination increased. However, the Ag embryos did not penetrate the Si substrate; instead, a clear SiO_x layer can be observed (Figure 8c). The above analysis indicates that under the current process conditions, the thickness of the n^+-poly-Si layer should be more than 40 nm.

The contact resistivity of the solar cells with varying n^+-poly-Si layer thicknesses is depicted in Figure 9. There is a remarkable change in ρ_c, which decreases from 40.4 to 7.8 mΩ·cm^2 with an increase in the layer thickness from 30 nm to 100 nm. Due to an increase in the total doping concentration, the width of the internal depletion region in a silicon wafer can be narrowed, enabling the quantum mechanical tunneling of charge carriers through Schottky barriers [47].

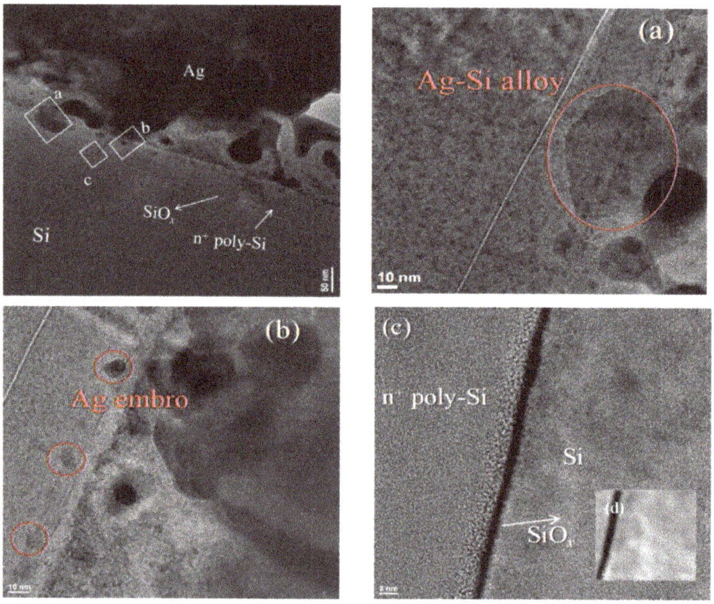

Figure 8. STEM bright-field cross-sectional images of Ag bulk/glass layer/50nm n^+ poly-Si/SiO_x layer contact; high-density of Ag embryos on SiN_x layer was observed in (**a**). Ag–Si alloy in the n^+ poly-Si layer (**b**), (**c**) Ag-embryo on n^+ poly-Si layer, and (**d**) selected area of tunneling layer SiO_x.

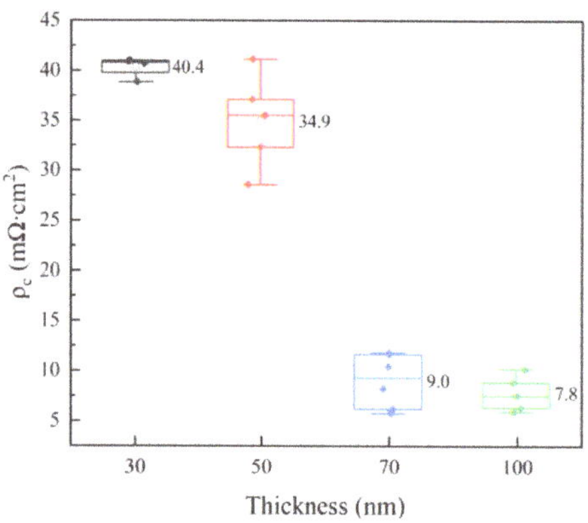

Figure 9. Plot of contact resistivity of solar cells.

3.2. *I–V Parameters*

Table 2 lists the *I–V* parameters of the cells as a function of the thickness of the n^+-poly-Si layer. The cells with 70 nm and 100 nm thick n^+-poly-Si layers have the same efficiency, approximately 25.45%, which is mainly due to the current gain compensating for the loss in V_{oc}. When the thickness of n^+-poly-Si layer is equal to 70 nm, the cells have a high $J_{0,metal}$ value, but the cells only exhibit a slight decrease in the V_{oc} value of 1 mV and have the best FF. This result is consistent with the aforementioned analysis of the ρ_c value. However, as shown in Figure 10, when the thickness of the n^+-poly-Si layer ranged from 30 to 50 nm, there was a sharp decline in the efficiency of the cells by over 0.2%. There is a slight decrease in the V_{oc} of 1 mV for the cells with 50 nm thick n^+-poly-Si layers, while there is a significant decrease in the V_{oc} of 8 mV for the cells with 30 nm thick

n^+-poly-Si layers. This occurred because the silver embryos penetrated the passivation layer of SiO_x. Meanwhile, the J_{sc} value of the cell does not exhibit any advantages with the decreased thickness of the poly-Si layer. The decrease in the passivation performance of the n^+-poly-Si layer is primarily attributed to Ag corrosion, which impacts the generation of photogenerated electrons. Based on the above data, it can be inferred that the optimal thickness of the n^+-poly-Si layer of cell for industrial production is 70 nm.

Table 2. *I–V* parameters of n-TOPCon solar cells.

Poly Thickness (nm)	E_{ff} (%)	V_{oc} (mV)	J_{sc} (mA/cm^2)	*FF* (%)	Cell Area (cm^2)
30	25.03	716.6	41.94	83.3	334.88
50	25.27	723.5	41.89	83.4	
70	25.47	723.5	42.01	83.8	
100	25.46	724.9	41.96	83.7	

E_{ff}: efficiency; V_{oc}: open circuit voltage; J_{sc}: short-circuit current density; *FF*: fill factor.

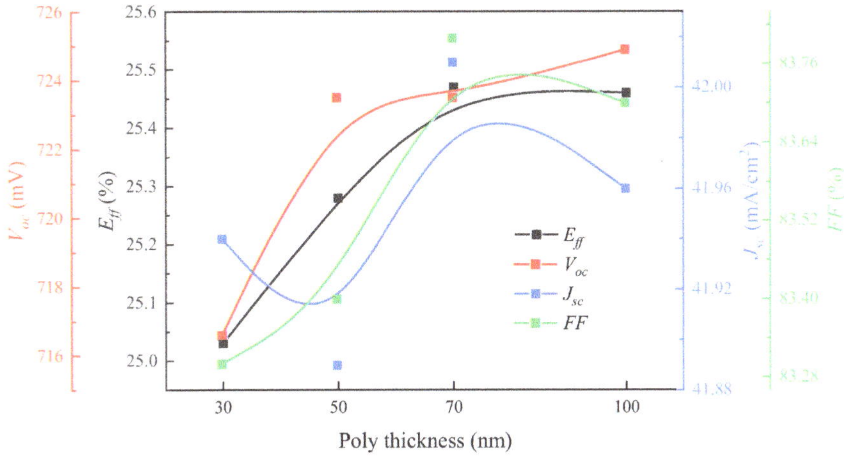

Figure 10. *I–V* parameters of n-TOPCon solar cells.

3.3. Failure Analysis

Current loss was investigated to understand the effects of the n^+-poly-Si layer thickness. As shown in Figure 11a, there is no significant difference in optical reflection between the 70 nm and 100 nm samples, as represented by a dashed curve. The solid line curve represents the internal quantum efficiency (IQE). A notable disparity in the IQE between 70 nm and 100 nm is observed at short wavelengths of <550 nm. The blue responses of these sizes were notably enhanced by the low total phosphorus doping concentration. The IQE for the 70 nm cell exhibited a good response at wavelengths < 900 nm. The curve shows that the values of the J_{sc} are increased by decreasing the thickness of the poly-Si layer owing to the low degree of parasitic absorption on the rear surface.

A current loss analysis of n-TOPCon cells with 70 nm and 100 nm thick n^+-poly-Si layers is shown in Figure 11b. The cells with an n^+-poly-Si layer thickness ranging from 70 nm to 100 nm have three advantages. The first is the "NIR parasitic absorption loss", which results in an increase of 0.14 mA/cm^2 in the J_{sc} value due to the thin n^+-poly-Si layer. The other advantages are "blue loss" and "base collection loss", which benefit from the low total dopant concentration of P. However, there is one disadvantage of "ARC reflectance" for the cells with a 70 nm thick n^+-poly-Si layer. This issue is related to the wraparound thin poly-Si layer, which is may damage the texture of the front surface and increase the reflectance [48].

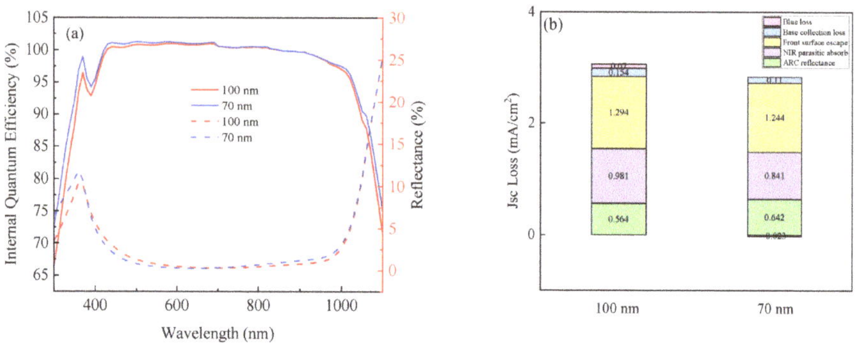

Figure 11. Solid line curves show internal quantum efficiency (IQE), and dashed line curves show optical refection (**a**) and current loss mechanisms of cells with 70 nm and 100 nm thick n^+ poly-Si layers (**b**).

After process optimization, the thin poly-Si layer can lead to an increase in J_{sc} value. However, it is essential to strike a balance between maximizing the J_{sc} value and minimizing the recombination effect of metallization on the etched depth of the poly-Si layer. Therefore, the development in metallization of industrialized poly-Si selective emitter technology will be the next focus of research.

4. Conclusions

J_0, $J_{0,metal}$, ρ_c, and microstructure analyses of contact formation were conducted, and the I–V parameters of solar cells were investigated as functions of the n^+-poly-Si layer thickness. We introduced an n^+-poly-Si layer with a thickness of 70 nm and a surface concentration of 5×10^{20} atoms/cm^3 to enhance the conversion efficiency to 25.65%. The results showed that the J_0 values of n^+-poly-Si layers exhibit a minor change of approximately 0.5 fA/cm^2 when the thickness ranges from 30 to 100 nm. However, the thickness of the n^+-poly-Si layer had a significant impact on $J_{0,metal}$ values, which increased from 304 to 545 fA/cm^2, and ρ_c values, which decreased from 9 to 40.4 mΩ·cm^2, when the thickness decreased from 70 to 30 nm. The analysis above indicates that the thickness of the n^+-poly-Si layer cannot be less than 40 nm under current process conditions. This limitation is primarily attributed to the presence of corroded silver (Ag) particles at a certain depth within the n^+-poly-Si layer. A reduction in the thickness of the poly-Si layer by 30–50 nm did not result in an increase in the short-circuit current density, resulting in an inability to effectively collect current. The ρ_c and $J_{0,metal}$ results, along with the I–V characteristics, indicate that the thickness of the n^+-poly-Si layer needs to exceed the depth of Ag particle corrosion to achieve a low ρ_c. Simultaneously, it is crucial to ensure that the $J_{0,metal}$ value is low. This optimized n^+-poly-Si layer process can be commercially applied in photovoltaics to reduce the processing time and thus lower costs.

Author Contributions: Q.W.: data curation, formal analysis, investigation, methodology, resources, software, writing—original draft, writing—review and editing, and supervision; B.G.: investigation, data curation, and methodology; W.W.: formal analysis, investigation, and writing—review and editing; W.H.: data curation, and formal analysis; K.G.: data curation, formal analysis, and investigation; J.D.: resources and writing—review and editing. All authors have read and agreed to the published version of the manuscript.

Funding: This work was partially supported by the National Natural Science Foundation Youth Project (62304199).

Institutional Review Board Statement: Not applicable.

Informed Consent Statement: Not applicable.

Data Availability Statement: Data is contained within the article.

Conflicts of Interest: Author Beibei Gao was employed by the company Jinko Solar Co., Ltd. The remaining authors declare that the research was conducted in the absence of any commercial or financial relationships that could be construed as a potential conflict of interest.

References

1. Ullah, H.; Czapp, S.; Szultka, S.; Tariq, H.; Qasim, U.B.; Imran, H. Crystalline Silicon (c-Si)-Based Tunnel Oxide Passivated Contact (TOPCon) Solar Cells: A Review. *Energies* **2023**, *16*, 715. [CrossRef]
2. Fischer, M.; Woodhouse, M.; Herritsch, S.; Trube, J. *International Technology Roadmap for Photovoltaic (ITRPV)*, 14th ed.; VDMA German Engineering Federation: Frankfurt am Main, Germany, 2023.
3. Jinko (n.d.). *High-Efficiency N-Type Monocrystalline Silicon Solar Cell Sets Our New Record with Maximum Conversion Efficiency of 26.4%*; Jinko Solar's: Zhejiang, China, 2023.
4. Hermle, M.; Feldmann, F.; Bivour, M.; Goldschmidt, J.C.; Glunz, S.W. Passivating contacts and tandem concepts: Approaches for the highest silicon-based solar cell efficiencies. *Appl. Phys. Rev.* **2020**, *7*, 021305. [CrossRef]
5. Nemeth, B.; Young, D.L.; Page, M.R.; LaSalvia, V.; Johnston, S.; Reedy, R.; Stradins, P. Polycrystalline silicon passivated tunneling contacts for high efficiency silicon solar cells. *J. Mater. Res.* **2016**, *31*, 671. [CrossRef]
6. Ng, K.K.; Card, H.C. Asymmetry in the SiO_2 tunneling barriers to electrons and holes. *J. Appl. Phys.* **1980**, *51*, 2153–2157. [CrossRef]
7. Chhetri, N.; Das, G.; Kole, A.; Saha, H.; Haldar, S.; Chatterjee, S. P-type polycrystalline silicon seed layer fabrication using RF-PECVD for thin film solar cells. *Mater. Today Proc.* **2022**, *58*, 665–667. [CrossRef]
8. Stöhr, M.; Aprojanz, J.; Brendel, R.; Dullweber, T. Firing-stable PECVD SiOxNy/n-Poly-Si surface passivation for silicon solar cells. *ACS Appl. Energy Mater.* **2021**, *4*, 4646–4653. [CrossRef]
9. Fırat, M.; Radhakrishnan, H.S.; Payo, M.R.; Choulat, P.; Badran, H.; van der Heide, A.; Govaerts, J.; Duerinckx, F.; Tous, L.; Hajjiah, A.; et al. Large-area bifacial n-TOPCon solar cells with in situ phosphorus-doped LPCVD poly-Si passivating contacts. *Sol. Energy Mater. Sol. Cells* **2022**, *236*, 111544. [CrossRef]
10. Stuckelberger, J.; Yan, D.; Phang, S.P.; Samundsett, C.; Wang, J.; Antognini, L.; Haug, F.J.; Wang, Z.; Yang, J.; Zheng, P.; et al. Pre-annealing for improved LPCVD deposited boron-doped poly-Si hole-selective contacts. *Sol. Energy Mater. Sol. Cells* **2023**, *251*, 112123. [CrossRef]
11. Mousumi, J.F.; Gregory, G.; Ganesan, J.P.; Nunez, C.; Provancha, K.; Seren, S.; Zunft, H.; Jurca, T.; Banerjee, P.; Kar, A.; et al. Process–structure–properties relationships of passivating, electron-selective contacts formed by atmospheric pressure chemical vapor deposition of phosphorus-doped polysilicon. *Phys. Status Solidi RRL—Rapid Res. Lett.* **2022**, *16*, 2100639. [CrossRef]
12. Li, S.; Pomaska, M.; Hoß, J.; Lossen, J.; Pennartz, F.; Nuys, M.; Hong, R.; Schmalen, A.; Wolff, J.; Finger, F.; et al. Poly-Si/SiO x /c-Si passivating contact with 738 mV implied open circuit voltage fabricated by hot-wire chemical vapor deposition. *Appl. Phys. Lett.* **2019**, *114*, 153901. [CrossRef]
13. Ingenito, A.; Allebé, C.; Libraro, S.; Ballif, C.; Paviet-Salomon, B.; Nicolay, S.; Leon, J.J.D. 22.8% full-area bifacial n-PERT solar cells with rear side sputtered poly-Si(n) passivating contact. *Sol. Energy Mater. Sol. Cells* **2023**, *249*, 112043. [CrossRef]
14. Nasebandt, L.; Min, B.; Hollemann, C.; Hübner, S.; Dippell, T.; Peibst, R.; Brendel, R. Sputtered phosphorus-doped poly-Si on oxide contacts for screen-printed Si solar cells. *Sol. RRL* **2022**, *6*, 2200409. [CrossRef]
15. Sedani, S.H.; Yasar, O.F.; Karaman, M.; Turan, R. Effects of boron doping on solid phase crystallization of in situ doped amorphous Silicon thin films prepared by electron beam evaporation. *Thin Solid Films* **2020**, *694*, 137639. [CrossRef]
16. Kafle, B.; Goraya, B.S.; Mack, S.; Feldmann, F.; Nold, S.; Rentsch, J. TOPCon-technology options for cost efficient industrial manufacturing. *Sol. Energy Mater. Sol. Cells* **2021**, *227*, 111100. [CrossRef]
17. Truong, T.N.; Yan, D.; Nguyen, C.P.T.; Kho, T.; Guthrey, H.; Seidel, J.; Al-Jassim, M.; Cuevas, A.; Macdonald, D.; Nguyen, H.T. Microstructure, and doping behaviour: A comparison between different deposition methods for poly-Si/SiOx passivating contacts. *Prog. Photovolt. Res. Appl.* **2021**, *29*, 857–868. [CrossRef]
18. Yan, D.; Cuevas, A.; Michel, J.I.; Zhang, C.; Wan, Y.; Zhang, X.; Bullock, J. Polysilicon passivated junctions: The next technology for silicon solar cells? *Joule* **2021**, *5*, 811–828. [CrossRef]
19. Chen, D.; Chen, Y.; Wang, Z.; Gong, J.; Liu, C.; Zou, Y.; He, Y.; Wang, Y.; Yuan, L.; Lin, W.; et al. 24.58% total area efffciency of screen-printed, large area industrial silicon solar cells with the tunnel oxide passivated contacts (i-TOPCon) design. *Sol. Energy Mater. Sol. Cells* **2020**, *206*, 110258. [CrossRef]
20. Chen, Y.; Chen, D.; Liu, C.; Wang, Z.; Zou, Y.; He, Y.; Wang, Y.; Yuan, L.; Gong, J.; Lin, W.; et al. Mass production of industrial tunnel oxide passivated contacts (i-TOPCon) silicon solar cells with average efffciency over 23% and modules over 345 W. *Prog. Photovolt. Res. Appl.* **2019**, *27*, 827–834. [CrossRef]
21. Stodolny, M.K.; Lenes, M.; Wu, Y.; Janssen, G.J.M.; Romijn, I.G.; Luchies, J.R.M.; Geerligs, L.J. n-Type polysilicon passivating contact for industrial bifacial n-type solar cells. *Sol. Energy Mater. Sol. Cells* **2016**, *158*, 24–28. [CrossRef]

22. Stodolny, M.K.; Anker, J.; Geerligs, B.L.; Janssen, G.J.; Van De Loo, B.W.; Melskens, J.; Santbergen, R.; Isabella, O.; Schmitz, J.; Lenes, M.; et al. Material properties of LPCVD processed n-type polysilicon passivating contacts and its application in PERPoly industrial bifacial solar cells. In *Energy Procedia*; Elsevier Ltd.: Amsterdam, The Netherlands, 2017; pp. 635–642.
23. Padhamnath, P.; Buatis, J.K.; Khanna, A.; Nampalli, N.; Nandakumar, N.; Shanmugam, V.; Aberle, A.G.; Duttagupta, S. Characterization of screen printed and ffre-through contacts on LPCVD based passivating contacts in monoPolyTM solar cells. *Sol. Energy* **2020**, *202*, 73–79. [CrossRef]
24. Padhamnath, P.; Khanna, A.; Nandakumar, N.; Nampalli, N.; Shanmugam, V.; Aberle, A.G.; Duttagupta, S. Development of thin polysilicon layers for application in monoPoly^{TM} cells with screen-printed and fired metallization. *Sol. Energy Mater. Sol. Cells* **2020**, *207*, 110358. [CrossRef]
25. Learn, A.J.; Foster, D.W. Deposition and electrical properties of in situ phosphorusdoped silicon fflms formed by low-pressure chemical vapor deposition. *J. Appl. Phys.* **1987**, *61*, 1898–1904. [CrossRef]
26. Meyerson, B.S.; Olbricht, W. Phosphorus-doped polycrystalline silicon via LPCVD: I. Process characterization. *J. Electrochem. Soc.* **1984**, *131*, 2361–2365. [CrossRef]
27. Meyerson, B.S.; Yu, M.L. Phosphorus-doped polycrystalline silicon via LPCVD: II. Surface interactions of the silane/phosphine/silic system. *J. Electrochem. Soc.* **1984**, *131*, 2366–2368. [CrossRef]
28. Mulder, J.G.M.; Eppenga, P.; Hendriks, M.; Tong, J.E. An industrial LPCVD process for in situ phosphorus-doped polysilicon. *J. Electrochem. Soc.* **1990**, *137*, 273–279. [CrossRef]
29. Ahmed, W.; Ahmed, E. An investigation of LPCVD and PECVD of in situ doped polycrystalline silicon for VLSI. *Adv. Mater. Opt. Electron.* **1992**, *1*, 255–259. [CrossRef]
30. Qiu, K.; Pomaska, M.; Li, S.; Lambertz, A.; Duan, W.; Gad, A.; Geitner, M.; Brugger, J.; Liang, Z.; Shen, H.; et al. Development of conductive SiCx:H as a new hydrogenation technique for tunnel oxide passivating contacts. *ACS Appl. Mater. Interfaces* **2020**, *12*, 29986–29992. [CrossRef] [PubMed]
31. Yang, Q.; Liu, Z.; Lin, Y.; Liu, W.; Liao, M.; Feng, M.; Zhi, Y.; Zheng, J.; Lu, L.; Ma, D.; et al. Passivating contact with phosphorus-doped polycrystalline silicon-nitride with an excellent implied open-circuit voltage of 745 mV and its application in 23.88% efffciency TOPCon solar cells. *Sol. RRL* **2021**, *5*, 2100644. [CrossRef]
32. Römer, U.; Peibst, R.; Ohrdes, T.; Lim, B.; Krügener, J.; Wietler, T.; Brendel, R. Ion implantation for poly-Si passivated back-junction back-contacted solar cells. *IEEE J. Photovolt.* **2015**, *5*, 507–514. [CrossRef]
33. Zheng, P.; Yang, J.; Wang, Z.; Wu, L.; Sun, H.; Chen, S.; Guo, Y.; Xia, H.; Phang, S.P.; Wang, E.C.; et al. Detailed loss analysis of 24.8% large-area screen-printed n-type solar cell with polysilicon passivating contact. *Cell Rep. Phys. Sci.* **2021**, *2*, 100603. [CrossRef]
34. Wu, W.F.; Tsai, K.C.; Chao, C.G.; Chen, J.C.; Ou, K.L. Novel multilayered Ti/TiN diffusion barrier for Al metallization. *J. Electron. Mater.* **2005**, *34*, 1150–1156. [CrossRef]
35. Yu, J.; Chen, Y.; He, J.; Bai, Y.; Su, R.; Cao, T.; Liu, W.; Chen, T. Enhancing poly-Si contact through a highly conductive and ultra-thin TiN layer for high-efficiency passivating contact silicon solar cells. *Sol. Energy Mater. Sol. Cells* **2023**, *260*, 112491. [CrossRef]
36. Morales-Vilches, A.B.; Larionova, Y.; Wietler, T.; Cruz, A.; Korte, L.; Peibst, R.; Brendel, R.; Schlatmann, R.; Stannowski, B. ZnO: Al/a-SiOx front contact for polycrystalline-silicon-on-oxide (POLO) solar cells. In *AIP Conference Proceedings*; AIP Publishing: Melville, NY, USA, 2018.
37. Adachi, D.; Terashita, T.; Uto, T.; Hernández, J.L.; Yamamoto, K. Effects of SiOx barrier layer prepared by plasma-enhanced chemical vapor deposition on improvement of long-term reliability and production cost for Cu-plated amorphous Si/crystalline Si heterojunction solar cells. *Sol. Energy Mater. Sol. Cells* **2017**, *163*, 204–209. [CrossRef]
38. Fellmeth, T.; Höffler, H.; Mack, S.; Krassowski, E.; Krieg, K.; Kafle, B.; Greulich, J. Laser-enhanced contact optimization on iTOPCon solar cells. *Prog. Photovolt. Res. Appl.* **2022**, *23630*, 1393–1399. [CrossRef]
39. Chaudhary, A.; Hoß, J.; Lossen, J.; Huster, F.; Kopecek, R.; van Swaaij, R.; Zeman, M. Influence of polysilicon thickness on properties of screen-printed silver paste metallized silicon oxide/polysilicon passivated contacts. *Phys. Status Solidi (a)* **2021**, *218*, 2100243. [CrossRef]
40. Meyer, F.; Ingenito, A.; Leon, J.J.D.; Niquille, X.; Allebé, C.; Nicolay, S.; Haug, F.J.; Ballif, C. Localisation of front side passivating contacts for direct metallisation of high-efficiency c-Si solar cells. *Sol. Energy Mater. Sol. Cells* **2022**, *235*, 111455. [CrossRef]
41. Çiftpınar, H.E.; Stodolny, M.K.; Wu, Y.; Janssen, G.J.; Löffler, J.; Schmitz, J.; Lenes, M.; Luchies, J.M.; Geerligs, L.J. Study of screen printed metallization for polysilicon based passivating contacts. *Energy Procedia* **2017**, *124*, 851–861. [CrossRef]
42. Wang, Q.; Wu, W.; Li, Y.; Yuan, L.; Yang, S.; Sun, Y.; Yang, S.; Zhang, Q.; Cao, Y.; Qu, H.; et al. Impact of boron doping on electrical performance and efficiency of n-TOPCon solar cell. *Sol. Energy* **2021**, *227*, 273–291. [CrossRef]
43. Wang, Q.; Peng, H.; Gu, S.; Guo, K.; Wu, W.; Li, B.; Li, L.; Yuan, N.; Ding, J. High-efficiency n-TOPCon bifacial solar cells with selective poly-Si based passivating contacts. *Sol. Energy Mater. Sol. Cells* **2023**, *259*, 112458. [CrossRef]
44. Wang, Q.; Yuan, L.; Li, L.; Peng, H.; Li, B.; Wang, A.; Zhang, L.; Wu, W.; Ding, J.; Yuan, N. Boron tube diffusion process parameters for high-efficiency n-TOPCon solar cells with selective boron emitters. *Sol. Energy Mater. Sol. Cells* **2023**, *253*, 112231. [CrossRef]
45. Liu, B.; Chen, Y.; Yang, Y.; Chen, D.; Feng, Z.; Altermatt, P.P.; Verlinden, P.; Shen, H. Improved evaluation of saturation currents and bulk lifetime in industrial Si solar cells by the quasi steady state photo conductance decay method. *Sol. Energy Mater. Sol. Cells* **2016**, *149*, 258–265. [CrossRef]

46. Yablonovitch, E. Statistical ray optics. *J. Opt. Soc. Am.* **1987**, *72*, 899–907. [CrossRef]
47. Allen, T.G.; Bullock, J.; Yang, X.; Javey, A.; De Wolf, S. Passivating contacts for crystalline silicon solar cells. *Nat. Energy* **2019**, *4*, 914–928. [CrossRef]
48. Wang, Q.; Wu, W.; Chen, D.; Yuan, L.; Yang, S.; Sun, Y.; Zhang, Q.; Cao, Y.; Qu, H.; Yuan, N.; et al. Study on the cleaning process of n^+-poly-Si wraparound removal of TOPCon solar cells. *Sol. Energy* **2020**, *211*, 324–335. [CrossRef]

materials

Article

Synthesis of Size-Adjustable $CsPbBr_3$ Perovskite Quantum Dots for Potential Photoelectric Catalysis Applications

Hang Li, Jiazhen He, Xiaoqian Wang, Qi Liu, Xuemin Luo, Mingwei Wang, Jinfeng Liu, Chengqi Liu and Yong Liu *

International School of Materials Science and Engineering (ISMSE), State Key Laboratory of Advanced Technology for Materials Synthesis and Processing, Wuhan University of Technology, Wuhan 430070, China; leehang@whut.edu.cn (H.L.); jiazhenhe0606@163.com (J.H.); 303568@whut.edu.cn (X.W.); liuq@whut.edu.cn (Q.L.); luoxuemin1123@163.com (X.L.); wmw1842591883@163.com (M.W.); liujinf990528@whut.edu.cn (J.L.); liuchengqi42@163.com (C.L.)

* Correspondence: liuyong3873@whut.edu.cn

Abstract: As a direct band gap semiconductor, perovskite has the advantages of high carrier mobility, long charge diffusion distance, high defect tolerance and low-cost solution preparation technology. Compared with traditional metal halide perovskites, which regulate energy band and luminescence by changing halogen, perovskite quantum dots (QDs) have a surface effect and quantum confinement effect. Based on the LaMer nucleation growth theory, we have synthesized $CsPbBr_3$ QDs with high dimensional homogeneity by creating an environment rich in Br^- ions based on the general thermal injection method. Moreover, the size of the quantum dots can be adjusted by simply changing the reaction temperature and the concentration of Br^- ions in the system, and the blue emission of strongly confined pure $CsPbBr_3$ perovskite is realized. Finally, optical and electrochemical tests suggested that the synthesized quantum dots have the potential to be used in the field of photocatalysis.

Keywords: $CsPbBr_3$ QDs; quantum confinement effect; blue emission; size-adjustable

1. Introduction

Over the past years, metal halide perovskite, a new type of semiconductor material, has attracted wide attention in the photovoltaic [1], luminescence [2] and detection [3] fields because of its excellent photoelectric properties. Compared with traditional semiconductors, perovskites have a special structure, and the general formula is ABX_3 [4]. This paper focuses on metallic lead halide perovskites with an A-site of MA^+, FA^+ or Cs^+; the B-site is occupied by metal cations such as Pb^{2+} or Sn^{2+}, and they have an X-site of Cl^-, Br^- or I^-. The B-site atoms are in the body center of the cube and form a regular octahedral structure of $BX_6{}^{4-}$ with the X-site atoms, the A-site atoms are in the vertex of the cube, and correspondingly, the X-site atoms are in the six face centers of the cube [5–9].

This material is a direct bandgap semiconductor with advantages such as high carrier mobility, long charge diffusion distance, high defect tolerance and a low-cost solution preparation process. Today, perovskite solar cells have achieved a photoelectric conversion efficiency of more than 25.5%, and perovskite quantum dots (QDs) combine the properties of perovskite block materials and traditional semiconductor quantum dots with excellent photoelectric properties, such as high fluorescence quantum yield, extremely narrow half-peak width, adjustable band gap, full-spectrum coverage in visible light, etc. These properties make perovskite quantum dots one of the most promising photoelectric materials at present; they are therefore widely used in light-emitting diodes, solar cells, photodetectors and other fields [10–13]. Meanwhile, there are relevant examples of the application of quantum dots in the field of photocatalysis [14]. Similarly, composite materials such as perovskite quantum dots combined with other materials to form heterojunctions also have the potential for photocatalytic applications [15]. Quantum dots are quasi-zero-dimensional particles composed of a small number of atoms whose dimensions are close to their Bohr radii [16]. In quantum dots, charge carriers are restricted in all directions and are sometimes referred to as artificial atoms [17]. Quantum dots can be thought of as bridges between molecular clusters and bulk crystals depending on their size. For semiconductors, an extremely important parameter is the gap separating the conduction band (CB) and the valence band (VB), which is fixed for macroscopic crystals and is determined only by the material properties. However, for quantum dots, when the size is reduced to the nanometer level, the electrons are limited by the boundary during movement, and the corresponding energy level will be adjusted to match the change in size. When a quantum dot is excited by a photon of higher energy, the electrons in the VB experience energy fluctuations and are excited to the CB, while a hole is left in the valence band, and the formed electron–hole pair (charge carriers) will be attracted to each other by electrostatic action; the average distance between the electron and the hole is regarded as the Bohr radius (The Bohr radius of $CsPbBr_3$ is 7 nm). The diffusion of charge carriers in the semiconductor lattice is related to the Bohr radius. When the size of the quantum dot is less than or close to its Bohr radius, the quantization of energy will be triggered, which will cause the electronic energy-level structure of the semiconductor material to change from a continuous energy level to a discrete energy level, and also cause the band gap between CB and VB to widen; this is the quantum-limited effect [18–20]. The quantum confinement effect can also be simply understood: the blue shift of the absorption and fluorescence spectra occurs as the size of the semiconductor decreases, and the blue shift of the spectrum becomes more obvious with a smaller size. In addition, a reduction in size will result in an increase in the proportion of atoms on the surface. Compared with bulk materials, quantum dots have a large specific surface area, and the surface atoms will form a large number of suspension bonds due to incomplete coordination [21,22]. The existence of suspension bonds may introduce defect levels in the semiconductor band gap and cause non-radiative recombination. At this time, the surface plays a decisive role in the structure and optical properties of the quantum dot [23–26]. In short, there are some properties distinguishing quantum dots compared to bulk materials.

Like traditional quantum dots, perovskite quantum dots can regulate the band gap through the quantum confinement effect to achieve tunable luminescence, and the absorption spectrum will gradually shift toward blue with the decrease in quantum dot size [27,28]. Another, lighter way is to adjust the luminescence through component engineering, that is, by adjusting the proportion of halogens. This is mainly because the CB of perovskite materials is contributed by the 6p orbital of Pb and the np orbital of halogens, and the 6p orbital contribution of Pb is a major part. The VB tops are mainly contributed by the 6s orbits of Pb and the np orbits of halogens [29]. For all metallic lead halide perovskites, the composition of lead is the same, so the band gap of perovskite can be continuously adjusted easily by adjusting the halogen composition, and, thus, the emission of perovskite in the entire visible band can be realized [30]. In addition, due to the soft ionic properties of halide perovskites, there is strong ionic interaction with surface ligands [31–33]. This makes it easy to exchange anions between perovskites with different halogens, and it is also easy and convenient to adjust the full spectrum of the visible light band of perovskites through composition engineering [34]. But, unfortunately, it is precisely because of this characteristic that the phase separation of perovskite mixed halogens occurs very easily in practical applications. Conversely, this problem can be fundamentally solved by regulating luminescence through quantum confinement effects. At the same time, how to prepare and control the size of quantum dots and ensure the uniformity of particle size has become a new problem to be solved urgently [35]. Based on previous work, we know that the synthesis of quantum dots follows LaMer nucleation growth theory. It mainly includes three parts; first, the monomer formation stage: because the precursor reacts under certain conditions, the monomer is formed quickly. Second, the nucleation stage: when the concentration of monomer exceeds the critical concentration of nucleation, it will be rapidly nucleated due to supersaturation, and a large amount of monomer will be consumed during nucleation, thereby reducing the monomer concentration to the critical concentration of nucleation again. Third, the growth stage: since the monomer concentration is lower than the critical concentration of nucleation, but still higher than the equilibrium concentration, within this range, QDs will carry out the growth stage [36]. Then, as the concentration of monomers in the solution is depleted, it enters the Ostwald ripening process. This process involves the dissolution of small-size QDs and the continued growth of large-size QDs, which will cause the widening of QDs [37]. The results show that accelerating the monomer formation rate can reduce the nucleation time of QDs, which is conducive to the formation of a more uniform QDs core. The homogeneous core can improve the uniformity of subsequent growth and inhibit the Ostwald ripening process, thus achieving the synthesis of QDs with a narrow size distribution.

Here, we add $ZnBr_2$ to create a Br^- ion-rich reaction system on the basis of the general hot injection method, and effectively regulate the synthesis of $CsPbBr_3$ QDs based on LaMer nucleation growth theory [38]. By simply adjusting the concentration of Br^- ions and changing the reaction temperature, $CsPbBr_3$ QDs of different sizes with ultra-high particle size uniformity can be obtained easily. These $CsPbBr_3$ QDs of similar size exhibit a certain self-assembly behavior due to their high homogeneity, which can be demonstrated by their XRD tests [39]. X-ray photoelectron spectroscopy (XPS) testing can also effectively prove that the Cs, Pb and Br elements in $CsPbBr_3$ QDs have a reasonable chemical state environment. In addition, the ultraviolet–visible absorption (UV-vis) spectrum and photoluminescence (PL) spectrum can prove that, based on the quantum confinement effect, the corresponding luminous band of $CsPbBr_3$ QDs is significantly blue-shifted with the reduction in their size [40]. Finally, optical and electrochemical tests show that the perovskite quantum dots have great potential for application in the field of photocatalysis.

2. Materials and Methods

2.1. Materials

Cesium carbonate (Cs_2CO_3, Aladdin, Shanghai China, 99.9%), lead bromide ($PbBr_2$, Aladdin, Shanghai China, 99.99%), zinc bromide ($ZnBr_2$, Aladdin, Shanghai China, 99%), oleic acid (OA, Aladdin, Shanghai China, AR), oleylamine (OAm, Aladdin, Shanghai China, 80~90%), 1-octadecene (ODE, Aladdin, Shanghai China, >90% (GC)), n-hexane (Aladdin, Shanghai China, >99%), acetone (Aladdin, Shanghai China, >99%).

2.2. Methods

2.2.1. Preparation of CsOA Precursors

The Cs-oleate precursor solution was prepared in a 50 mL three-necked round-bottomed flask by dissolving Cs_2CO_3 (0.4 g) in a mixture of oleic acid (OA, 5 mL) and 1-octadecene (ODE, 15 mL). After drying with a vacuum pump for 1 h at 120 °C, it was heated under N_2 protection to 140 °C until all Ca_2CO_3 dissolved to form Cs-oleate. The Cs-oleate precursor solution was preheated to 120 °C before injection.

2.2.2. Synthesis of $CsPbBr_3$ Quantum Dots (QDs)

ODE (10 mL), $PbBr_2$ (0.138 g) and $ZnBr_2$ were loaded in a 50 mL 3-necked flask and vacuum-dried for 1 h at 120 °C, and then oleylamine (2 mL) and OA (2 mL) were injected under a N_2 atmosphere. After the $PbBr_2$ and $ZnBr_2$ dissolved, the temperature was set to a different temperature, and the Cs-oleate solution (0.8 mL, 0.125 M in ODE) was quickly injected in. The reaction mixture was cooled by an ice-water bath after reacting for 30 s for the crystal growth. The product was collected by centrifugation. The amount of $ZnBr_2$ added depended on the molar ratio of bromine/lead. When the molar ratio of Br:Pb = 6:1, $ZnBr_2$ (0.17 g) was required, and when the molar ratio of Br:Pb = 10:1, twice as much was required.

2.2.3. Purification of $CsPbBr_3$ Quantum Dots (QDs)

For the product in the centrifugation, the supernatant was left on the bench top under ambient conditions for ~3 h until the unreacted salts precipitated. Then the mixture was centrifuged to get the clear supernatant. With a proper amount of the clarified solution, ~5 mL of acetone was slowly added until the mixture became turbid to avoid decomposition of the QDs. Then the QDs were centrifuged at 10,000 rpm for 10 min and the precipitate was collected and dissolved in ~500 μL of hexane.

2.3. *Characterization*

2.3.1. Characterization via Transmission Electron Microscopy (TEM)

Images captured via transmission electron microscopy (TEM) were generated utilizing a JEOL JEM-1400 Plus microscope (Beijing, China). This instrument, featuring a thermionic emission source, was set to function at an acceleration power of 120 kV. For the examination, quantum dot (QD) specimens underwent a precise preparation process, involving the deposition of a thinned QD mixture in n-hexane onto grids coated with carbon on copper. Following this, the analysis of the QDs' size distribution was conducted through the application of Nano Measurer software (1.2), bearing version number 1.02.0005.

2.3.2. Characterization via X-ray Diffraction (XRD)

X-ray diffraction (XRD) measurements were conducted utilizing a Bruker D8 ADVANCE diffractometer, from Karlsruhe, Germany, employing Cu-Kα radiation at a wavelength of 1.540598 Å. Data collection covered a 2θ range from 10° to 60°, utilizing a step increment of 0.05° and a dwell time of 0.5 s per step. The instrument was set to a voltage of 40 kV and a current of 40 mA. To prepare for XRD, a dense suspension of quantum dots (QDs) in n-hexane was drop-cast onto silicon dioxide/silicon (SiO_2/Si) supports. Analysis of the XRD patterns was subsequently performed with Jade 6.5 software.

2.3.3. X-ray Photoelectron Spectroscopy (XPS) Characterization

Spectra from X-ray photoelectron spectroscopy (XPS) of perovskite quantum dots were captured with an AXIS SUPRA spectrometer from Kratos Analytical, part of the Shimadzu Group, located in Manchester, UK. This device features a Thermo Scientific Kα spectrometer and utilizes a monochromatic Al-Kα X-ray for illumination. Data collected were then transformed into VGD format and analyzed using Avantage software (599-31), release 5.9922. Calibration of the binding energy scale was aligned using the C 1 s peak, attributing a binding energy of 284.8 eV specifically to the carbon–carbon (C–C) bond.

2.3.4. Fluorescence Spectrum Characterization

Optical absorption spectra in the UV–visible range were obtained using a Shimadzu UV-1800 spectrophotometer (Kyoto, Japan), which was operated with UVProbe 2.52 software. Concurrently, the emission characteristics were assessed through steady-state photoluminescence (PL) spectra, recorded using a Shimadzu RF-6000 spectrophotometer (Kyoto, Japan) equipped with LabSolutions RF software (1.11). For $CsPbBr_3$, the excitation wavelength (λ_{ex}) utilized was 400 nm. The preparation for optical evaluation involved the dilution of quantum dot (QD) solutions in n-hexane, followed by their allocation into quartz cuvettes, each with a path length of 10 mm, to ensure accurate measurements.

2.3.5. Electrochemistry (Photo) Measurements

Photocurrent and electrochemical impedance analyses on halide perovskite quantum dots (QDs) were conducted using a Zennium workstation, supplied by Zahner Company of Kronach, Germany. This investigation utilized a tri-electrode configuration comprising a sample-based working electrode, a counter-electrode of platinum disk, and a Ag/AgCl reference electrode (in saturated KCl solution) with a potential of +0.1989 V relative to the Normal Hydrogen Electrode (NHE) within an electrolyte solution of 0.5 M Na_2SO_4. Illumination for the photocurrent assessments was provided by a 405 nm wavelength

LED, with its intensity quantified using a Newport photometer (Irvine, CA, USA). The electrochemical impedance spectra were acquired at a potential of −0.1 V versus NHE, with application of a 10 mV signal over a frequency range from 100 mHz to 20 kHz, in a 0.5 M Na_2SO_4 solution.

3. Results

The traditional synthesis of $CsPbBr_3$ nanocrystals involves simple dissolution of $PbBr_2$ and ligands in a solvent and hot injection of cesium oleate to obtain the corresponding halogen perovskite nanocrystals. Although monodisperse nanocrystals have been synthesized successfully, their sizes are not uniform, and effective size control methods are lacking. Based on the LaMer nucleation growth theory, it can be seen that this is an extremely rapid explosive nucleation process, accompanied by the Oswald maturation process in parallel with crystal growth. Therefore, a BR^--rich reaction system was constructed by introducing $ZnBr_2$, which can improve the formation of perovskite monomer and shorten the nucleation time. The homogeneous core is conducive to inhibiting the Oswald ripening process in the later stage of growth, avoiding size widening and obtaining a narrow size distribution of quantum dots. In addition, the size of the perovskite QDs can be adjusted by changing the reaction temperature during hot injection. As the size of QDs decreases, we obtain strongly confined $CsPbBr_3$ QDs. Due to the great quantum confinement effect, the luminous wavelength of $CsPbBr_3$ QDs shifts from green to blue. Compared with the component engineering of Cl/Br mixed halogens, the blue light formed based on dimension engineering fundamentally solves the problem of easy phase separation of mixed halogens. At the same time, with the addition of excessive Br^- ions, the Br vacancy defects on the surface of perovskite are greatly passivated [41]. In bromine-rich reaction systems, Pb^{2+} and Br^- ions are combined to form a complex of $[PbBr_3]^-$ firstly, and then the complex is transformed into an octahedron. Once the octahedron is formed, sufficient nucleation sites are provided for rapid nucleation with the addition of Cs-OA. We obtained $CsPbBr_3$ QDs of different sizes at different reaction temperatures (150 °C, 130 °C, 110 °C, 90 °C) (Scheme 1). Based on the weak-to-strong quantum confinement effect, the continuously tunable emission of $CsPbBr_3$ QDs from green light to blue light was realized.

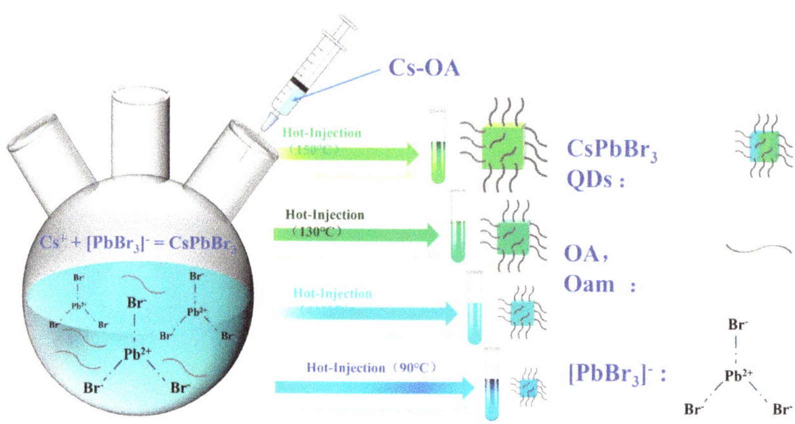

Scheme 1. Illustrative representation of the $CsPbBr_3$ QDs synthesis process.

Shown in Figure 1A–D are TEM images of $CsPbBr_3$ QDs corresponding to reaction temperatures of 150 °C, 130 °C, 110 °C and 90 °C, respectively. The size of $CsPbBr_3$ QDs can be visually observed through TEM images. It can be clearly seen from the TEM images that the size of the obtained $CsPbBr_3$ QDs decreased with the gradual decrease in the reaction temperature during hot injection. In addition, we calculated the sizes of hundreds of quantum dots, made a bar chart, and placed it under the TEM images of quantum dots we obtained, arranged by the corresponding reaction temperatures (Figure 1E–H). According to the information in the bar chart, the average size and particle size distribution of $CsPbBr_3$ QDs obtained, from high to low reaction temperature, are d = 5.7 ± 0.1 nm, d = 4.9 ± 0.05 nm, d = 3.9 ± 0.1 nm and d = 3.5±0.11 nm, respectively. In addition, comparing Figures S1 and 1, it can be found that as the molar ratio of bromine to lead increases from Br: Pb = 6:1 to Br: Pb = 10:1, the sizes of QDs in the system with higher concentrations of Br^- ions at the same reaction temperature become smaller. This shows that in addition to the temperature condition, the change in Br^- ion content in the system also really affects the size of the quantum dot. At the same time, in an environment with high-energy particle impacts such as TEM, there are no white spots, which indicates the destruction of the structure, and small black patches, which indicate the collapse of the structure, leading to the precipitation of $PbBr_2$. Studies indicate that increasing the concentration of halogen ions can improve the radiation resistance of halide perovskite QDs [42]. This may be attributed to the fact that with the increase in Br^- ion content in the system, more nucleation sites are formed while the surface defects are passivated. The increase in nucleation sites greatly accelerates the nucleation process and, thus, effectively inhibits Oswald ripening in some aspects, which means that the sizes of the perovskite quantum dots with extremely high size uniformity obtained by the improved method in this paper can be controlled simply and effectively. What is more interesting is that, compared with the monodisperse $CsPbBr_3$ QDs synthesized by the general method, they show a chaotic arrangement (Figure S2). The improved $CsPbBr_3$ QDs synthesized in this paper are arranged in an orderly manner on the substrate, and show a trend of self-assembly. It should be clear that this trend of self-assembly has high requirements for the dimensional uniformity of the assembled units, which is also evidence that the synthesized $CsPbBr_3$ QDs have ultra-high dimensional uniformity. This high dimensional uniformity and size adjustability will also greatly facilitate the installation of $CsPbBr_3$ QDs in channel materials, further developing their potential in areas such as catalysis.

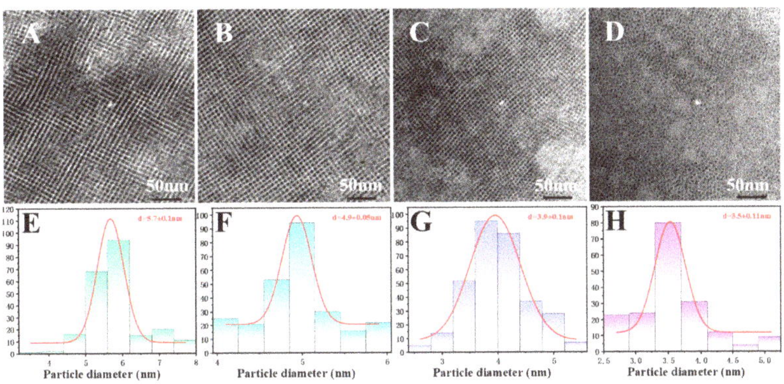

Figure 1. (**A–D**) TEM images of $CsPbBr_3$ QDs at different reaction temperatures, respectively (scale bar: 50 nm) and (**E–H**) their size-of-particle-diameter distribution diagrams.

It is noteworthy that the X-ray diffraction (XRD) pattern exhibits only two highly intense peaks, near the diffraction angles 2θ = 15° and 2θ = 30°. These peaks correspond to the (001) and (002) crystal faces of $CsPbBr_3$ perovskite, respectively. The presence of these specific peaks, in accordance with the ICSD code 029073, confirms the cubic crystal structure of $CsPbBr_3$ QDs (Figure S3). Just like the extremely ordered arrangement of quantum dots in the TEM images, $CsPbBr_3$ QDs are arranged along the preferred orientation of the (001) crystal plane, with structural coherence, so that there are extremely strong diffraction peaks at the peaks of the (001) crystal plane and the (002) crystal plane in the XRD pattern. The special XRD pattern of the synthesized $CsPbBr_3$ QDs is in good agreement with the special arrangement in the TEM images. Figures 2B–D and S4 show the XPS spectrum of $CsPbBr_3$ QDs. According to the curve of the spectrum, the chemical valence states of Cs, Pb and Br were analyzed [43]. Figure 2B shows the binding energy curve of the Cs 3d orbit, where the peaks corresponding to Cs $3d_{5/2}$ and Cs $3d_{3/2}$ are located near 724 eV and 738 eV, respectively. Similarly, Figure 2C shows the binding energy curve of the Pb 4f orbit, where the peaks corresponding to Pb $4f_{7/2}$ and Pb $4f_{5/2}$ are located near 138 eV and 143 eV, respectively. In addition, in Figure 2D, the coupling peaks near 68 eV and 69 eV come from the Br 3d orbit, representing Br $3d_{5/2}$ and Br $3d_{3/2}$, respectively. In the above peak location, the peak location of the Pb 4f track is similar to that recorded in reference [44].

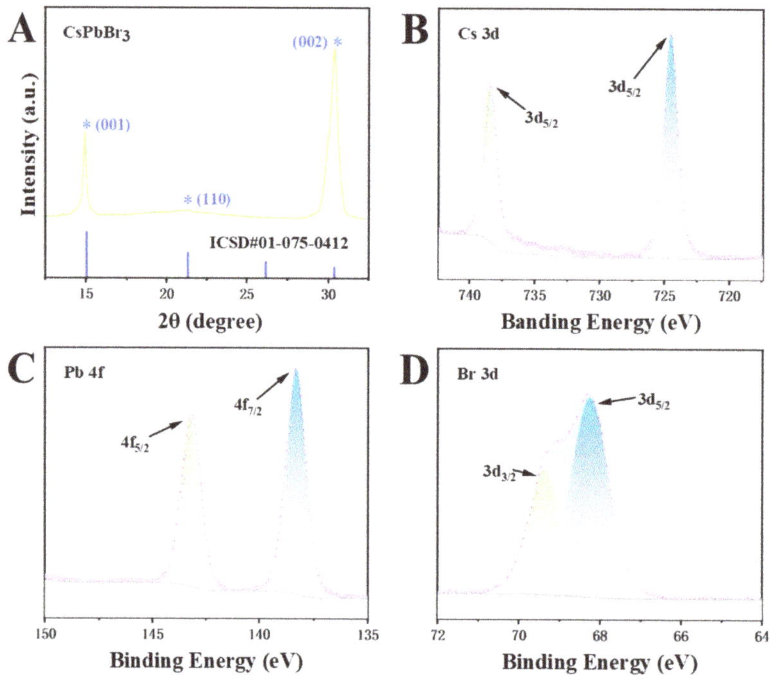

Figure 2. (**A**) XRD pattern and (**B–D**) XPS spectra of $CsPbBr_3$ QDs.

Subsequently, we performed optical characterization of $CsPbBr_3$ quantum dot samples at various reaction temperatures, including UV-vis and photoluminescence (PL) spectra. As the average particle size of $CsPbBr_3$ QDs decreases, the quantum confinement effect changes from weak to strong, the continuous electronic energy levels split into discrete energy levels, and the energy levels are constantly widened. Shown in Figure 3A–D are $CsPbBr_3$ quantum dot samples corresponding to reaction temperatures of 150 °C, 130 °C, 110 °C and 90 °C, respectively. We dispersed them in hexane and obtained the transition of excitation light from green to blue light under the excitation of an ultraviolet lamp. The corresponding UV-vis absorption peak and PL emission peak also gradually transited from a green band to a deep blue band, indicating that the band gap gradually widened. According to the detailed information shown in the UV-vis spectra and PL spectra in Figure 3, the emission peak positions of PL shift from 494 nm, 484 nm and 474 nm to 464 nm (deep blue light), while the corresponding absorption peak positions shift from 485 nm, 474 nm and 459 nm to 450 nm. It should be noted that the intrinsic PL excitation band of $CsPbBr_3$ QDs should be around 520 nm. For details, see the illustration in Figure S5. The solution of $CsPbBr_3$ QDs dispersed with larger sizes appears bright green under ultraviolet irradiation. The excited light color of the $CsPbBr_3$ QDs solution obtained at the reaction temperature of 150 °C under the UV lamp is more like the transition color between green and blue, because the average particle size of the sample at this time was d ≈ 5.7 nm < 7 nm. At this time, an obvious quantum confinement effect appeared, and the original continuous electronic energy levels split into discrete energy levels. As the energy level widened, the band gap increased and the PL peak position shifted to blue. The quantum confinement effect can also increase the exciton binding energy of the crystal; the exciton binding energy in the halogen perovskite film is about 25 meV, and the excitons in the perovskite film will decompose into free electrons and holes at room temperature. For perovskite quantum dots, the exciton binding energy can reach hundreds of meV, and the increase in exciton binding energy increases the possibility of radiation recombination, which can significantly improve luminous performance. In addition, the adjustable band gap of $CsPbBr_3$ QDs also has great potential in the field of photocatalysis. By adjusting the band gap of $CsPbBr_3$ QDs and other materials to form heterostructures (Figure S7), it is easier to form a type II band structure, effectively realize the separation of electrons and holes, and improve the catalytic performance.

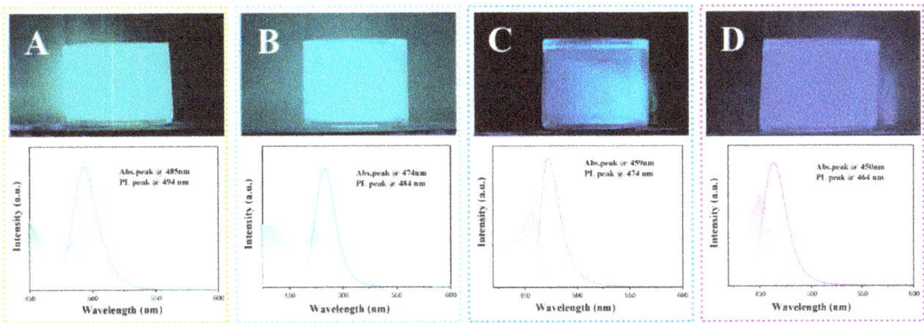

Figure 3. (**A–D**) images of $CsPbBr_3$ QDs at different reaction temperatures under ultraviolet lamp excitation and the corresponding UV-vis and PL spectra.

$CsPbBr_3$ quantum dots (QDs), fabricated at a temperature of 150 °C, were subjected to assessments of photocurrent response and electrochemical impedance spectroscopy (EIS), as illustrated in Figure 4. The equivalent circuit analysis involved C_{PE} as the notation for double-layer capacitance, with R_S representing the resistance across the electrolyte between the reference and working electrodes, and R_C indicating the resistance related to charge transfer at the electrode interface. The EIS patterns for the $CsPbBr_3$ QDs showcased a distinct semicircle, a marker for charge transfer resistance. To accurately measure the photocurrent response exclusive of external effects, measurements were systematically taken with the initiation of the light source. Remarkably, when the illumination was discontinued, the photocurrent density approached zero, but surged immediately when the light was reinstated after a 30 s interval. Such periodic fluctuation in current, with light modulation every 30 s, underscores the $CsPbBr_3$ perovskite's adeptness at carrier transport. This efficiency in carrier migration underscores the potential of $CsPbBr_3$ QDs for future photoelectrocatalysis applications.

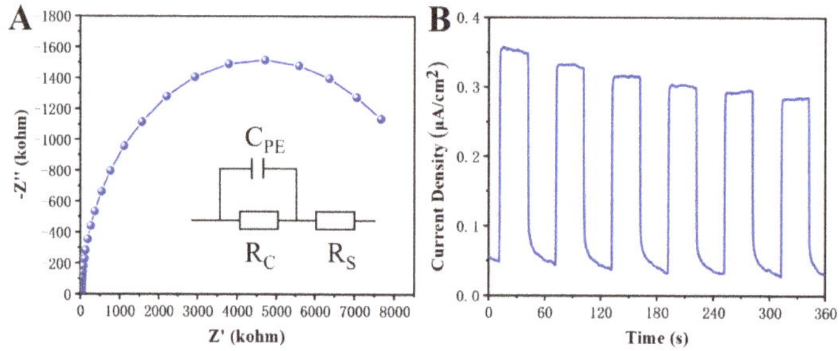

Figure 4. (**A**) Electrochemical impedance spectroscopy (Nyquist plot) of $CsPbBr_3$ QDs synthesized at a reaction temperature of 150 °C. The inset illustrates the equivalent circuit model. (**B**) Instantaneous photocurrent response of $CsPbBr_3$ QDs in neutral water (0.5 M Na_2SO_4) at −0.1 V vs. NHE.

4. Conclusions

Based on the synthesis of perovskite by the hot injection method, we introduced more Br^- ions by adding $ZnBr_2$ to create a bromine-rich reaction environment to regulate the nucleation and growth process of perovskite, and finally synthesized monodisperse, extremely uniform and adjustable $CsPbBr_3$ QDs. Due to the presence of excessive Br^- ions in the solution, the reaction system had abundant nucleation sites, which greatly sped up the explosive nucleation process in the hot injection stage, and effectively inhibited the Oswald ripening process in the subsequent crystal growth. Thus, $CsPbBr_3$ QDs with an extremely uniform size were obtained. We can regulate the size of $CsPbBr_3$ QDs by adjusting the reaction temperature during hot injection and the amount of Br^- ions added. The general rule is as follows: when the content of Br^- ions in the system is constant, the size of the obtained quantum dots becomes smaller and smaller as the reaction temperature gradually decreases. Similarly, when the temperature is constant, the size of the quantum dot is negatively correlated with the Br^- ion content in the system to a certain extent. In addition, because the size is uniform enough, there is an interesting self-assembly arrangement, which can be clearly seen in TEM images and is to some extent supported by XRD. In addition, based on the quantum confinement effect, the size changes of $CsPbBr_3$ QDs also bring special optical properties. When the size of the quantum dot enters the critical value of the Boer exciton radius (7 nm), with a further reduction in the average size, its photoluminescence wavelength shifts to a deep blue band. The results of spectral and structural characterization show that the emission of $CsPbBr_3$ perovskite, from green light to blue light, can be continuously adjusted by adjusting the sizes of quantum dots. In general, on the one hand, from the perspective of structure, $CsPbBr_3$ QDs can achieve more efficient catalytic performance by attaching to other materials, and the adjustable size property greatly facilitates this method. Furthermore, based on the change in band gap width caused by the size effect, the formation of a type II band structure with other materials also has the potential to be applied in the field of catalysis.

Supplementary Materials: The following supporting information can be downloaded at: https://www.mdpi.com/article/10.3390/ma17071607/s1, Figure S1: (A–D) TEM images of $CsPbBr_3$ QDs at different reaction temperatures (scale bar: 50 nm) and (E,F) their size-of-particle-diameter distribution diagrams; Figure S2: Image of $CsPbBr_3$ QDs at different reaction temperatures under ultraviolet lamp excitation and the corresponding UV-vis and PL spectra; Figure S3: (a) HR-TEM image and its magnified local view; (b) the reciprocal space image after FFT; (c) the lattice image obtained by inverse FFT and the reciprocal space image; Figure S4: The XPS spectrum of $CsPbBr_3$ QDs; Figure S5: Image of $CsPbBr_3$ QDs under ultraviolet lamp excitation and the corresponding PL spectra; Figure S6: (a) TEM image and (b) EDS spectra of $CsPbBr_3$ QDs; Figure S7: The relationship between absorbance and photon energy. The band gap (B_g) of $CsPbBr_3$ QDs of different sizes.

Author Contributions: H.L. and Y.L. conceived projection and material synthetic design. H.L., X.W. and J.H. performed the experiment, sample testing, data collection and analysis. H.L., Q.L. and X.L.

performed data discussion and formal analyses. H.L., J.L., C.L. and M.W. wrote, reviewed and edited the manuscript. All authors discussed the results. All authors have read and agreed to the published version of the manuscript.

Funding: This work was financially supported by the National Natural Science Foundation of China (52072281, Y.L.), the Major Program of the National Natural Science Foundation of China (22293021, Y.L.), the National Innovation and Entrepreneurship Training Program for College Students (No. S202210497011, Y.L.), Wuhan University of Technology Doctoral Research Initiation Fee (40120543, Y.L.), and Hubei Province's state-allocated dual-class research platform fee (40120623, Y.L.).

Institutional Review Board Statement: Not applicable.

Informed Consent Statement: Not applicable.

Data Availability Statement: Data are available in a publicly accessible repository that does not issue DOIs or upon request from the corresponding author.

Acknowledgments: We are grateful to the School of Chemistry and Chemical and Life Sciences of Wuhan University of Technology for providing UV-vis and PL characterization and to the Center for Materials Research and Analysis of Wuhan University of Technology Materials for XPS and HR-TEM testing.

Conflicts of Interest: The authors declare no conflicts of interest.

References

1. Yuan, J.; Hazarika, A.; Zhao, Q.; Ling, X.; Moot, T.; Ma, W.; Luther, J.M. Metal Halide Perovskites in Quantum Dot Solar Cells: Progress and Prospects. *Joule* **2020**, *4*, 1160–1185. [CrossRef]
2. Solari, S.F.; Kumar, S.; Jagielski, J.; Shih, C.J. Monochromatic LEDs based on perovskite quantum dots: Opportunities and challenges. *J. Soc. Inf. Disp.* **2019**, *27*, 667–678. [CrossRef]
3. Li, X.; Wang, Y.; Chen, T.; Xu, T.; Ye, S.; Liu, R.; Sun, Z.; Tan, C.; Lv, X.; Yang, J.; et al. Unencapsulated $CsPbClBr_2$ Film Photodetectors Grown by Thermal Vacuum Deposition Exhibit Exceptional Environmental Stability in High-Humidity Air. *ACS Appl. Energy Mater.* **2022**, *5*, 8709–8716. [CrossRef]
4. Akkerman, Q.A.; Manna, L. What Defines a Halide Perovskite? *ACS Energy Lett.* **2020**, *5*, 604–610. [CrossRef] [PubMed]
5. Ceratti, D.R.; Tenne, R.; Bartezzaghi, A.; Cremonesi, L.; Segev, L.; Kalchenko, V.; Oron, D.; Potenza, M.A.C.; Hodes, G.; Cahen, D. Self-Healing and Light-Soaking in $MAPbI_3$: The Effect of H_2O. *Adv. Mater.* **2022**, *34*, 2110239. [CrossRef] [PubMed]
6. Li, Y.; Ding, T.; Luo, X.; Tian, Y.; Lu, X.; Wu, K. Synthesis and Spectroscopy of Monodispersed, Quantum-Confined $FAPbBr_3$ Perovskite Nanocrystals. *Chem. Mater.* **2019**, *32*, 549–556. [CrossRef]
7. Saparov, B.; Mitzi, D.B. Organic–Inorganic Perovskites: Structural Versatility for Functional Materials Design. *Chem. Rev.* **2016**, *116*, 4558–4596. [CrossRef] [PubMed]
8. Seth, S.; Ahmed, T.; De, A.; Samanta, A. Tackling the Defects, Stability, and Photoluminescence of $CsPbX_3$ Perovskite Nanocrystals. *ACS Energy Lett.* **2019**, *4*, 1610–1618. [CrossRef]
9. Zhang, J.; Zhong, Y. Origins of p-Doping and Nonradiative Recombination in $CsSnI_3$. *Angew. Chem. Int. Ed.* **2022**, *61*, 106800.
10. Fakharuddin, A.; Gangishetty, M.K.; Abdi-Jalebi, M.; Chin, S.H.; bin Mohd Yusoff, A.R.; Congreve, D.N.; Tress, W.; Deschler, F.; Vasilopoulou, M.; Bolink, H.J. Perovskite light-emitting diodes. *Nat. Electron.* **2022**, *5*, 203–216. [CrossRef]
11. Jing, H.; Peng, R.; Ma, R.M.; He, J.; Zhou, Y.; Yang, Z.; Li, C.Y.; Liu, Y.; Guo, X.; Zhu, Y.; et al. Flexible Ultrathin Single-Crystalline Perovskite Photodetector. *Nano Lett.* **2020**, *20*, 7144–7151. [CrossRef] [PubMed]
12. Kim, J.Y.; Lee, J.W.; Jung, H.S.; Shin, H.; Park, N.G. High-Efficiency Perovskite Solar Cells. *Chem. Rev.* **2020**, *120*, 7867–7918. [CrossRef] [PubMed]
13. Tao, S.; Yang, D.; He, G.; Guo, D.; Yang, L.; Zheng, J.; Li, J.; Chen, J.; Ma, D. Photomultiplication-type perovskite photodetectors base on air-processed perovskite films. *Org. Electron.* **2023**, *118*, 106800. [CrossRef]
14. Kannan, R.; Kim, A.R.; Eo, S.K.; Kang, S.H.; Yoo, D.J. Facile one-step synthesis of cerium oxide-carbon quantum dots/RGO nanohybrid catalyst and its enhanced photocatalytic activity. *Ceram. Int.* **2017**, *43*, 3072–3079. [CrossRef]
15. Liu, W.; Liu, J.; Wang, X.; He, J.; Li, Y.; Liu, Y. Synthesis of Asymmetrical $CsPbBr_3/TiO_2$ Nanocrystals with Enhanced Stability and Photocatalytic Properties. *Catalysts* **2023**, *13*, 1048. [CrossRef]
16. Rossetti, R.; Nakahara, S.; Brus, L.E. Quantum size effects in the redox potentials, resonance Raman spectra, and electronic spectra of CdS crystallites in aqueous solution. *J. Chem. Phys.* **1983**, *79*, 1086–1088. [CrossRef]
17. Kagan, C.R.; Murray, C.B. Charge transport in strongly coupled quantum dot solids. *Nat. Nanotechnol.* **2015**, *10*, 1013–1026. [CrossRef] [PubMed]
18. Halperin, W.P. Quantum size effects in metal particles. *Rev. Mod. Phys.* **1986**, *58*, 533–606. [CrossRef]
19. Feldheim, L.; Keating, C.D. Self-assembly of single electron transistors and related devices. *Chem. Soc. Rev.* **1998**, *27*, 1–12. [CrossRef]

20. Zhou, C.; Lin, H.; Tian, Y.; Yuan, Z.; Clark, R.; Chen, B.; van de Burgt, L.J.; Wang, J.C.; Zhou, Y.; Hanson, K.; et al. Luminescent zero-dimensional organic metal halide hybrids with near-unity quantum efficiency. *Chem. Sci.* **2018**, *9*, 586–593. [CrossRef]
21. Ioannou, D.; Griffin, D.K. Nanotechnology and molecular cytogenetics: The future has not yet arrived. *Nano Rev.* **2010**, *1*, 5117. [CrossRef] [PubMed]
22. Xu, Y.F.; Yang, M.Z.; Chen, B.X.; Wang, X.D.; Chen, H.Y.; Kuang, D.B.; Su, C.Y. A $CsPbBr_3$ Perovskite Quantum Dot/Graphene Oxide Composite for Photocatalytic CO_2 Reduction. *J. Am. Chem. Soc.* **2017**, *139*, 5660–5663. [CrossRef] [PubMed]
23. Li, P.; Wei, Z.; Wu, T.; Peng, Q.; Li, Y. Au−ZnO Hybrid Nanopyramids and Their Photocatalytic Properties. *J. Am. Chem. Soc.* **2011**, *133*, 5660–5663. [CrossRef] [PubMed]
24. Petronella, F.; Pagliarulo, A.; Striccoli, M.; Calia, A.; Lettieri, M.; Colangiuli, D.; Curri, M.L.; Comparelli, R. Colloidal Nanocrystalline Semiconductor Materials as Photocatalysts for Environmental Protection of Architectural Stone. *Crystals* **2017**, *7*, 30. [CrossRef]
25. Pincik, E.; Matsumoto, T. Progress in applied surface, interface and thin film science 2017. *Appl. Surf. Sci.* **2018**, *461*, 1–2. [CrossRef]
26. Ball, P.; Garwin, L. Science at the atomic scale. *Nature* **1992**, *355*, 761–764. [CrossRef]
27. Fu, P.; Shan, Q.; Shang, Y.; Song, J.; Zeng, H.; Ning, Z.; Gong, J. Perovskite nanocrystals: Synthesis, properties and applications. *Sci. Bull.* **2017**, *62*, 369–380. [CrossRef] [PubMed]
28. Protesescu, L.; Yakunin, S.; Bodnarchuk, M.I.; Krieg, F.; Caputo, R.; Hendon, C.H.; Yang, R.X.; Walsh, A.; Kovalenko, M.V. Nanocrystals of Cesium Lead Halide Perovskites ($CsPbX_3$, X = Cl, Br, and I): Novel Optoelectronic Materials Showing Bright Emission with Wide Color Gamut. *Nano Lett.* **2015**, *15*, 3692–3696. [CrossRef]
29. Endres, J.; Egger, D.A.; Kulbak, M.; Kerner, R.A.; Zhao, L.; Silver, S.H.; Hodes, G.; Rand, B.P.; Cahen, D.; Kronik, L.; et al. Valence and Conduction Band Densities of States of Metal Halide Perovskites: A Combined Experimental–Theoretical Study. *J. Phys. Chem. Lett.* **2016**, *7*, 2722–2729. [CrossRef]
30. Liu, Y.; Li, F.; Li, Q.; Yang, K.; Guo, T.; Li, X.; Zeng, H. Emissions at Perovskite Quantum Dot/Film Interface with Halide Anion Exchange. *ACS Photonics* **2018**, *5*, 4504–4512. [CrossRef]
31. Akkerman, Q.A.; D'Innocenzo, V.; Accornero, S.; Scarpellini, A.; Petrozza, A.; Prato, M.; Manna, L. Tuning the Optical Properties of Cesium Lead Halide Perovskite Nanocrystals by Anion Exchange Reactions. *J. Am. Chem. Soc.* **2015**, *137*, 10276–10281. [CrossRef]
32. Jiang, Y.; Sun, C.; Xu, J.; Li, S.; Cui, M.; Fu, X.; Liu, Y.; Liu, Y.; Wan, H.; Wei, K.; et al. Synthesis-on-substrate of quantum dot solids. *Nature* **2022**, *612*, 679–684. [CrossRef]
33. Uddin, M.A.; Glover, J.D.; Park, S.M.; Pham, J.T.; Graham, K.R. Growth of Highly Stable and Luminescent $CsPbX_3$ (X = Cl, Br, and I) Nanoplates via Ligand Mediated Anion Exchange of $CsPbCl_3$ Nanocubes with AlX_3. *Chem. Mater.* **2020**, *32*, 5217–5225. [CrossRef]
34. Wang, X.; Liu, W.; He, J.; Li, Y.; Liu, Y. Synthesis of All-Inorganic Halide Perovskite Nanocrystals for Potential Photoelectric Catalysis Applications. *Catalysts* **2023**, *13*, 1041. [CrossRef]
35. Shekhirev, M.; Goza, J.; Teeter, J.D.; Lipatov, A.; Sinitskii, A. Synthesis of Cesium Lead Halide Perovskite Quantum Dots. *J. Chem. Educ.* **2017**, *94*, 1150–1156. [CrossRef]
36. Chu, D.B.K.; Owen, J.S.; Peters, B. Nucleation and Growth Kinetics from LaMer Burst Data. *J. Phys. Chem. A* **2017**, *121*, 7511–7517. [CrossRef]
37. Huang, X.; Guo, Q.; Yang, D.; Xiao, X.; Liu, X.; Xia, Z.; Fan, F.; Qiu, J.; Dong, G. Reversible 3D laser printing of perovskite quantum dots inside a transparent medium. *Nat. Photonics* **2020**, *14*, 82–88. [CrossRef]
38. Dong, Y.; Qiao, T.; Kim, D.; Parobek, D.; Rossi, D.; Son, D.H. Precise Control of Quantum Confinement in Cesium Lead Halide Perovskite Quantum Dots via Thermodynamic Equilibrium. *Nano Lett.* **2018**, *18*, 3716–3722. [CrossRef]
39. Rainò, G.; Becker, M.A.; Bodnarchuk, M.I.; Mahrt, R.F.; Kovalenko, M.V.; Stöferle, T. Superfluorescence from lead halide perovskite quantum dot superlattices. *Nature* **2018**, *563*, 671–675. [CrossRef] [PubMed]
40. Hu, H.; Guan, W.; Xu, Y.; Wang, X.; Wu, L.; Chen, M.; Zhong, Q.; Xu, Y.; Li, Y.; Sham, T.-K.; et al. Construction of Single-Atom Platinum Catalysts Enabled by $CsPbBr_3$ Nanocrystals. *ACS Nano* **2021**, *15*, 13129–13139. [CrossRef] [PubMed]
41. Wang, H.; Ye, F.; Sun, J.; Wang, Z.; Zhang, C.; Qian, J.; Zhang, X.; Choy, W.C.H.; Sun, X.W.; Wang, K.; et al. Efficient $CsPbBr_3$ Nanoplatelet-Based Blue Light-Emitting Diodes Enabled by Engineered Surface Ligands. *ACS Energy Lett.* **2022**, *7*, 1137–1145. [CrossRef]
42. Li, F.; Liu, Y.; Wang, H.; Zhan, Q.; Liu, Q.; Xia, Z. Postsynthetic Surface Trap Removal of $CsPbX_3$ (X = Cl, Br, or I) Quantum Dots via a ZnX_2/Hexane Solution toward an Enhanced Luminescence Quantum Yield. *Chem. Mater.* **2018**, *30*, 8546–8554. [CrossRef]
43. Kim, H.; Bae, S.R.; Lee, T.H.; Lee, H.; Kang, H.; Park, S.; Jang, H.W.; Kim, S.Y. Enhanced Optical Properties and Stability of CsPbBr3 Nanocrystals Through Nickel Doping. *Adv. Funct. Mater.* **2021**, *31*, 2102770. [CrossRef]
44. Liu, A.; Bi, C.; Li, J.; Zhang, M.; Cheng, C.; Binks, D.; Tian, J. High Color-Purity and Efficient Pure-Blue Perovskite Light-Emitting Diodes Based on Strongly Confined Monodispersed Quantum Dots. *Nano Lett.* **2023**, *23*, 2405–2411. [CrossRef]

materials

Article

Performance of Large Area n-TOPCon Solar Cells with Selective Poly-Si Based Passivating Contacts Prepared by PECVD Method

Zhaobin Liu [1,2], Chunlin Guo [3], Ya Liu [1,2], Jianhua Wang [1,2], Xuping Su [1,2,*] and Qinqin Wang [4,*]

1 School of Materials Science and Engineering, Changzhou University, Changzhou 213164, China
2 Jiangsu Collaborative Innovation Center of Photovoltaic Science and Engineering, Changzhou University, Changzhou 213164, China
3 Institute of Microelectronics of the Chinese Academy of Sciences, Beijing 100029, China
4 Institute of Technology for Carbon Neutralition, Yangzhou University, Yangzhou 225009, China
* Correspondence: sxping@cczu.edu.cn (X.S.); wangqinqin@yzu.edu.cn (Q.W.)

Abstract: Selective emitter (SE) technology significantly influences the passivation and contact properties of n-TOPCon solar cells. In this study, three mask layers (SiO_x, SiN_x, and SiO_xN_y) were employed to fabricate n-TOPCon solar cells with phosphorus (P)-SE structures on the rear side using a three-step method. Additionally, phosphosilicon glass (PSG) was used to prepare n-TOPCon solar cells with P-SE structure on the rear side using four-step method, and the comparative analysis of electrical properties were studied. The SiO_x mask with a laser power of 20 W (O2 group) achieved the highest solar cell efficiency (E_{ff}, 24.85%), The open-circuit voltage (V_{oc}) is 2.4 mV higher than that of the H1 group, and the fill factor (*FF*) is 1.88% higher than that of the L1 group. Furthermore, the final E_{ff} of solar cell is 0.17% higher than that of the L1 group and 0.20% higher than that of the H1 group. In contrast, using the four-step method and a laser power of 20 W (P2 group), a maximum E_{ff} of 24.82% was achieved. Moreover, it exhibited an V_{oc}, which is elevated by 3.2 mV compared to the H1 group, and *FF* increased by 1.49% compared to the L1 group. Furthermore, the overall E_{ff} of the P2 group outperforms both the L1 and H1 groups by approximately 0.14% and 0.17%, respectively. In the four-step groups, the E_{ff} of each laser condition group was improved compared with the L1 group and H1 group, The stability observed within the four-step method surpassed that of the three-step groups. However, in terms of full-scale electrical properties, the three-step method can achieve comparable results as those obtained from the four-step method. This research holds significant guiding implications for upgrading the n-TOPCon solar cell rear-side technology during mass production.

Keywords: n-TOPCon; mask; laser; P-SE

1. Introduction

The TOPCon (tunnel oxide passivated contact) solar cell with a surface area of 4 cm^2 was proposed and developed by the Fraunhofer Institute for Solar Energy Systems in Germany in 2013, achieving an impressive efficiency of 23.7% [1]. The efficiency (E_{ff}) was enhanced to 25.7% in 2017 by optimizing the village resistivity and process flow at the Fraunhofer Institute for Solar Energy Systems in Germany [2]. The efficiency of the POLO-IBC solar cell with the TOPCon structure combined with IBC reached 26.1% [3] in 2018. Due to their excellent passivation performance, TOPCon solar cells have been extensively studied in recent years [4–14]. There are many factors that affect the efficiency of solar cells, including surface recombination rate [15] and contact resistance [16]. The surface recombination rate primarily relies on the passivation quality of the surface, while the contact resistance is influenced by the doping concentration in the contact region. As the efficiency of TOPCon solar cells continues to improve, the reduction in frontal contact area has emerged as a predominant factor constraining solar cell performance [17]. A commonly employed technique involves the fabrication of a selective emitter (SE) on the front side [18]. This approach serves two purposes: firstly, it mitigates carrier recombination in the emitter region and enhances surface passivation; secondly, it reduces contact resistivity (ρ_c) in the contact region, thereby augmenting the *FF* [19]. Currently, laser treatment has been shown to increase V_{oc} by 5–8 mV [20]. The maturity and widespread adoption of positive B-SE technology in industrial production, coupled with the advancement of laser-enhanced contact optimization (LECO) technology, offer potential for further enhancement of V_{oc} and *FF*, thereby augmenting the E_{ff} of solar cells. Additionally, the rear-side surface is passivated using SiO_x/Poly-Si(n^+) [21]. Despite the improved passivation and contact effects, due to the high P atoms doping concentration leading to high recombination, it is still possible to achieve SE through laser technology reducing the P atoms doping concentration in the non-contact region while increasing it in the contact region. Consequently, optimal rear surface passivation and superior rear contact performance are attained.

The principle of laser doping involves focusing a laser beam on the surface of crystalline silicon, resulting in the generation of high temperatures exceeding 1414 °C, followed by a rapid quenching process (~735 °C/s) [22]. Sufficient energy can induce rapid melting and recrystallization in the local area of the wafer surface, with minimal heat generated outside of the laser region. The solubility of doped elements in liquid Si is an order of magnitude higher than that in the solid state. By adjusting laser power, wavelength, and pulse parameters, diffusion or activation of dopants can be achieved [23]. Due to the significant recombination introduced by the frontal contact of n-TOPCon solar cells [17], B-SE technology has garnered increased attention and been extensively investigated by scholars [11]. Employing laser irradiation on the metallized pattern region enables the attainment of selective emitter characteristics, with low doping concentration in the undoped region and high doping concentration in the doped region. Nevertheless, it is worth noting that laser-induced defects, contamination, and discontinuities in selective emitters may also lead to a degradation in solar cells' performance [24,25]. As is widely recognized, the conventional SE technology involves the laser processing of boron-rich boron-silicate glass (BSG) or phosphate-rich phosphorus-silicon glass (PSG) as precursors, followed by high-temperature annealing and oxidation to form a protective film on the surface. Traditional B diffusion typically requires over 80 min at temperatures ranging from 900 to 1000 °C [26]. Similarly, P annealing oxidation also necessitates approximately 50 min at temperatures between 800 and 900 °C [27]. Consequently, the conventional SE method exhibits prolonged process duration and significant energy loss.

Currently, the P-SE of n-TOPCon solar cells prepared using the PECVD (plasma-enhanced chemical vapor deposition) method can be accomplished through four steps: firstly, preparing SiO_x/a-Si(n^+) and SiO_x; secondly, depositing the P source to form PSG; thirdly, performing laser treatment (SE); and finally, conducting annealing and oxidation. Alternatively, it can be achieved in three steps: first, by the PECVD preparation of SiO_x/a-Si(n^+) and a mask layer; then, by laser treatment (opening film and SE); and lastly by annealing oxidation followed by deposition of P diffusion and oxidation. The two laser methods have different effects as the three-step laser involves opening film and SE effects, while the four-step laser only includes the SE effect. The LPCVD (low pressure chemical vapor deposition) method can only be completed using the four-step method due to its inability regarding in situ doping. Investigating various approaches to realize P-SE is highly significant for enhancing the E_{ff} of solar cells and reducing energy consumption.

2. Experiment and Characterization

2.1. Experiment

In this experiment, 182 × 182 × 0.14 mm^3 n-type Cz silicon wafers (LONGi Green Energy Technology Co., Ltd., Xi'an, China) with a resistivity of 1.0–1.2 Ω.cm were used to fabricate n-TOPCon solar cells. The preparation process and variables for the n-TOPCon solar cell test samples are illustrated in Figure 1. The three-step method to achieve the SE is shown in Figure 1a. Alkali texturing was used to create a pyramid-like surface structure, enhancing light trapping. The height of the pyramids was controlled to be less than 2 μm, and the reflectivity was maintained at 9.5 ± 0.2%. B diffusion is used to form the PN junction with a sheet resistance of 140 Ω/sq and junction depth of 0.8 μm. The rear side and surrounding BSG layers were removed through HF chain cleaning, followed by polishing with a mixed aqueous solution containing additives and alkali to eliminate the B wrap-round on the rear and edges simultaneously, and in this step, the rear-side reflectance was controlled to be 42.0 ± 0.2%. Subsequently, the tunneling oxidation passivation structure of a-Si(n)/SiO_x was prepared using PECVD (CETC, Changsha, China). An electric field was applied in the reactor, relying on radio frequency induction to ionize the target material source gas and generate plasma, increasing reactant activity for the low temperature deposition (≤450 °C) process. After evacuating the wafers into the PECVD equipment cavity, leakage detection was performed at 420 °C. The entire preparation process was carried out at 420 °C. Firstly, the tunneling oxide layer, SiO_x, was prepared with a flow rate of N_2O set at 8000 sccm, with the pressure maintained at 300 Pa, power set to 10,000 W, and pulse on/off ratio adjusted to 2 ms/200 ms. Then, an a-Si(n) layer was fabricated for a total duration of 1300 s. The PH_4 flow rate was adjusted to 1200 sccm, while PH_3 and H_2 were introduced at rates of 1000 sccm and 3500 sccm, respectively. The process operated under a pressure of 300 Pa with a power input of 14,000 W, employing a pulse on/off ratio of 4 ms/50 ms. Finally, the mask layer deposition took place over a period lasting for 100 s. During this step, the flow rate of SiH_4 was 1100 sccm, the flow rate of N_2O was 6000 sccm, the pressure was 180 Pa, the power was 9000 W, and the pulse on/off was 5 ms/100 ms. After completion of the steps, the tube underwent two cycles of pumping out and nitrogen filling to restore normal pressure, and then the tube was released.

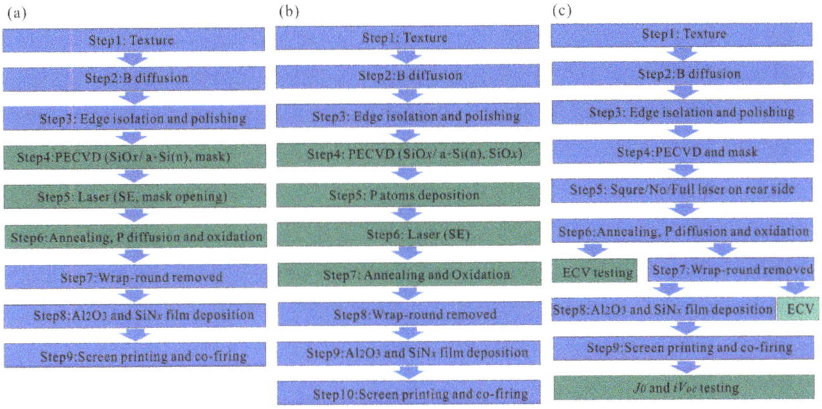

Figure 1. Experimental flow chart: (**a**) three-step SE process; (**b**) four-step SE process; (**c**) sample test flow chart.

A green nanosecond laser was employed for film opening as well as selective emitter formation, and the spot size of the laser is 120 × 120 μm^2. The wafers were annealed, P atoms diffused and oxidized; The PSG on the edge was removed through front chain HF cleaning, while a portion of the BSG was retained on the front. The wrap-round was removed through alkali polishing. Alumina passivation was applied to the front, along with silicon nitride passivation on both sides. Metal electrodes were printed and fired to form the final solar cells in this study. The deposition times of SiO_x, SiN_x, and SiO_xN_y were set at 200 s to prevent P diffusion in the lightly doped region after laser treatment. The power, source amount, and deposition temperature for all three types of masks deposited by PECVD were kept consistent. The laser SE powers used were 15 W, 20 W, and 25 W.

Figure 1b illustrates the four-step P-SE process of the n-TOPCon solar cell preparation process, wherein oxidation was required post laser to generate a SiO_x protective layer safeguarding the SE region when the wrap-round was removed. Despite the realization of a selective emitter in the four-step method, only one step corresponds to an effective selective emitter. In the three-step method, although no P atoms exist in the surface mask layer, laser played a role by driving P atoms from the shallow surface layer deeper. Following laser treatment, the surface mask layer was etched away and subsequently, annealing, P deposition, and drive in were conducted. However, due to the blocking effect of the non-laser areas' mask layers, it was challenging for P atoms to infiltrate poly. Conversely, on the laser-treated region's surface, P atoms were easily driven into poly as part of forming a selective emitter, thus, constituting the second step involved in this process. Although there are merely three steps involved here, two of them contribute towards establishing a selective emitter.

The test wafers prepared during the experiment are shown in Figure 1c. The laser area of Step 5 was square, full-face, and laser free. The laser block film after Step 6 was utilized for sheet resistance (R_{sheet}) and the ECV test, while the ECV test was conducted on the wafer after the wrap-round was removed to confirm the cleaning window and compare it with the wafer without the wrap-round removed. J_0 and iV_{oc} were tested using both full laser and non-laser wafers. The difference between the J_0 of the non-laser wafer and J_0 of the full-face laser wafer at the rear side represents the compound J_0 caused by laser damage. By considering the width of the gate laser slot and number of gate lines, we can calculate the increase in J_0 brought about by rear-side lasers.

Figure 2 shows the schematic diagram of the high temperature diffusion process of the two laser routes. Step 6 of the three-step method raised the temperature to 920 °C first for nitrogen protection and crystallization for 45 min, after which $POCl_3$ was fed for the next 7 min, then the temperature was lowered to 850 °C, and then the temperature was lowered after 30 min of oxygen constant temperature. Figure 2b shows Step 5 of the four-step method, where the P atoms were deposited for 30 min at a low temperature of 750 °C; Figure 2c shows Step 7 of the four-step method, where the temperature was raised to 920 °C, and oxygen was injected under constant temperature for 45 min. Figure 3 illustrates a schematic diagram of laser film opening. Table 1 presents all the experiments along with their respective experimental conditions.

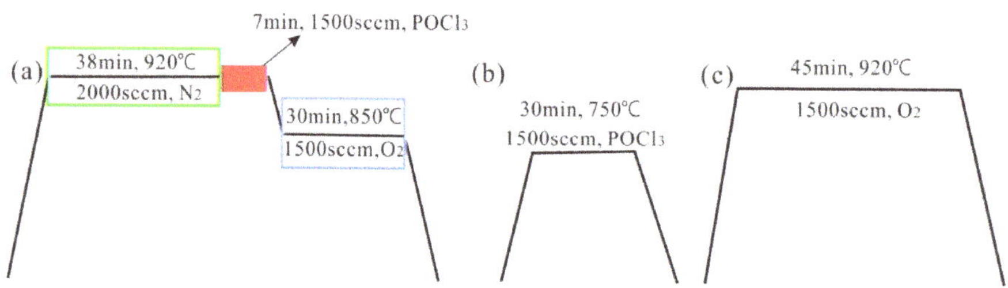

Figure 2. (**a**) Step 6 of the three-step method; (**b**) Step 5 of the four-step laser method; (**c**) Step 7 of the four-step method.

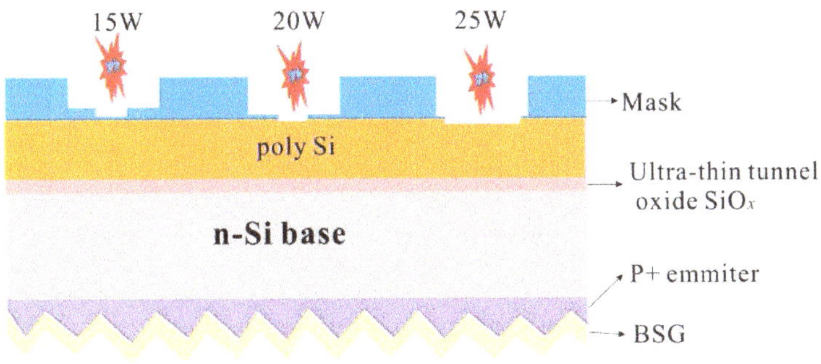

Figure 3. Laser diagram of P-SE process.

Table 1. Different experimental groups of P-SE process.

Groups	L1	H1	O1	O2	O3	N1	N2	N3	S1	S2	S3	P1	P2	P3
Laser/w	—	—	15	20	25	15	20	25	15	20	25	15	20	25
mask	—	—		SiO_x			SiN_x			SiO_xN_y		PSG	PSG	PSG
Steps	—	—					Three						Four	

2.2. Characterization

The rear-side morphologies were examined using an Olympus optical microscope (DSX-HRSU, Olympus Optical Co., Ltd., Tokyo, Japan) under different experimental conditions. Additionally, the R_{sheet} on the sample's rear side was measured using a Napson test instrument (RG-2000PV, Napson, Wuxi, Chaina), and the P atoms doping curve was determined under various experimental settings using a WEP Wafer Profiler (CVP21, Saratoga Technology International, Saratoga, CA, USA). The solar cell's open circuit voltage (V_{oc}), short circuit current density (J_{sc}), fill factor (FF), series resistance (R_s), and efficiency (E_{ff}) were evaluated using an IV test instrument (Delta Electronics, Inc., Shanghai, China) under standard testing conditions of 1000 w/m^2 illumination, temperature of 25 $\pm$ 1 °C, 70% ambient humidity, and simulated am 1.5 spectral distribution.

In order to assess the minority carrier lifetime (τ_{eff}) and implied open circuit voltage (iV_{oc}), we conducted relevant experiments using the WCT-120 [28] instrument (Sinton Instruments, Sinton, TX, USA). Additionally, the specific contact resistivity of the grid line was measured using the transmission line model (TLM) and GP-4 [29] probe (GP Solar GmbH, Memmingen, Germany) transmission line model method. The *EQE* test was performed utilizing the PVE300-IV test equipment (Enli Technology Co., Ltd., Shenzhen, China), while the current loss analysis relied on the V1.4 calculation tool provided by the Singapore Solar Energy Research Institute [30].

3. Results and Discussions

3.1. Morphologies Analysis

After laser treatment under different experimental conditions, the spot morphologies are depicted in Figure 4. By manipulating the laser processing speed and pulse frequency, the laser overlap ratio can be adjusted [31]. In the mass production process, we regulated the slight overlap of spots due to differential cooling rates between the spot's edge and its middle. Under high power conditions, micro-textures may manifest in the overlapping region [32], leading to a reduction in the minority carrier lifetime of the solar cell. We know that in the mass production process of n-TOPCon prepared using the low-pressure chemical vapor deposition (LPCVD) method, some manufacturers judge the thickness of poly silicon by the color of the front of the wafer after P atoms diffusion and annealing, because the mask layer deposited on different poly silicon thicknesses show different colors. Following the deposition of diverse mask layers through the PECVD method, a light gray hue akin to the wafers is observed on their surfaces. The localized high temperature generated by laser formation induces crystallization of a-Si(n$^+$) into Poly-Si(n$^+$), which shows a pink coloration on the masked layers, while cyan coloration emerges at spot junctions, particularly noticeable at 15 W power. The 15 W laser produces the darkest color in each mask group, while the 25 W laser produces the lightest color. Additionally, the SiO_xN_y mask group's 25 W laser creates the lightest color. Due to the presence of the mask layer, Poly-Si with different thicknesses show different colors. The colors disappear after the mask layer is removed, and there are some areas at the edge of the spot where the film is completely opened. Higher power lasers can create cracks or even grooves in open areas [33]. This creates more damage, which reduces the lifetime of a carrier. Previous studies have demonstrated that noticeable defect formations occur only when laser irradiation surpasses the silicon melting threshold, with the recrystallization rate being a crucial parameter for defect generation. Specifically, the surface electrical and optical properties of solar cells are influenced by dislocation density and oxygen incorporation [34,35]. Generally, the types of defects commonly observed depend largely on two factors: (1) initial concentration of external impurities (carbon, oxygen, nitrogen, dopants); and (2) structural defects in the silicon crystal, which directly impact its melting threshold [36].

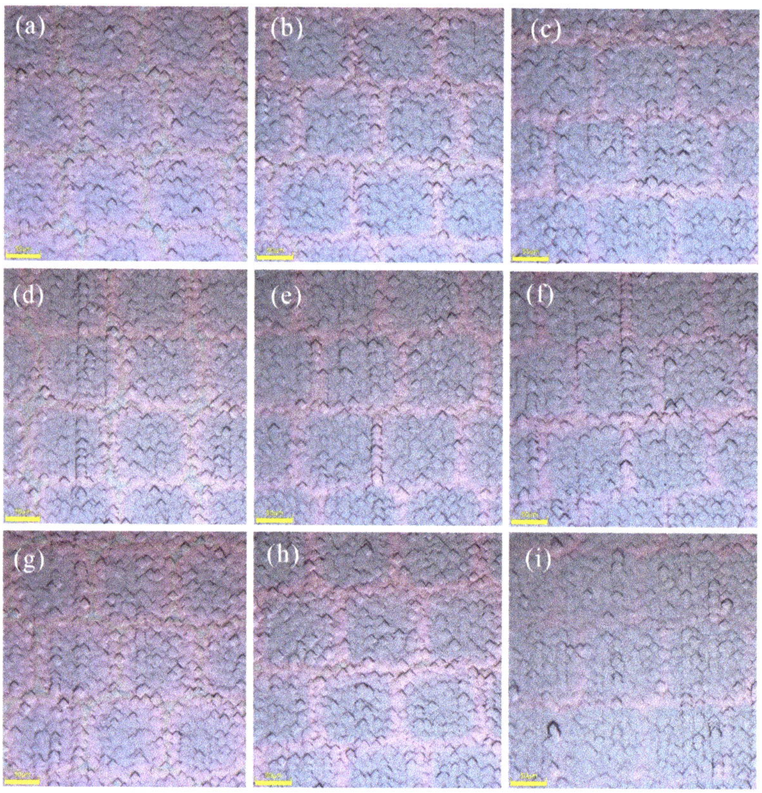

Figure 4. Microscope image of the laser spot on n-TOPCon cell: (**a**) O1; (**b**) O2; (**c**) O3; (**d**) N1; (**e**) N2; (**f**) N3; (**g**) S1; (**h**) S2; (**i**) S3.

In the three-step process, the microscopic morphology damage of the spot after laser treatment with different masks is not readily discernible, whereas in the four-step process, a more distinct visualization of the state of the surface film layer opening can be observed. As depicted in Figure 5, at a power of 15 W, the laser spot is barely discernible on the surface. However, upon increasing the laser power to 20 W, it becomes evident that the surface film layer Is removed, thereby exposing the wafer directly. Notably, the initial removal by the laser occurs precisely at the center of the spot. Subsequently, raising the laser power to 25 W leads to a further expansion of the PSG removal area. The PSG serves as an effective reservoir for doping diffusion, resulting in highly doped emitters with significant junction depths [37]. When moved from the interstitial position to the substituted position, the inactive phosphorus atoms in the emitter become electrically active [38]. If excessively high laser power is selected, the ablation of the surface is initiated, leading to severe silicon damage that can be visualized under microscopy [39]. As depicted in red circle position of Figure 5c, excessive heat provided by lasers causes the ablation of mask layers and melting on transmitter surfaces, resulting in debris upon solidification on wafer surfaces [40]. Furthermore, extensive dislocations along with micro-twins and swirling defects have been identified [41–44]. Additionally, SEM imaging has revealed other defects within melting zones when continuous wave (CW) lasers are employed [45]. Other defects in the melting region have been also observed using SEM imaging, and the trap state defects are mainly associated with oxygen vacancy and vacancy [44–47].

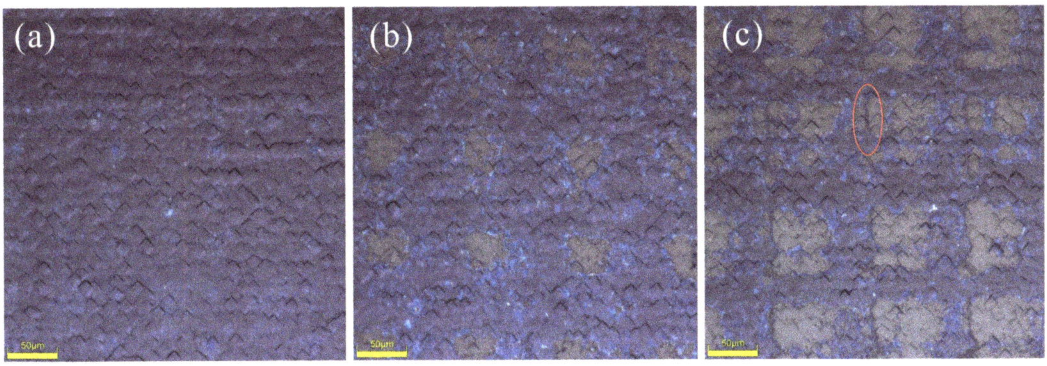

Figure 5. Microscope image after four-step SE laser: (**a**) P1, (**b**) P2, (**c**) P3.

The microscope images after the wrap-round removal and cleaning are presented in Figure 6. We show the SiO_x mask groups in the three-step method, and the images of the four steps after the wrap-round removal and cleaning. As the laser energy increases, the extent of the damage becomes more pronounced, with Figure 6a,b showing slight damage in the center area of the laser spot, while Figure 6c provides a clearer depiction. The laser area of the four-step method may be due to the thick PSG, and the laser spot is indistinct after cleaning. To achieve a higher laser doping concentration through the P-SE process, reliance on PSG as a P source is necessary but limited; thus, increasing the P atoms content within PSG becomes imperative for augmenting the SE's P source. Augmenting the P content also leads to an increase in the PSG thickness, which subsequently lowers the damage layer during the application of the laser SE technique. Nevertheless, Figure 6f clearly demonstrates the recrystallization occurring on the silicon wafer post melting due to the surface tension effects between liquid phases [25], resulting in a blurred division line at the pyramid base.

The four-step process of P diffusion annealing after the laser is mainly used for three reasons: firstly, it facilitates the conversion of a-Si(n^+) into Poly-Si(n^+); secondly, it enables the deposition and promotion of P atoms to enhance the P concentration in the contact region; thirdly, it facilitates the formation of a protective PSG film. During the deposition process of P atoms and oxidation, BPSG tends to wrap round the front surface of the wafer, as illustrated in Figure 7a. Chain single-sided HF cleaning is required before the wrap-round removal and cleaning, and the corrosion rate of BPSG is much higher than that of PSG. However, a critical aspect of this step is that the corrosion time of BPSG should be longer than that of PSG. If the deposition of P atoms and oxidation takes too long, an increased amount of BPSG is formed, leading to a comparable or shorter corrosion time for BPSG compared to PSG. Consequently, there is no available window for the wrap-round removal and cleaning. In such cases, if the front part of the BPSG is retained during the wrap-round removal and cleaning, it becomes impossible to clean the Poly-Si(n^+) around the wafer (as shown in Figure 7b), resulting in a pyramid base size on the side remaining unchanged after the wrap-round removal and cleaning. Our observation indicates that after the wrap-round removal and cleaning, an increase in the pyramid base size from its original range of 8–9 μm to more than 13 μm (as depicted in Figure 7c) is required for the complete removal/cleaning of Poly-Si(n^+)/SiO_x. It should be noted that a certain thickness of a PSG protective layer needs to be prepared in order to ensure Poly-Si corrosion by alkali solution during the wrap-round removal and cleaning. The oxidation time cannot be reduced here, and only the time for deposition of P atoms can be reduced. But if the amount of deposition of P atoms is insufficient, it is impossible to pull apart the concentration difference between the heavily doped region and the lightly doped region, and then a good selective emitter cannot be formed, so the time for P deposition needs to be strictly controlled. We verified that the best P deposition time is 7 min. However, when the deposition time exceeds 10 min, no window is observed for the wrap-round removal and cleaning. Additionally, if the deposition time of P atoms is less than 4 min, the concentration difference between the heavily doped and lightly doped regions falls below 1×10^{20} atoms/cm^3, rendering it incapable of forming a satisfactory selective emitter.

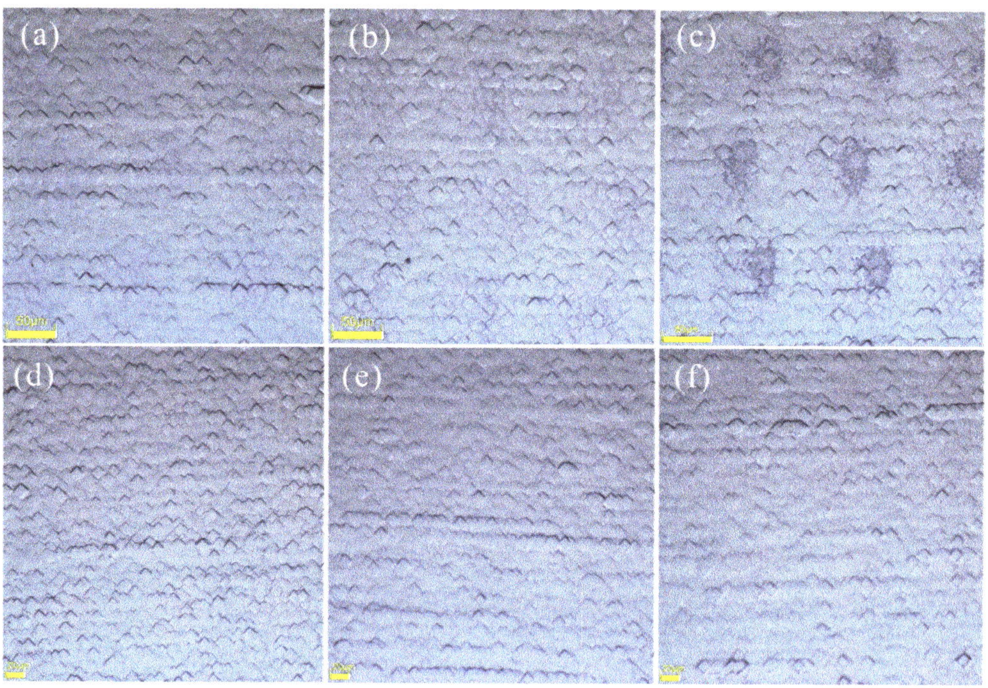

Figure 6. Morphologies of laser region after wrap-round removal and cleaning: (**a**) O1; (**b**) O2; (**c**) O3; (**d**) P1; (**e**) P2; (**f**) P3.

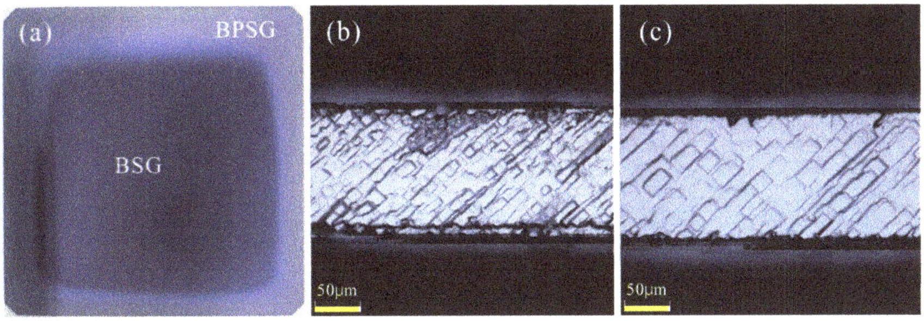

Figure 7. (**a**) BPSG wrap-round; (**b**) Unqualified side; (**c**) Qualified side.

3.2. ECV Profiles of Experimental Groups

The inhibitory effects of various mask structures and mask deposition durations in the three-step laser method on the advancement of P atoms deposition (ECV profiles) are illustrated in Figure 8, while Table 2 presents the peak concentrations of P atoms doping. The measured P atoms concentration after direct annealing in the lightly doped zone is determined to be 2.30×10^{20} atoms/cm^3. It is observed that both SiO_x depositions for 100 s and 150 s exhibit weak blocking effects on P atoms during annealing deposition in the lightly doped zone, failing to effectively maintain the P concentration within this region. However, a good blocking effect against P atoms entering the wafer is achieved with a SiO_x deposition time of 200 s. The peak concentration of P atoms only increases by merely 4.3×10^{19} atoms/cm^3 for this case, whereas for SiN_x at a deposition time of 150 s, it increases by just 5.4×10^{19} atoms/cm^3, and for SiO_xN_y, it increases by merely 3.6×10^{19} atoms/cm^3. The deposited SiN_x and SiO_xN_y layers at a deposition time of 150 s demonstrate an effective blocking effect on P atoms driving in, while even stronger blocking effects are observed by their respective depositions at 200 s. Considering the imperative for efficient mass production, it is crucial to minimize the process time of the mask layer to ideally within 200 s. To ensure comparability in our study, we standardized the deposition time of the mask layer to 200 s. According to the differential blocking effects of various mask layers on P atoms, SiO_x exhibits the weakest blocking effect, followed by SiN_x and SiO_xN_y. Both SiN_x and SiO_xN_y demonstrate similar blocking effects. Considering the mask layer structure, deposition mask time, and laser power, we designed the experimental groups as presented in Table 2. Additionally, we designed the L1 group with a low doping concentration of non-SE and the H1 group with a high doping concentration of non-SE, respectively. To compare the SE groups using the four-step methods, we also designed the SE groups using the four-step methods with the laser powers for F1–F3 were set at 15 W, 20 W, and 25 W, respectively.

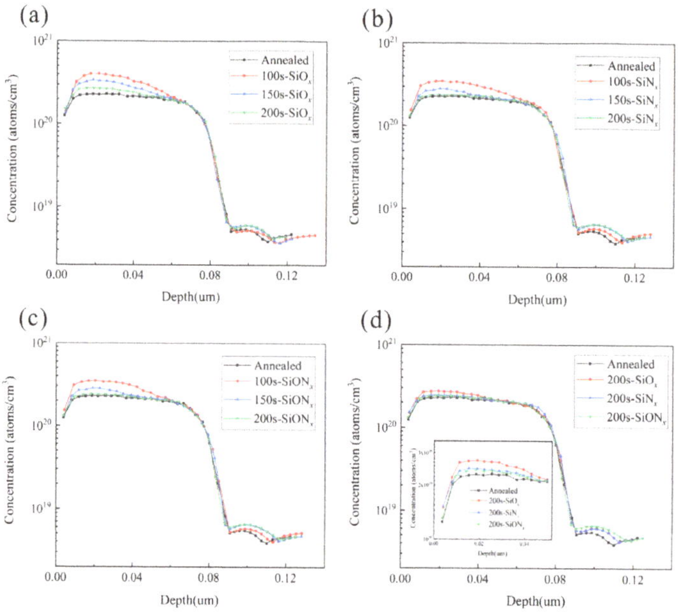

Figure 8. ECV profiles of light diffusion zones with different masks and times: (**a**) SiO_x groups; (**b**) SiN_x groups; (**c**) SiO_xN_y groups; (**d**) 200 s for three mask layers.

Table 2. Peak concentrations of light doping regions in different mask layer groups.

Mask	100 s (Atoms/cm^3)	150 s (Atoms/cm^3)	200 s (Atoms/cm^3)
SiO_x	4.05×10^{20}	3.42×10^{20}	2.73×10^{20}
SiN_x	3.46×10^{20}	2.69×10^{20}	2.47×10^{20}
SiO_xN_y	3.49×10^{20}	2.84×10^{20}	2.42×10^{20}

In order to facilitate the investigation of doping concentration changes in the laser region P atoms, we employed a 4×4 cm^2 square laser for experimentation. To avoid discrepancies between wafers, we put the laser square in the position adjacent to the same piece of wafer. The ECV profiles obtained after annealing in different SE regions are depicted in Figure 9. It is observed that varying laser powers significantly influence ECV profiles types, with higher power lasers resulting in steeper emitter profiles and increased doping depths [38]. Elevating the laser power leads to the additional diffusion of phosphorus atoms from the phosphorus emission layer into silicon, thereby forming a highly doped region [38]. The depth of diffusion is typically correlated with the mass of the SiO_x film and the annealing temperature [19]. In the three-step method, the concentration of the shallow surface is driven in by the laser. After opening the mask layer, the wafer is deposited and subsequently driven in. Due to oxidation at 850 °C, the P atoms in the shallow surface precipitate out, resulting in the ECV profiles becoming a parabola. The higher the laser power, the higher the peak concentration.

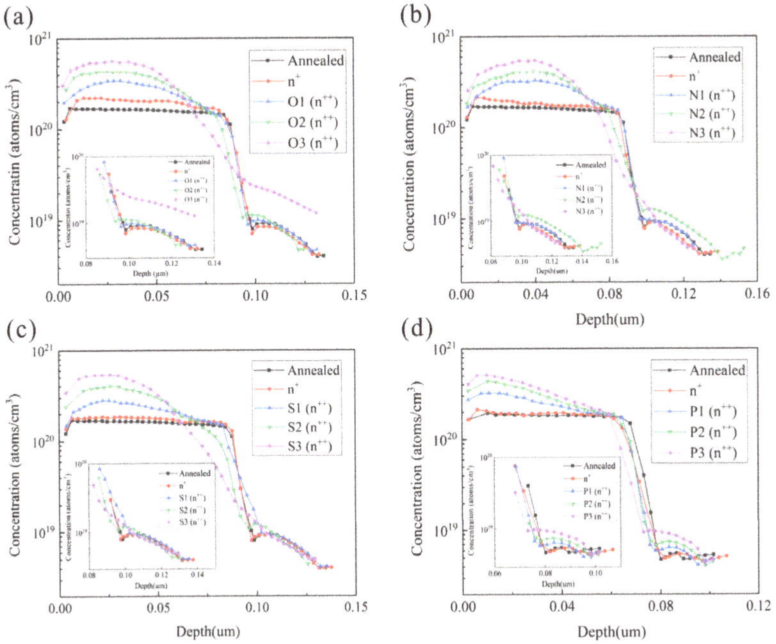

Figure 9. ECV profiles of (**a**) SiO_x groups, (**b**) SiO_xN_y groups, (**c**) SiN_x groups, and (**d**) four-step method groups.

Laser not only affects peak concentration but also significantly influences poly thickness and ECV inflection point concentration near the wafer. In the three-step method, increasing laser power gradually raises the lower inflection point concentration while weakening the tunneling oxide layer blocking effect. At 25 W laser power, no obvious inflection point is observed in the ECV profiles, and instead it appears smooth. This indicates that the tunneling oxide layer fails to effectively block P atoms from entering into the wafer, and its passivation ability weakens accordingly. Additionally, at 25 W laser power level, the partial ablation of poly occurs making it impossible to determine Poly-Si layer thickness from ECV profiles alone. Figure 9d illustrates the ECV profiles of the four-step method, where peak concentration closely aligns with surface due to mutual diffusion between PSG and Poly-Si [48].

The peak concentrations and inflection point concentrations of various experimental conditions are presented in Table 3. It is observed that the inflection point concentration exhibits an increasing trend with the augmentation of laser power. However, when the laser power reaches 25 W, the inflection point becomes less discernible. The primary cause lies in the excessive infiltration of phosphorus atoms into the wafer through the tunnelling oxide layer, thereby compromising the passivation of SiO_x/Poly-Si(n^+) and subsequently exerting a direct impact on the solar cells' electrical performance. In the four-step method, the peak concentration of 15 W is increased from 2.15×10^{20} atoms/cm^3 to 3.25×10^{20} atoms/cm^3, resulting in a net increase of 1.10×10^{20} atoms/cm^3 in concentration. Furthermore, when the laser power is raised to 20 W, an additional increase of 1.13×10^{20} atoms/cm^3 is observed. However, due to limited P atom availability in PSG, increasing the laser power to 25 W only leads to a marginal increment of 7.5×10^{19} atoms/cm^3 compared to that at 20 W.

Table 3. Peak concentrations of ECV profiles at different conditions.

Mask	n^+ (Atoms/cm^3)	15 W (Atoms/cm^3)	20 W (Atoms/cm^3)	25 W (Atoms/cm^3)
SiO_x	2.24×10^{20}	3.44×10^{20}	4.35×10^{20}	5.66×10^{20}
SiN_x	2.18×10^{20}	3.28×10^{20}	4.19×10^{20}	5.50×10^{20}
SiO_xN_y	1.87×10^{20}	2.82×10^{20}	4.06×10^{20}	5.43×10^{20}
PSG	2.15×10^{20}	3.25×10^{20}	4.38×10^{20}	5.13×10^{20}

3.3. R_{sheet} of Each Experimental Group

Since the formation of a selective emitter enhances the current collection characteristics of the PN junction, the improved *FF* in SE can be attributed to an enhancement in R_s resulting from improved electrode-emitter contact and reduced R_{sheet} [49]. The R_{sheet} of each experimental group is illustrated in Figure 10. In the lightly doped region, the R_{sheet} of the SiO_x mask decreased from 56.3 Ω/sq to 54.4 Ω/sq, followed by the SiN_x group and then the SiO_xN_y group. The blocking ability of the 200 s deposited mask layer against P atoms deposition and diffusion was found to be in the order of $SiO_x < SiN_x < SiO_xN_y$. These findings are consistent with our ECV testing results in the lightly doped region. However, the R_{sheet} of the four-step method is considered to be constant due to the simultaneous oxygenation and deposition of P atoms using low temperature precipitation at 750 °C. At this temperature, we believe that there is no driving in effect, and P atoms are solely deposited on the surface. SiO_x can maintain continuity at 750 °C [50]. However, due to the PSG's blocking action and the phosphorus absorption effect, it becomes challenging for P atoms to drive in or separate out.

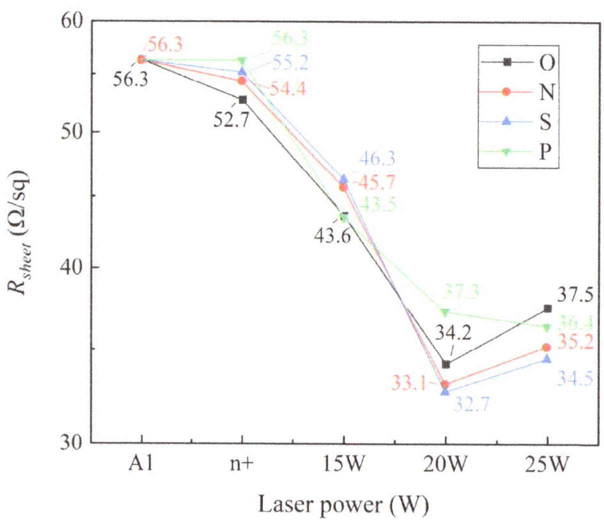

Figure 10. R_{sheet} of different experimental groups.

When employing 15 W laser power, the three-step laser method only reduces the thickness of the mask layer in the laser region, resulting in a certain level of blocking effect. Conversely, the four-step laser method directly introduces low-temperature deposited P-source into the wafer with a weaker blocking effect. Consequently, at 15 W power, the R_{sheet} decline is most pronounced for the four-step laser method compared to the three-step laser method. However, as the laser power increases to 20 W, the phenomenon of partial laser film opening to poly is visible in the three-step laser method. As the deposition rate of SiO_x was the slowest, the laser was switched on the most, and partial poly was dissolved. Although the square resistance decreased, the decrease rate was smaller than that of the SiN_x and SiO_xN_y groups. There is no significant difference between the SiN_x and SiO_xN_y

groups. From 15 W to 20 W, the wafer R_{sheet} of the SiO_x group decreased by 9.4 Ω/sq, the R_{sheet} of the SiN_x group decreased by 10.5 Ω/sq, and the R_{sheet} of the SiO_xN_y group decreased by 13.6 Ω/sq. When the laser power increases from 15 W to 20 W, the R_{sheet} continues to decrease with the increase of laser power in the four-step laser method, but the rate of decrease is less than that of the three-step laser method. After the increase of laser power in the three-step laser method, more P atoms are driven into the wafer after the mask is opened, but the amount of P atoms in the PSG was limited. As the laser power increases to 25 W, the R_{sheet} of the four-step method continues to decrease to 36.4 Ω/sq, slightly surpassing the R_{sheet} observed with 20 W laser power in the three-step method. Moreover, the 25 W laser power of the three-step method ablates part of the a-Si(n), so the R_{sheet} of the three-step method shows an increasing trend, and the increase of amplitude is similar. Considering that the SiO_x, SiO_xN_y, and SiN_x mask methods adhere to the same principles under varying power conditions, we exclusively selected the S2 and N2 groups for a comparative analysis.

3.4. J_0, iV_{oc} and τ_{eff}

In order to compare the conventional process with P-SE, we prepared two experimental groups of concentrations for comparison. The first group is a low concentration group (L1 group) with a peak concentration of approximately 2.0×10^{20} atoms/cm^3. The second group consisted of high concentration samples (H1 group) with a peak concentration of 4.0×10^{20} atoms/cm^3. The rear side of the wafer has 130 grid lines with an actual printed line width of 35 $\pm$ 3 μm. Our laser has a spot width of 120 μm, and the grid graphic incorporates a half-piece design to prevent breakage of the grid lines. Based on the laser used for the grid graphic, we calculated that the laser area covers approximately 8.6%. We prepared the non-SE group (SE group light doping zone concentration), L1 group, H1 group, and full-surface laser group following the experimental process in Figure 11b. Before printing and then firing, a J_0 test was carried out. The results are shown in Figure 11.

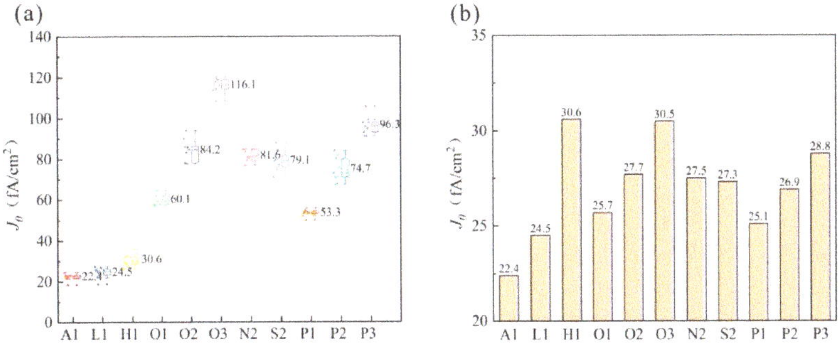

Figure 11. (**a**) J_0 of the full-face laser wafer; (**b**) J_0 calculated from the average value.

The J_0 is the current density of recombination, which reflects the strength of the recombination. In order to improve the efficiency of the solar cell, it is necessary to reduce the value of J_0 as much as possible. The J_0 value of the entire wafer after laser treatment is denoted as $J_{0t,laser}$. The J_0 value of the front and rear sides of the full-face laser on the rear side of the wafer is represented as $J_{0t,full}$. The sum of the J_0 values for both the front and rear sides of the wafer after direct annealing is expressed as $J_{0t,annealed}$. The J_0 value of the heavily doped region on the rear side of the wafer is indicated by $J_{0,finger}$.The lightly doped region on the rear side of the wafer is denoted as $J_{0,light}$. The front area's J_0 value is represented by $J_{0,front}$. S_{laser} represents the J_0 value in the laser area, while S_{wafer} denotes its corresponding film area. For calculating different experimental lasers' respective values of $J_{0t,laser}$, the following formula is used:

$$J_{0t,laser} = J_{0,finger} + J_{0,light} + J_{0,front} = (J_{0t,full} - J_{0t,annealed}) \times (S_{laser}/S_{wafer}) + J_{0t,annealed}$$

After calculation, the $J_{0t,laser}$ values are shown in Figure 11b. J_0 increases with the increase of doping concentration, whether it is the three-step method or the four-step method; J_0 increases with the increase of laser power. In laser groups, due to an only 8.6% laser region and a 91.4% light doping region, the comprehensive J_0 is between the L1 and H1 groups. The J_0 of the three masks is very close. Under the same laser power conditions, the J_0 of the four-step method is lower than that of the three-step method.

We conducted direct firing and testing of the wafers prior to printing in order to obtain iV_{oc} and τ_{eff} (effective carrier lifetime), as illustrated in Figure 12. Both iV_{oc} and τ_{eff} exhibit consistent trends. The L1 group achieved the highest iV_{oc} at 731 mV, surpassing all the other groups. In contrast, the highly concentrated H1 group exhibited a lower iV_{oc} of only 720 mV. In the laser groups, the iV_{oc} of the O3 group with only 25 W is lower than that of the H1 group, while all the other groups exhibit higher iV_{oc} values compared to the H1 group. Although the peak concentrations in the doped region of the O2, S2, and N2 groups is higher than that in the H1 group, due to its limited coverage (8.6% of the rear-side surface), the lightly doped region exhibits a lower concentration (91.4% of the rear-side surface) compared to that in the H1 group. Additionally, the mask layer in the lightly doped region fails to completely obstruct the P atoms, resulting in a similar concentration of the lightly doped region in the SE group compared to the L1 groups. Consequently, laser damage occurs within this region, serving as a recombination center for carriers. As a result, both iV_{oc} and τ_{eff} are lower in the SE group than those observed in the L1 group with a larger reduction range.

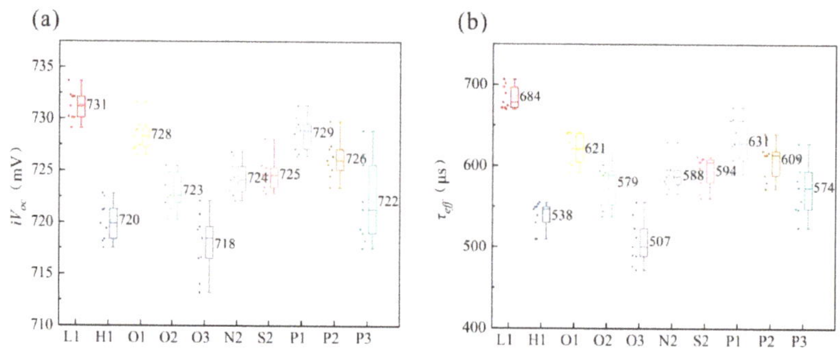

Figure 12. (**a**) iV_{oc} and (**b**) τ_{eff} of the precursors.

In order to enhance the concentration of the contact region, the F1–F3 groups employed a four-step laser method to deposit P atoms ahead of the laser, while utilizing a four-step laser method resulted in a thicker PSG mask layer, thereby minimizing damage compared to the three-step laser method. Under identical laser power conditions, iV_{oc} and τ_{eff} exhibit higher values than those achieved with the three-step laser method. Increasing the laser power led to elevated recombination rates, indicating that high-power lasers can induce defect formation (such as cracks and displacements). However, when operating at powers below 20 W, even graphics with less than 9% coverage display favorable iV_{oc} and τ_{eff}.

3.5. Specific Contact Resistivity

The significant increase in *FF* can be attributed to the presence of a heavily doped surface beneath the silver-metal contacts, which effectively drives a limited number of carriers by reducing surface recombination through field-effect passivation [49]. The specific contact resistivity of the rear side for each experimental group is illustrated in

Figure 13. As the concentration of the contact area increases, there is a gradual decrease in the specific contact resistivity of the rear side. Among all groups, the O3 group exhibits superior performance with a remarkable specific contact resistivity value of only 0.25 mΩ/cm^2. Notably, the concentration of P atoms within the contact area emerges as a critical factor influencing the observed specific contact resistivity. The specific contact resistivity decreases as the P atoms doping concentration in the contact region increases. Under the 20 W laser condition, both the three-step and four-step laser methods outperform the H1 high doping contacts. By combining iV_{oc} and τ_{eff} in Figure 12, it is observed that at 20 W, the iV_{oc} and three-step laser method exhibit superior performance compared to the high-doping concentration group, thus achieving a selective emitter effect in this study. The low doping concentration group achieved a higher iV_{oc} but compromised on contact quality.

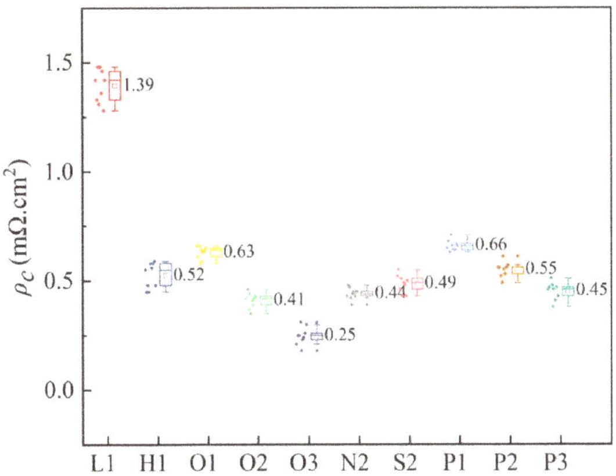

Figure 13. Specific contact resistivity of the rear side for each experimental group.

3.6. Electrical Performance

The electrical performance data of each experimental group are presented in Table 4. The low doping concentration group (L1) exhibits the highest V_{oc}, albeit at the expense of a significant reduction in *FF*, resulting in a solar cell efficiency of 24.68%. On the other hand, the high doping concentration group (H1) achieved higher *FF* by sacrificing V_{oc}, leading to an efficiency of 24.65%, which is marginally lower than that of the L1 group by 0.03%. Among the different laser power groups of SiO_x, the O1 group exhibits a higher V_{oc}, which is 5.2 mV greater than that of the H1 group. Additionally, it demonstrates significantly improved *FF* compared to the L1 group and achieves a solar cell efficiency that is 0.04% higher than that of the L1 group. Despite Group O2 having a slightly compromised V_{oc} compared to Group O1, it showcases further enhancement in *FF* and attains the highest solar cell efficiency of 24.85%, surpassing Group L1 by 0.17%. The implementation of a selective emitter enhances the current collection characteristics of the PN junction, while the enhancement in *FF* can be attributed to an improvement in series resistance (R_s) [49]. However, when the laser power is increased to 25 W, Group O3 exhibits a significant decrease in *FF* and J_{sc}, resulting in a solar cell efficiency of only 23.72%, which is 0.96% lower than that of Group L1. This decline can primarily be attributed to severe damage inflicted on Poly-Si(n$^+$) due to excessive laser power, leading to substantial surface damage and consequently causing notable reductions in V_{oc} and J_{sc}. It can be seen from Figure 9a that the laser power of 25 W reduced the thickness of Poly-Si, which resulted in the decrease of *FF*. At the same time, too many P atoms entered the Si bulk through the tunnel oxide layer, the tunneling effect of the solar cell weakened, and a large recombination was brought about, all of which resulted in the decrease of V_{oc} and J_{sc}. As shown in Figure 12a, iV_{oc} decreased significantly, and there was a corresponding relationship between iV_{oc} and V_{oc}. From the efficiency formula of the solar cell, $E_{ff} = V_{oc} \times J_{sc} \times FF/1000 \times 100\%$, it can be seen that the E_{ff} of the solar cell is proportional to V_{oc}, J_{sc}, and FF, and the V_{oc}, J_{sc}, and FF of the O3 group are significantly reduced than those of the O2 group, bringing about a significant decline in E_{ff}. The laser power of the S2 and N2 groups are equivalent; however, there exists a slight disparity in the ECV profiles. The discrepancies in V_{oc}, J_{sc}, and FF are minimal, with only a 0.02% variance observed in the efficiency of solar cells. Furthermore, the difference in efficiency of solar cells between the N2 group and O2 group is merely 0.1%. Notably, the V_{oc} of the O2, S2, and N2 groups gradually diminishes primarily due to light doping within this region. Additionally, distinct mask layers exhibit varying abilities to impede P diffusion into Poly-Si. The four-step method adheres to a consistent rule, with the P2 group exhibiting superior efficiency. As a result of direct laser doping, the concentration of the non-contact region remains unchanged and is lower compared to that achieved by the three-step method. Consequently, higher V_{oc} and J_{sc} values can be attained in comparison to those obtained using the three-step method. We found that the FF improvement in the O2 group is larger than that in the P2 group. This could potentially be attributed to a higher P atoms doping concentration in the O2 group as opposed to the P2 group, primarily driven by externally deposited P atoms and further facilitated by elevated temperatures, thereby enabling the attainment of a higher FF in Group O2. The surface damage in Group P3 occurred due to the high laser power, resulting in a slight decrease in V_{oc} and J_{sc}. However, the improvement of FF is insufficient to compensate for this decline, leading to an overall reduction in the efficiency of solar cells. Finally, the efficiency of the P3 group is close to that of the P1 group.

Table 4. Electrical performance parameters of each experimental group.

Groups	V_{oc} (mV)	J_{sc} (mA/cm^2)	R_s (mΩ)	FF (%)	E_{ff} (%)
L1	711.4	41.87	5.37	82.86	24.68
H1	703.3	41.34	3.48	84.77	24.65
O1	708.5	41.76	4.63	83.56	24.72
O2	705.7	41.56	3.93	84.74	24.85
O3	693.5	41.22	3.65	82.96	23.72
N2	705.8	41.64	4.01	84.45	24.82
S2	706.3	41.68	4.26	84.38	24.84
P1	708.7	41.78	4.87	83.42	24.70
P2	706.5	41.65	4.12	84.35	24.82
P3	703.4	41.41	3.76	84.76	24.69

4. Discussion

The rear-side laser has a significant impact on the V_{oc}, J_{sc}, and FF, which has an impact on the efficiency of solar cells. The main factors affecting the V_{oc} are as follows:

$$Voc = \frac{kT}{q}\ln\frac{Jsc + J01}{J01} \approx \frac{kT}{q}\ln\frac{Jsc}{J01}$$

In this equation, $J_{sc} \gg J_{01}$, a strong correlation between V_{oc} and the dark current is observed, encompassing reverse saturation current, thin layer leakage current, and bulk leakage current. During the process of achieving a selective emitter on the rear side, certain damages are formed that act as recombination centers by trapping electrons and holes for subsequent recombination. The V_{oc} of the SiO_x, SiN_x, and SiO_xN_y groups exhibit minimal variation. As shown in Figure 14, the V_{oc} gradually decreases with increasing laser power for the three-step and four-step SE groups. In the 15 W laser group (O1), only a partial vaporization of the mask layer occurred without any damage to the poly layer. The main gasification of the mask layer still occurred in the 20 W laser group (O1), while causing some destruction to the Poly-Si layer. However, when the laser power was increased to 25 W (O3 group), a further decrease in V_{oc} was observed, primarily due to more significant damage inflicted by the laser on the poly layer. Consequently, compared to the O1 group, the O2 group exhibited a smaller reduction in V_{oc}, whereas the O3 group demonstrated a greater decrease than that of the O2 group. In the four-step method, despite an increase in laser power, the declining trend of V_{oc} remained consistent. This can be attributed to the relatively thicker PSG layer compared to the SiO_x mask layer and the minimal damage caused by a 25 W laser on Poly-Si.

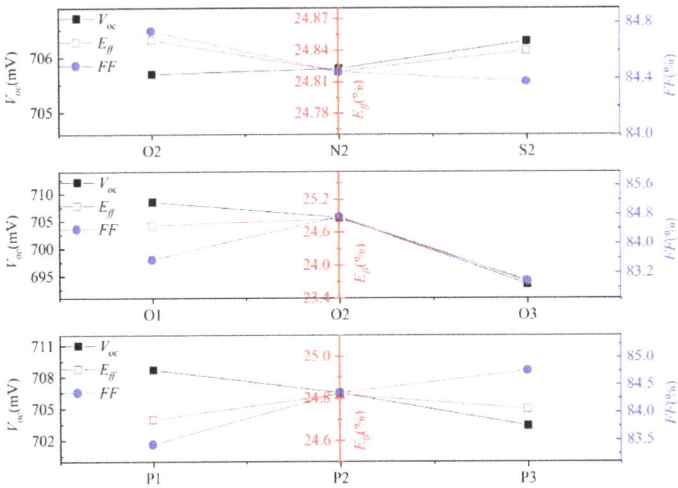

Figure 14. Comparison of V_{oc}, *FF* and E_{ff} of three masks groups, three-step groups, and four-step groups.

The *FF* has a direct relationship with contact resistance, and the contact resistance is directly affected by the concentration of P atoms in the contact region [49]. Selective emitter formation enhances the current collection characteristics of PN junctions. The P atoms doping concentrations of the SiO_x, SiN_x, and SiO_xN_y mask layers are comparable. Hence, there is no significant difference in the *FF* values. Moreover, due to similar V_{oc} values among these three groups, their efficiencies also exhibit close proximity.

In the three-step selective emitter groups, the V_{oc} exhibits a gradual decline with increasing laser power, while the *FF* and E_{ff} display an initial increase followed by a subsequent decrease. Amongst these groups, the 20 W O2 group demonstrates the highest efficiency, whereas the 25 W O3 group exhibits significantly lower efficiency compared to both the O1 and O2 groups.

In the four-step selective emitter groups, the *FF* shows a progressive increase with increasing power; however, the overall E_{ff} also showcases an initial rise followed by a subsequent decline due to the combined influence of V_{oc}, J_{sc}, and *FF*. The P2 group achieves the highest efficiency, while the P1 group has the lowest efficiency. Notably, there is minimal disparity in the efficiency of solar cells among the three laser groups indicating their higher tolerance towards laser irradiation as compared to that observed in the four-step method.

The increase in annealing temperatures leads to a higher kinetic energy in dopant atoms, while SiO_x is degraded at elevated temperatures, resulting in an augmentation of pinhole formation and increased bulk recombination [51] and thereby reducing passivation efficiency. The annealing temperature not only influences the density and integrity of SiO_x but also significantly impacts the crystallization rate of Poly-Si (χc). The χc value of silicon can be increased by raising the annealing temperature [52]. However, it should be noted that χc does not exhibit a linear relationship with temperature due to its susceptibility to various factors such as film composition and stress release. Moreover, achieving optimal passivation results requires a delicate balance between multiple mechanisms. Therefore, a higher Poly-Si value of χc does not necessarily guarantee superior passivation performance.

The current loss analysis of each experimental groups is presented in Figure 15. The influence mechanism of a short-circuit current is relatively complex, which is comprehensively affected by surface area recombination, metal recombination, contact properties, etc. The n-TOPCon solar cells mitigate compound loss in the base layer by tunneling through the oxide layer and implementing Poly-Si bulk passivation.

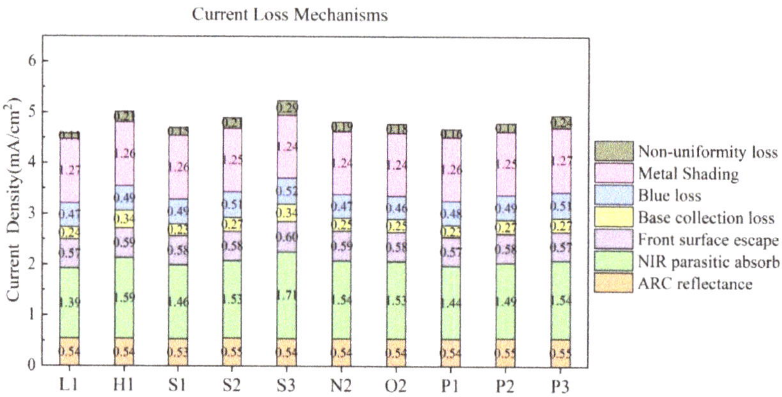

Figure 15. Current loss analysis of each experimental group.

The primary differences among the groups are attributed to NIR parasitic absorption, base collection loss, and non-uniformity loss. Near-infrared light is partially reflected by the rear, and some of it is absorbed and lost, resulting in rear parasitic absorption loss. Different laser conditions lead to varying degrees of passivation in the laser region, which affect near infrared parasitic absorption losses. Higher laser power results in greater NIR parasitic absorption loss. The parasitic absorption loss of near-infrared is greater at high doping concentrations compared to low doping concentrations. The near-infrared parasitic absorption loss follows the same trend as the base collection loss and non-uniform loss, which are also influenced by the quality of rear-side passivation. Despite an increase in laser power leading to a decrease in specific contact resistance and an improvement in carrier collection ability, there is a simultaneous increase in base collection loss. Amongst all groups, the H1 group exhibits the highest base collection loss primarily due to its high doping concentration on the rear side. On the other hand, the O3 group experiences decreased bulk passivation mass resulting from the high laser energy causing some P atoms to penetrate through the tunnel oxidation layer into the wafer.

Both the three-step and four-step laser methods can enhance the efficiency of n-TOPCon solar cells, with the former reducing one step in industrial production. The various mask structures of the three-step method effectively block the light-doped region P atoms to achieve a selective emitter. The three-step method is simpler and more effective but not more stable. Because the P atoms of the three-step method are deposited at 920 °C for 7 min, the high temperature increases the diffusion rate of P atoms, which forms a severe wrap-round of P atoms. The severe wrap-round of P atoms forms BPSG with BSG at the front, which is difficult to clean. The four-step method only needs to deposit the P atoms at a low temperature because there is also a step of oxidation annealing for 45 min, which is enough to form good PSG that is protective of the wafer. The corrosion in the cleaning process needs to be strictly controlled, and the four-step method provides better control. The large-scale production additives and cleaning processes currently focus on SiO_x and PSG, with better coordination of the SiO_x mask layer. However, the mask layer containing N also presents an environmental issue of N emission, so, we recommend the SiO_x mask. The efficiency of SiO_x is the highest, which is mainly due to the best matching of 20 W laser power, 200 s SiO_x, the amount of deposited P atom, etc. Due to the better blocking effect of SiN_x, the optimal power matching condition may be above 20W but below 25 W. The same effect can be achieved in the four-step method.

The precise control of laser power during mask opening yields excellent comprehensive electrical properties without damaging Poly-Si. Although laser technology is a very attractive approach for a large number of applications in silicon photovoltaic manufacturing, its application requires a significant amount of fine-tuning of laser parameters to reduce the harmful effects that accompany laser processing. In particular, laser processing can adversely affect device performance by reducing minority carrier lifetime [41,53–55], increasing leakage current [41,42,56], and reducing carrier transport [57]. All of these depend on the surface preparation, the pre-existing dielectric pile, and the laser conditions (pulse duration and energy density) of the application. The optimization direction of the rear side of n-TOPCon solar cells can be achieved through the experimental method employed in this study, enabling the preparation of a-Si without doping or with a lower concentration of doping to achieve a poly-finger structure on the rear side. Currently, mass-produced LECO technology [58] utilizes full-face laser treatment, which not only enhances contact on the front side but also improves contact performance on the rear side. This approach enables excellent contact performance at low doping concentrations and allows for lower doping concentrations in non-contact areas on the rear side, resulting in higher overall electrical properties. The future optimization direction of n-TOPCon solar cells also encompasses the reduction in poly layer thickness to mitigate near parasitic absorption losses [59,60]. Furthermore, the prospective technological advancements include double-sided TOPCon technology, poly-finger technology, and ultimately integrating BC with TBC technology.

5. Conclusions

The rear-side selective emitter technique significantly impacts the passivation and contact properties of n-TOPCon solar cells. Both the three-step and the four-step laser methods can effectively reduce the specific contact resistivity at the rear-side surface. In the three-step method, among the three mask layers deposited for 200 s, the blocking effect on phosphorus in the non-contact region followed the order: SiO_x < SiN_x < SiO_xN_y. Considering current mass production cleaning conditions, SiO_x is recommended as the optimal choice for mass production without introducing nitrogen elements. In terms of stability, the four-step method exhibits higher performance compared to the three-step method. However, in terms of comprehensive electrical properties, the three-step method can achieve comparable results to the four-step method. The O2 group demonstrated the highest efficiency (24.85%), with a comprehensive V_{oc} of 2.4 mV higher than that of the high doping concentration group (H1 group). Additionally, it exhibits an increase in *FF* by 1.88% compared to the low doping concentration group (L1 group), resulting in a final efficiency that is 0.17% higher than the L1 group and 0.20% higher than the H1 group. Compared with Group O2 (20 W), its V_{oc} decreases by 12.2 mV, leading to a substantial decline in the solar cell's electrical performance with an efficiency of only 23.72%. In the four-step method, the group with a laser power of 20 W (P2 group) achieved the highest efficiency (24.82%). The V_{oc} is 3.2 mV higher than that of the H1 group, while the *FF* is 1.49% higher than that of the L1 group. Moreover, the overall efficiency is found to be 0.14% higher than that of the L2 group and 0.17% higher than that of the H1 group. Notably, in comparison to both the L1 and H1 groups, each laser condition group demonstrates significant improvements within this four-step method study, thereby providing valuable guidance for enhancing n-TOPCon solar cells' back technology during mass production.

Author Contributions: Z.L.: Conceptualization, Methodology, Writing—Original draft preparation; C.G.: Data curation, Writing—Original draft preparation; Y.L.: Investigation, Visualization; J.W.: Validation, Writing—Reviewing and Editing; X.S.: Supervision and Editing; Q.W.: Supervision and Editing. All authors have read and agreed to the published version of the manuscript.

Funding: This research was funded by the Priority Academic Program Development of Jiangsu higher education institutions is gratefully acknowledged. Additionally, this work also has received partial funding from the National Natural Science Foundation youth project (62304199) and the National Key Research and Development Program of China (2023YFB4202600).

Institutional Review Board Statement: Not applicable.

Informed Consent Statement: Not applicable.

Data Availability Statement: Data are contained within the article.

Conflicts of Interest: The authors declare that they have no known competing financial interests or personal relationships that could have appeared to influence the work reported in this paper.

References

1. Feldmann, F.; Bivour, M.; Reichel, C.; Hermle, M.; Glunz, S.W. *A Passivated Rear Contact for High-Efficiency n-Type Si Solar Cells Enabling High Voc's and FF > 82%*; EU PVSEC: Paris, France, 2013; pp. 988–992.
2. Richter, A.; Benick, J.; Feldmann, F.; Fell, A.; Hermle, M.; Glunz, S.W. N-Type Si solar cells with passivating electron contact: Identifying sources for efficiency limitations by wafer thickness and resistivity variation. *Sol. Energy Mater. Sol. Cells* **2017**, *173*, 96–105. [CrossRef]
3. Haase, F.; Hollemann, C.; Schäfer, S.; Merkle, M.; Rienäcker, M.; Krügener, J.; Brendel, R.; Peibst, R. Laser contact openings for local poly-Si-metal contacts enabling 26.1%-efficient POLO-IBC solar cells. *Sol. Energy Mater. Sol. Cells* **2018**, *186*, 184–193. [CrossRef]
4. Wang, Q.; Peng, H.; Gu, S.; Guo, K.; Wu, W.; Li, B.; Li, L.; Yuan, N.; Ding, J. High-efficiency n-TOPCon bifacial solar cells with selective poly-Si based passivating contacts. *Sol. Energy Mater. Sol. Cells* **2023**, *259*, 112458. [CrossRef]
5. Lohmuller, E.; Baliozian, P.; Gutmann, L.; Kniffki, L.; Beladiya, V.; Geng, J.; Wang, L.; Dunbar, R.; Lepert, A.; Hofmann, M.; et al. Thermal laser separation and high-throughput layer deposition for edge passivation for TOPCon shingle solar cells. *Sol. Energy Mater. Sol. Cells* **2023**, *258*, 112419. [CrossRef]
6. Khokhar, M.Q.; Hussain, S.Q.; Kim, Y.; Dhungel, S.K.; Yi, J. A novel passivating contact approach for enhanced performance of crystalline silicon solar cells. *Mater. Sci. Semicond. Process.* **2023**, *155*, 107231. [CrossRef]
7. Ma, S.; Liao, B.; Qiao, F.Y.; Ding, D.; Gao, C.; Li, Z.P.; Tong, R.; Kong, X.Y.; Shen, W.Z. 24.7% industrial tunnel oxide passivated contact solar cells prepared through tube PECVD integrating with plasma-assisted oxygen oxidation and in-situ doped polysilicon. *Sol. Energy Mater. Sol. Cells* **2023**, *257*, 112396. [CrossRef]
8. Chen, W.; Liu, W.; Yu, Y.; Chen, P.; Wan, Y. A study of activated phosphorus distribution within silicon substrate for polysilicon passivating contacts based on an in-line PVD system. *Sol. Energy* **2023**, *259*, 375–380. [CrossRef]
9. Cheng, H.; Liu, W.; Liu, Z.; Yang, Z.; Ma, D.; Du, H.; Luo, J.; Xing, H.; Liao, M.; Zeng, Y.; et al. Emitter formation with boron diffusion from PECVD deposited boron-doped silicon oxide for high-efficiency TOPCon solar cells. *Sol. Energy Mater. Sol. Cells* **2022**, *240*, 111713. [CrossRef]
10. Liao, B.; Ge, J.; Wu, X.; Wang, Q.; Yeo, R.J.; Du, Z. Unlocking the potential of boronsilicate glass passivation for industrial tunnel oxide passivated contact solar cells. *Prog. Photovolt.* **2022**, *30*, 310–317. [CrossRef]
11. Hong, J.; Liu, X.; Ge, J.; Sun, J.; Liu, S.; Yang, W.; Jaffer, S.; Lu, J.; Qin, T.; Chen, R.; et al. Superb improvement of boron doping in selective emitter for TOPCon solar cells via boron-doped silicon paste. *Sol. Energy* **2022**, *247*, 115–122. [CrossRef]
12. Firat, M.; Radhakrishnan, H.S.; Singh, S.; Duerinckx, F.; Payo, M.R.; Tous, L.; Poortmans, J. Industrial metallization of fired passivating contacts for n-type tunnel oxide passivated contact (n-TOPCon) solar cells. *Sol. Energy Mater. Sol. Cells* **2022**, *240*, 111692. [CrossRef]
13. Xu, J.; Chen, C.; Liu, C.; Chen, J.; Liu, Z.; Yuan, X.; Li, H. High efficiency TOPCon solar cells with micron/nano-structured emitter for a balance of light-trapping and surface passivation. *Sol. Energy Mater. Sol. Cells* **2022**, *238*, 111606. [CrossRef]
14. Ma, Z.Q.; Wang, Y.L.; Lan, Z.X.; Zhao, L.; Xu, F.; Xu, J. On the presence of oxyphosphorus bonds in TOPCon solar cell polycrystalline silicon films. *Sol. Energy Mater. Sol. Cells* **2022**, *246*, 111910. [CrossRef]
15. Hilali, M.M.; Nakayashiki, K.; Ebong, A.; Rohatgi, A. Investigation of high-efficiency screen-printed textured si solar cells with high sheet-resistance emitters. In Proceedings of the Conference Record of the Thirty-first IEEE Photovoltaic Specialists Conference, Lake Buena Vista, FL, USA, 3–7 January 2005; pp. 1185–1188.
16. Meier, D.L.; Schroder, D.K. Contact resistance: Its measurement and relative importance to power loss in a solar cell. *IEEE Trans. Electron Devices* **1984**, *31*, 647–653. [CrossRef]
17. Ding, D.; Lu, G.; Li, Z.; Zhang, Y.; Shen, W. High-efficiency n -type silicon PERT bifacial solar cells with selective emitters and poly-Si based passivating contacts. *Sol. Energy* **2019**, *193*, 494–501. [CrossRef]
18. Dubé, C.E.; Tsefrekas, B.; Buzby, D.; Tavares, R.; Zhang, W.; Gupta, A.; Low, R.J.; Skinner, W.; Mullin, J. High efficiency selective emitter cells using patterned ion implantation—ScienceDirect. *Energy Procedia* **2011**, *8*, 706–711. [CrossRef]
19. Xiao, M.; Yang, Z.; Liu, Z.; Du, H.; Lin, N.; Wei, H.; Xing, H.; Wu, Q.; Liu, W.; Liao, M.; et al. SiO_x/polysilicon selective emitter prepared by PECVD-deposited amorphous silicon plus one-step firing enabling excellent $J_{0,met}$ of <235 fA/cm^2 and ρ_c of <2 mΩ·cm^2. *Sol. Energy* **2023**, *262*, 111887.
20. Lin, W.; Chen, D.; Liu, C.; Wang, Y.; He, Y.; Zou, Y.; Yuan, L.; Gong, J.; Yang, Y.; Feng, Z.; et al. Green-laser-doped selective emitters with separate BBr3 diffusion processes for high-efficiency n-type silicon solar cells. *Sol. Energy Mater. Sol. Cells* **2020**, *210*, 110462. [CrossRef]

21. Kafle, B.; Goraya, B.S.; Mack, S.; Feldmann, F.; Nold, S.; Rentsch, J. TOPCon—Technology options for cost efficient industrial manufacturing. *Sol. Energy Mater. Sol. Cells* **2021**, *227*, 111100. [CrossRef]
22. Van-Driel, H.M. Physics of pulsed laser processing of semiconductors. In *Semiconductors Probed by Ultrafast Laser Spectroscopy*; Academic Press: Cambridge, MA, USA, 1984; Volume 2, pp. 57–94.
23. Vaqueiro-Contreras, M.; Hallam, B.; Chan, C. Review of Laser Doping and its Applications in Silicon Solar Cells. *IEEE J. Photovolt.* **2023**, *13*, 373–384. [CrossRef]
24. Hameiri, Z.; Puzzer, T.; Mai, L.; Sproul, A.B.; Wenham, S.R. Laser induced defects in laser doped solar cells. *Prog. Photovolt.* **2011**, *19*, 391–405. [CrossRef]
25. Molpeceres, C.; Sánchez-Aniorte, M.I.; Morales, M.; Muoz, D.; Martín, I.; Ortega, P.; Colina, M.; Voz, C.; Alcubilla, R. In Influence of wavelength on laser doping and laser-fired contact processes for c-Si solar cells. *SPIE* **2012**, *8473*, 61–71.
26. Basnet, R.; Phang, S.P.; Samundsett, C.; Yan, D.; Liang, W.; Sun, C.; Armand, S.; Einhaus, R.; Degoulange, J.; Macdonald, D. 22.6% Efficient Solar Cells with Polysilicon Passivating Contacts on n-type Solar-Grade Wafers. *Solar RRL* **2019**, *3*, 1900297. [CrossRef]
27. Li, H.; Ma, F.J.; Hameiri, Z.; Wenham, S.; Abbott, M. On elimination of inactive phosphorus in industrial $POCl_3$ diffused emitters for high efficiency silicon solar cells. *Sol. Energy Mater. Sol. Cells* **2017**, *171*, 213–221. [CrossRef]
28. Sinton, R.A.; Cuevas, A. *A Quasi-Steady-State Open-Circuit Voltage Method for Solar Cell Characterization*; EU PVSEC: Glasgow, UK, 2000; Volume 2, pp. 1152–1155.
29. Guo, S.; Gregory, G.; Gabor, A.M.; Schoenfeld, W.V.; Davis, K.O. Detailed investigation of TLM contact resistance measurements on crystalline silicon solar cells. *Sol. Energy* **2017**, *151*, 163–172. [CrossRef]
30. Wong, J.; Raj, S.; Ho, J.W.; Wang, J.; Lin, J. Voltage Loss Analysis for Bifacial Silicon Solar Cells: Case for Two-Dimensional Large-Area Modeling. *IEEE J. Photovolt.* **2017**, *6*, 1421–1426. [CrossRef]
31. Wan, H.; Min, J.; Lin, J.; Carlson, B.E.; Surender, M.; Sun, C. Effect of laser spot overlap ratio on surface characteristics and adhesive bonding strength of an Al alloy processed by nanosecond pulsed laser. *J. Manuf. Process.* **2021**, *62*, 555–565. [CrossRef]
32. Kittiboonanan, P.; Mcwilliams, J.; Taechamaneesatit, P.; Ratanavis, A. Effects of Laser Pulse Overlap using Nanosecond Fiber Lasers for Structuring of AA6061-T6 Surfaces. *Lasers Manuf. Mater. Process.* **2023**, *10*, 190–230. [CrossRef]
33. Wang, S.S.; Ho, J.J.; Liou, J.J.; Ho, J.S.; Hsu, W.C.; Lu, W.H.; Tsai, S.Y.; Hung, H.S.; Wang, K.L. Performance Improvements of Selective Emitters by Laser Openings on Large-Area Multicrystalline Si Solar Cells. *Int. J. Photoenergy* **2014**, *2014*, 291904. [CrossRef]
34. Nguyen, H.T.; Han, Y.; Ernst, M.; Fell, A.; Franklin, E.; Macdonald, D. Dislocations in laser-doped silicon detected by micro-photoluminescence spectroscopy. *Appl. Phys. Lett.* **2015**, *107*, 22101. [CrossRef]
35. Han, Y.J.; Franklin, E.; Fell, A.; Ernst, M.; Nguyen, H.T.; Macdonald, D.; Macdonald, D. Low-temperature micro-photoluminescence spectroscopy on laser-doped silicon with different surface conditions. *Appl. Phys. A* **2016**, *122*, 420. [CrossRef]
36. Mesli, A.; Muller, J.C.; Siffert, P. Analysis and origin of point defects in silicon after liquid phase transient annealing. *J. Phys.* **1983**, *44*, 281–295. [CrossRef]
37. Poulain, G.; Boulord, C.; Blanc, D.; Kaminski, A.; Gauthier, M.; Dubois, C.; Semmache, B.; Lemiti, M. Direct laser printing for high efficiency silicon solar cells fabrication. *Appl. Surf. Sci.* **2011**, *257*, 5241–5244. [CrossRef]
38. Velichko, O.I.; Aksenov, V.V.; Anufriev, L.P.; Golubev, N.F.; Komarov, A.F. Modeling high-concentration phosphorus diffusion in crystalline silicon. *J. Eng. Phys. Thermophys.* **2013**, *86*, 667–675. [CrossRef]
39. Li, J.; Zhang, W.; Zhou, Y.; Zhang, X.; Jin, G. Experimental study of the damage morphology induced by the millisecond-nanosecond combined-pulse laser with different pulse delay on silicon. *Optik* **2021**, *226*, 165904. [CrossRef]
40. Zhang, D.; Gökce, B.; Sommer, S.; Streubel, R.; Barcikowski, S. Debris-free rear-side picosecond laser ablation of thin germanium wafers in water with ethanol. *Appl. Surf. Sci.* **2016**, *367*, 222–230. [CrossRef]
41. Sun, Z.; Gupta, M.C. A study of laser-induced surface defects in silicon and impact on electrical properties. *J. Appl. Phys.* **2018**, *124*, 223103. [CrossRef]
42. Fairfield, J.M.; Schwuttke, G.H. Silicon diodes made by laser irradiation. *Solid State Electron.* **1968**, *11*, 1175-IN6. [CrossRef]
43. Besi-Vetrella, U.; Salza, E.; Pirozzi, L.; Noel, S.; Slaoui, A.; Muller, J.C. Selective doping of silicon by rapid thermal and laser assisted processes. *Mater. Sci. Semicond. Process.* **1999**, *1*, 325–329. [CrossRef]
44. Amer, M.S.; Dosser, L.; Leclair, S.; Maguire, J.F. Induced stresses and structural changes in silicon wafers as a result of laser micro-machining. *Appl. Surf. Sci.* **2002**, *187*, 291–296. [CrossRef]
45. Kray, D.; Fell, A.; Hopman, S.; Mayer, K.; Willeke, G.P.; Glunz, S.W. Laser Chemical Processing (LCP)-A versatile tool for microstructuring applications. *Appl. Phys. A* **2008**, *93*, 99–103. [CrossRef]
46. Tan, H.S.; Ng, S.C.; Woon, H.S.; Hultquist, G. Transient capacitance measurements of laser radiation-induced defects in silicon. *Semicon. Sci. Technol.* **1990**, *5*, 657–662. [CrossRef]
47. Kachurin, G.A.; Nideav, E.V.; Popov, A.I. Laser annealing of radiation defects, studied by the capacitance spectroscopy method. *Sov. Phys. Semicond. USSR* **1982**, *16*, 13–15.
48. Jäger, U.; Okanovic, M.; Hörtheis, M.; Grohe, A.; Preu, R. *Selective Emitter by Laser Doping from Phosphosilicate Glass*; EU PVSC: Hamburg, Germany, 2009; pp. 1740–1743.
49. Ju, M.; Park, J.; Cho, Y.H.; Kim, Y.; Lim, D.; Cho, E.C.; Yi, J. A Novel Method to Achieve Selective Emitter Using Surface Morphology for PERC Silicon Solar Cells. *Energies* **2020**, *13*, 5207. [CrossRef]

50. Yang, X.; Bullock, J.; Xu, L.; Bi, Q.; Surve, S.; Ernst, M.; Weber, K. Passivated contacts to laser doped p+ and n+ regions. *Sol. Energy Mater. Sol. Cells* **2015**, *140*, 38–44. [CrossRef]
51. Yang, Z.; Liu, Z.; Cui, M.; Sheng, J.; Chen, L.; Lu, L.; Guo, W.; Yang, X.; Zhao, Y.; Yang, W.; et al. Charge Carrier Dynamics for Silicon Oxide Tunneling Junctions Mediated by Local Pinholes. *Cell Rep. Phys. Sci.* **2021**, *2*, 100667. [CrossRef]
52. Choi, S.; Min, K.H.; Jeong, M.S.; Lee, J.I.; Kang, M.G.; Song, H.E.; Kang, Y.; Lee, H.S.; Kim, D.; Kim, K.H. Structural evolution of tunneling oxide passivating contact upon thermal annealing. *Sci. Rep.* **2017**, *7*, 12853. [CrossRef] [PubMed]
53. Tang, K.; Øvrelid, E.J.; Tranell, G.; Tangstad, M. Thermochemical and Kinetic Databases for the Solar Cell Silicon Materials. *Cryst. Growth Si Sol. Cells* **2009**, *14*, 219–251.
54. Abbott, A.; Cousins, P.; Chen, F.; Cotter, J.E. *Laser-Induced Defects in Crystalline Silicon Solar Cells*; IEEE PVSC: Lake Buena Vista, FL, USA, 2005; pp. 1241–1244.
55. Ametowobla, M.; Bilger, G.; Kohler, J.R.; Werner, J.H. Laser induced lifetime degradation in p-type crystalline silicon. *J. Appl. Phys.* **2012**, *111*, 114515. [CrossRef]
56. Young, R.T.; Wood, R.F.; Narayan, J.; White, C.W.; Christie, W.H. Laser processing for high-efficiency Si solar cells. *J. Appl. Phys.* **1982**, *53*, 1178–1189. [CrossRef]
57. Ernst, M.; Huyeng, J.D.; Walter, D.; Fong, K.C.; Blakers, A. Unravelling the Origins of Contact Recombination for Localized Laser-Doped Contacts. In Proceedings of the 2018 IEEE 7th World Conference on Photovoltaic Energy Conversion (WCPEC) (A Joint Conference of 45th IEEE PVSC, 28th PVSEC & 34th EU PVSEC), Waikoloa, HI, USA, 10–15 June 2018; pp. 2195–2199.
58. Krassowski, E.; Jaeckel, B.; Zeller, U.; Pander, M.; Schenk, P.; Hofmueller, E.; Hanifi, H. Reliability Evaluation of Photovoltaic Modules Fabricated from Treated Solar Cells by Laser-Enhanced Contact Optimization Process. *Solar RRL* **2022**, *6*, 2100537. [CrossRef]
59. Zhou, Y.; Tao, K.; Liu, A.; Jia, R.; Qu, H. Screen-printed n-type industry solar cells with tunnel oxide passivated contact doped by phosphorus diffusion. *Superlattices Microstruct.* **2020**, *148*, 106720. [CrossRef]
60. Padhamnath, P.; Nampalli, N.; Nandakumar, N.; Buatis, J.K.; Duttagupta, S. Optoelectrical properties of high-performance low-pressure chemical vapor deposited phosphorus-doped polysilicon layers for passivating contact solar cells. *Thin Solid Films* **2020**, *699*, 137886. [CrossRef]

materials

MDPI

Article

The Design of PAN-Based Janus Membrane with Adjustable Asymmetric Wettability in Wastewater Purification

Yuehui Wang [1,†], Jun Huang [1,†], Ye Zhang [1], Shiwen Zhang [1], Lili Li [1,*] and Xuan Pang [2,*]

[1] Key Laboratory of Automobile Materials, Ministry of Education, College of Materials Science and Engineering, Jilin University, Changchun 130022, China; 15161110960@163.com (Y.W.); 17843334500@163.com (J.H.); 15124698738@163.com (Y.Z.); shiwenuknow@163.com (S.Z.)

[2] Key Laboratory of Polymer Ecomaterials, Changchun Institute of Applied Chemistry, Chinese Academy of Sciences, Changchun 130022, China

* Correspondence: lilyllee@jlu.edu.cn (L.L.); xpang@ciac.ac.cn (X.P.)

† These authors contributed equally to this work.

Abstract: In this paper, an environmentally friendly polyacrylonitrile-based (PAN-based) composite membrane with a Janus structure for wastewater treatment was successfully fabricated. To achieve the optimum adsorption of PAN-based Janus composite membrane, the asymmetric wettability was regulated through electrospinning, resulting in TiO_2 modifying PAN as the hydrophilic substrate layer, and PCL gaining a different thickness as the hydrophobic layer. The prepared Janus composite membrane (PAN/TiO_2-PCL20) showed excellent oil/water separation performance for diverse surfactant-stabilized oil-in-water emulsions. For n-hexane-in-water emulsion, the permeate flux and separation efficiency reached 1344 L m^{-2} h^{-1} and 99.52%, respectively. Even after 20 cycles of separation, it still had outstanding reusability and the separation efficiency remained above 99.15%. Meanwhile, the PAN/TiO_2-PCL20 also exhibited an excellent photocatalytic activity, and the removal rate for RhB reached 93.2%. In addition, the research revealed that PAN/TiO_2-PCL20 possessed good mechanical property and unidirectional water transfer capability. All results indicated that PAN/TiO_2-PCL20 with photocatalysis and oil/water separation performance could be used for practical complex wastewater purification.

Keywords: Janus composite membrane; PCL; oil/water separation; photocatalysis

1. Introduction

In recent years, water pollution problems from frequent oil spills and discharge of industrial oily wastewater have posed a potential threat to the ecological environment and human health [1,2]. The wastewater containing organic oils and dyes from the process of industrial production needs to be settled urgently due to its toxicity, non-biodegradability, carcinogenicity and mutagenicity [3–6]. Therefore, it is necessary to develop efficient and environmentally-friendly technology to remove organic contaminants from the water environment.

Traditional methods for treating oil-contaminated wastewater, such as biological treatment, membrane separation, flocculation, air flotation, gravity separation and adsorption etc., have been reported [7–15]. Among them, the membrane separation method attracts tremendous attention in oil/water separation because of its high separation efficiency, wide application range and easy operation [16–18]. Han et al. fabricated a titanium dioxide-acrylonitrile-butadiene-styrene composite membrane (TiO_2-ABS) via 3D printing. This TiO_2-ABS composite membrane demonstrated exceedingly high flux (1.8×10^5 L m^{-2} h^{-1}) and oil rejection rate (99.5%) [19].

The conventional membrane separation method for water purification is not efficient enough. Recently, some researchers retained the filtration properties and fabricated membranes used in an integrated filtration-adsorption process [20,21]. Photocatalytic degradation is considered to be an effective and green approach for traditional organic dye adsorption and removal [22,23]. Zhao et al. prepared a n-p type $Bi_2WO_6/AgInS_2$ S-scheme heterojunction for organic pollutants degradation. The $Bi_2WO_6/AgInS_2$ exhibited much higher photocatalytic activity, achieving degradation efficiency of 97% for RhB under visible light for 60 min [24]. Recently, the utilizing of both photocatalysis and membrane adsorption technologies have become a research hotpot for attempts to remove organic dyes in wastewater [25,26]. Baig et al. fabricated superhydrophobic/superoleophilic photocatalytic membranes (CeO_2 nanoparticles coated membrane) that exhibited simultaneously high removal efficiency of dyes (99.96%) and oils (99.95%) [27].

The name "Janus membrane" comes from Janus, the two-faced god in ancient Roman mythology [28]. Janus membrane is a general term for membranes with opposite properties, such as composition, morphology, wettability, surface charge, and so on. Layer-by-layer electrospinning, surface coating, unilateral deposition and chemical modification are often used to prepare Janus membranes [29–32]. In recent years, Janus membranes with a unique hierarchical structure of asymmetric hydrophilicity/hydrophobicity have attracted more and more attention [33,34]. Meanwhile, Janus membranes are widely used as an ideal material for water treatment [35,36]. Chen et al. fabricated a Janus wood (JW) through a unidirectional vacuum impregnation method that had a unidirectional transport of water, and presented high flux (3700 L m^{-2} h^{-1}) and separation efficiency (99.6%) [37]. Sun et al. fabricated a Janus membrane via hydrophilic ZnO nanowires modified hydrophobic polyvinylidene fluoride (PVDF) nanofiber. The membrane showed excellent oil/water separation fluxes (1210 L m^{-2} $h^{-1)}$ for heavy oil-water mixtures; 7653 L m^{-2} h^{-1} for light oil-water mixtures; and good photocatalytic degradation performance (95%) [38].

Herin, PAN and PCL were used as a matrix to design a multifunctional oil-water separator with a Janus membrane structure. In order to improve its oil-water separation efficiency and endow it with multifunctionality for water remediation, TiO_2 was added to modify the hydrophilic PAN layer, which was expected to improve oil-water separation performance by modifying asymmetric wettability, and further endowing the multifunction (such as photocatalytic degradation and unidirectional water transfer capability) of the Janus membranes. The morphology, chemical structure, and wettability of all the as-prepared PAN/TiO_2-PCLs were characterized by SEM, LSCM, FT-IR and contact angle measurement. The oil/water separation performance of the PAN/TiO_2-PCLs were operated by a homemade experiment device. The photocatalytic degradation performance of the PAN/TiO_2-PCL20 was investigated under simulated solar irradiation. The mechanical properties and unidirectional water transfer capability were also tested.

2. Materials and Methods

2.1. Materials

Polyacrylonitrile (PAN, Mw = 150,000), Polycaprolactone (PCL, Mw = 50,000), TiO_2 nanoparticles (anatase-type nano powder, particle size: 5~10 nm), dichloromethane, coloring agent: Rhodamine B (RhB), Sudan III and methylene blue were provided by Aladdin Industrial Corporation, China. Sodium dodecyl sulfate (SDS, ACS reagent) was purchased from Saan Chemical Technology (Shanghai) Co., Ltd., China. Chloroform (CF), n-hexane, glycerin and N,N-dimethylformamide (DMF) were obtained from Beijing Chemical Works, Beijing, China. Diesel oil was supplied from National Petroleum Co., Ltd., China. Olive oil was purchased from China Oil & Foodstuffs Corporation (COFCO) (Beijing, China). Benzyl benzoate was obtained from Xilong Chemical Technology Co., Ltd., Guangzhou, China. 1,1,2,2-Tetrabromoethane was purchased from Jiangshun Chemical Technology Co., Ltd., Guangzhou, China. All chemical reagents and drugs used in this study were analytical reagents without further purification.

2.2. *Fabrication of PAN-Based Janus Membrane*

Fabrication of PAN/TiO_2-PCL Janus Membrane

1 g PAN and 0.3 g TiO_2 were dissolved in 10 mL DMF by magnetic stirring for 12 h at room temperature, and then PAN/TiO_2 precursor solution was obtained. At the same time, 2 g PCL was added into 10 mL CF/DMF (4:1, v/v) mixed solvent to obtain 20 wt% PCL precursor solution. Both precursor solutions were placed into 5 mL syringes. The PAN/TiO_2 substrate layer was fabricated by electrospinning at 16 kV applied voltage and 1 mL h^{-1} feeding speed for 2 h. Then the PCL fiber membranes with different thickness were electrospun on PAN/TiO_2 substrate layer with an applied voltage of 12 kV and a flow rate of 2 mL h^{-1} for different electrospinning times (10 min, 20 min and 30 min, respectively). The prepared Janus membranes were named PAN/TiO_2-PCLx, where x represented the electrospinning time of PCL. The manufacturing process of PAN/TiO_2-PCLx was schematically displayed in Scheme 1. The PAN-PCLx Janus membranes were prepared by the same procedure and more details were referred to the Supplementary Materials.

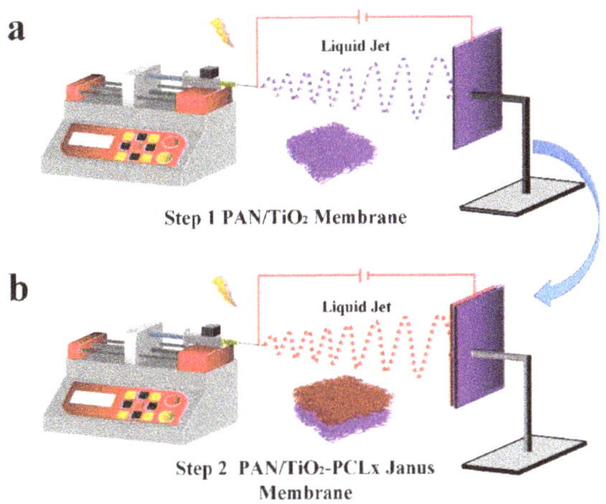

Scheme 1. The fabrication process of (**a**) electrospinning PAN/TiO_2 substrate; (**b**) electrospinning PCL layer on PAN/TiO_2 substrate.

2.3. *Characterizations*

The morphologies of all prepared nanofiber membranes were observed by a cold-field emission-scanning electron microscope (SEM, JSM-6700F, JEOL, Tokyo, Japan) at 8 kV acceleration voltage under the vacuum condition. The nanofiber diameters were assisted by Image J software (http://cnij.imjoy.io/, accessed on 17 April 2023). Viscometer (NDJ-1, Shanghai Precision Instruments Corporation, Shanghai, China) was used to measure the viscosity of the solution, and conductivity of the solution was measured by using a conductivity meter (DDS-11A, Shanghai Precision Instruments Corporation, Shanghai, China). The specific surface area and porosity of all prepared membranes were measured according to the Brunauer-Emmett-Teller analyzer (BET, Autosorb-iQ, Quantachrome Instruments, Boynton Beach, FL, USA). The surface morphology and roughness of membranes were detected through a Confocal Laser Scanning Microscope (LSCM, OLS3000, Olympus Corporation, Tokyo, Japan). The chemical compositions and functional groups of all samples were analyzed through a Fourier-transform-infrared spectrometer (FT-IR, FTIR-4100, JASCO, Tokyo, Japan). The mechanical strength was tested on an electronic universal testing machine (XQ-1C, Shanghai New Fiber Instrument Co., Ltd., Shanghai, China). The underwater oil contact angles (UOCAs) were tested by using a droplet shape analyzer (DSA100, Cruise, Berlin, Germany). The droplet size distribution of oil-water emulsion, both before and after filtration, was measured through dynamic light scattering (DLS, Nano ZS90, Malvern Instruments Ltd., Malvern, UK) and optical microscopy (TE2000-U, Nikon, Tokyo, Japan).

The photocatalytic performance of the membranes was studied by using an ultraviolet spectrophotometer (UV-6100s, MAPADA, Shanghai, China).

2.4. Oil/Water Separation Experiments

2.4.1. Separation of Immiscible Oil-Water Mixtures

The separation experiments of immiscible oil-water mixtures were performed under gravity. The immiscible oil-water mixtures were prepared by mixing water (methylene blue dyed) and oil (1:1, v/v). The separation effective area was 12.96 cm^2. During separation, water was collected after penetrating the membrane, while oil was rejected on the membrane. After separation, the device was left to stand for 10 min to ensure that all water droplets permeated through. The separation efficiency (%) was determined by the Equation (1):

$$Efficiency(\%) = (V/V_i) \times 100 \tag{1}$$

where V_i (L) and V (L) represented the volume of water before and after oil/water separation, respectively. In addition, the water permeate flux (L m^{-2} h^{-1}) was calculated according to the following Equation (2):

$$Flux = \frac{V}{S \times \Delta t} \tag{2}$$

where V (L) was the volume of water permeated through the membrane, S (m^2) was the effective filtration area of membrane, and Δt (h) was the total filtration time.

2.4.2. Separation of Oil-in-Water Emulsions

For the preparation of the oil-in-water emulsions, SDS was chosen as the emulsifier. A total of 5 mL oil and 500 mL water were taken into a beaker, and then added 0.1 g Span 80 at 1200 rpm for 24 h by stirring. Ultimately, the surfactant-stabilized emulsion was obtained. N-hexane, diesel oil, olive oil, benzyl benzoate, 1,1,2,2-tetrabromoethane and glycerol were used as oils. The separation efficiencies and permeate fluxes were calculated by Equations (1) and (2), respectively.

The reusability of the PAN/TiO_2-PCL20 was evaluated by separation of the n-hexane-in-water emulsion for 20 cyclic experiments. After each separation, the PAN/TiO_2-PCL20 was washed with ethanol and dried at 40 °C for 3 h in an oven to remove the residual solvent.

2.5. Adsorption and Photocatalytic Experiments

To evaluate the adsorption process, PAN/TiO_2-PCL20 was immersed in 50 mL RhB aqueous solution (12 mg L^{-1}) and magnetically stirred under dark conditions for 60 min. 3.5 mL solution was extracted from the RhB solution at 15 min intervals to evaluate the adsorption capacity. The adsorption capacities were calculated by Equation (3).

$$q = \frac{C_i - C_t}{m} \times V \tag{3}$$

where C_i (mg L^{-1}) and C_t (mg L^{-1}) were the initial concentration and the equilibrium concentration of RhB at time t. V (L) represented the volume of dye solution and m (mg) was the adsorbent dosage.

The photocatalytic activity of PAN/TiO_2-PCL20 for RhB was measured through the following experiments. The PAN/TiO_2-PCL20 was performed for RhB degradation under simulated solar irradiation for 80 min. A total of 3.5 mL liquid was extracted at 10 min intervals to evaluate removal rate. The removal rates (R%) of RhB were calculated by Equation (4).

$$R(\%) = \frac{C_0 - C_t}{C_0} \times 100 \tag{4}$$

where C_0 (mg L^{-1}) and C_t (mg L^{-1}) represented the initial and equilibrium concentration of the RhB organic dye, respectively.

3. Results and Discussion

3.1. FT-IR Analysis

The functional group characteristics of the PAN, PAN/TiO_2, PAN/TiO_2-PCL20 were investigated by FT-IR (Figure 1). The characteristic peaks of PAN were as follows: 1452 cm^{-1} and 2939 cm^{-1} (-CH_2 symmetric and asymmetric stretching vibration), 2242 cm^{-1} (-C≡N stretching vibration) and 1631 cm^{-1} (-C=N stretching vibration) [39]. Compared with the FT-IR spectra of PAN, the broad band in the range of 3000–3500 cm^{-1} in PAN/TiO_2 spectra corresponded to the stretching vibration of -OH, which was attributed to the introduction of TiO_2 [40]. At the same time, the adsorption band at 450–900 cm^{-1} was attributed to the Ti-O bonds of TiO_2 [41]. For PAN/TiO_2-PCL20, the new peak that appeared at 1730 cm^{-1} belonged to the C=O stretching vibration, and the peak at 1157 cm^{-1} contributed to the C-O stretching vibration of PCL [42].

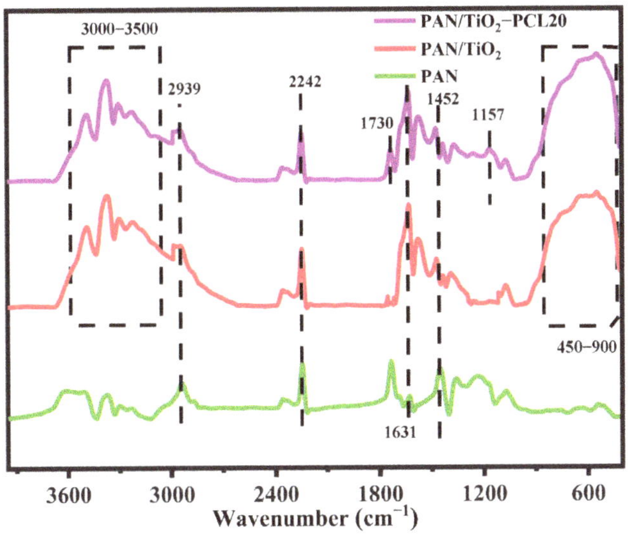

Figure 1. FTIR spectra of PAN, PAN/TiO_2 and PAN/TiO_2-PCL20. (Two dashed boxes represented the range of characteristic peaks).

3.2. Surface Morphology Analysis

As depicted in Figure 2, the surface morphologies of the PAN/TiO_2-PCL20 were characterized by SEM images. The surface of PCL layer was relatively rough and the average fiber diameter was 553.7 ± 4.3 nm (Figure 2a,e). The PAN substrate layer had a smooth surface with an average fiber diameter of 158.4 ± 2.4 nm (Figure 2b,f). After the incorporation of TiO_2 into PAN fibers, a large number of beaded structures appeared in PAN/TiO_2 fibers and the PAN/TiO_2 fiber diameter decreased to 153.8 ± 1.7 nm (Figure 2c,g). This was attributed to an increase in the viscosity and conductivity of the PAN/TiO_2 precursor solution (Table 1).

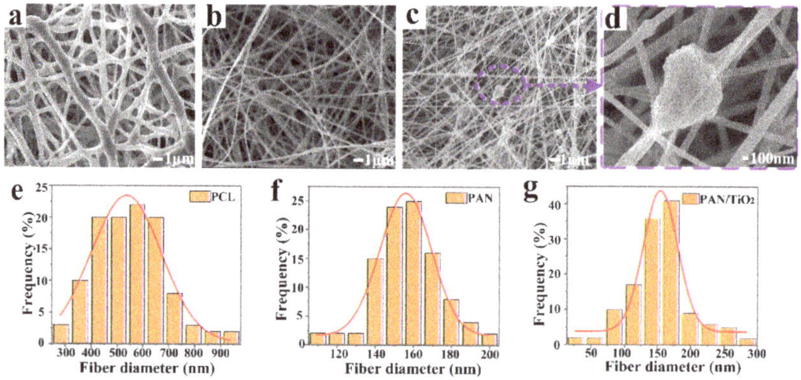

Figure 2. SEM of (**a**) PCL top layer, (**b**) PAN and (**c**,**d**) PAN/TiO_2 substrate layer; Diameter distribution of (**e**) PCL top layer, (**f**) PAN and (**g**) PAN/TiO_2 substrate layer.

Table 1. The physical parameters of electrospinning solutions.

Sample	Viscosity (mPa s)	Conductivity (μS cm^{-1})
PCL	360.2	0.351
PAN	85.5	34.2
PAN/TiO_2	125.4	67.8

3.3. BET Analysis

The Nitrogen (N_2) adsorption-desorption isotherms for PAN, PAN/TiO_2, PAN/TiO_2-PCLx are shown in Figure S1, while the BET results are analyzed in Figure 3. The specific surface area and pore volume of pure PAN were 45.198 m^2 g^{-1} and 0.220 cc g^{-1}, respectively. Compared with pure PAN, the specific surface area and pore volume of PAN/TiO_2 increased to 113.901 m^2 g^{-1} and 0.251 cc g^{-1}. This indicated that PAN/TiO_2 had a smaller pore size and larger specific surface area, which further enhanced the hydrophilicity of PAN substrate [43]. After the incorporation of the PCL layer, the specific surface area and pore volume of PAN/TiO_2-PCLx increased significantly. The PAN/TiO_2-PCL20 had the largest specific surface area (147.377 m^2 g^{-1}) and pore volume (0.326 cc g^{-1}). Compared with PAN, the PAN/TiO_2-PCL20 Janus membrane appeared to have a large number of pore diameters smaller than 50 nm—this was confirmed by BET analysis, which showed the advantages for the separation of oil-water emulsions and the adsorption of RhB [44,45].

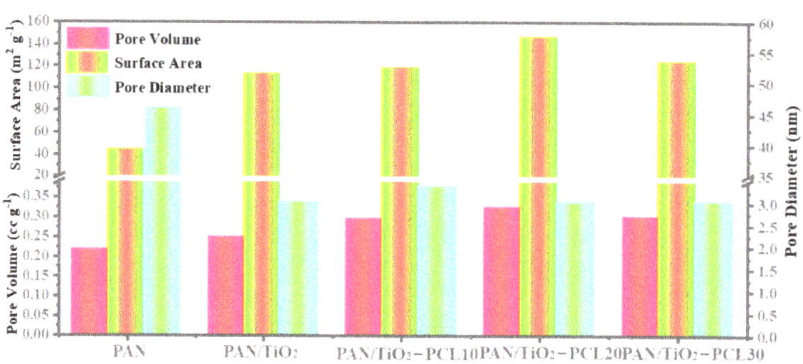

Figure 3. BET analysis of the PAN, PAN/TiO_2 and PAN/TiO_2-PCLx.

3.4. LSCM Analysis

The average surface roughness (Ra) of the PCL layer, and the PAN and PAN/TiO_2 substrate layers is analyzed in Figure 4. The Ra of the PCL layer and PAN substrate layer was 1.547 ± 0.023 μm and 1.113 ± 0.047 μm, respectively (Figure 4a,b). Compared with the PAN substrate layer, the Ra of PAN/TiO_2 substrate layer increased to 1.716 ± 0.073 μm (Figure 4c). It could be observed that the surface roughness of the PAN/TiO_2 substrate layer improved significantly. Owing to the impact of PAN on the hydrophilic layer, the larger roughness for PAN/TiO_2 could lead to a further improvement in hydrophilicity [46], which could further increase the asymmetric wettability of the Janus membrane when combined with BET analysis.

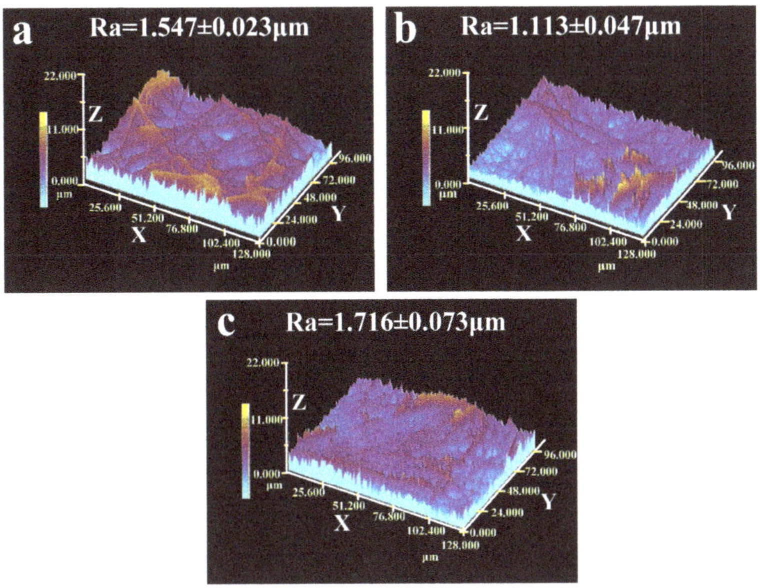

Figure 4. LSCM images of the (**a**) PCL top layer; (**b**) PAN layer; and (**c**) PAN/TiO_2 substrate layer. (Ra: arithmetic mean height; X, Y, Z: the axis of 3D images).

3.5. Surface Wettability

The surface wettability of Janus composite membranes was studied by measuring the underwater oil contact angle (UOCA). Figure S2 showed the dynamic wetting behaviors of water droplets on the PAN, PAN/TiO_2 and PAN/TiO_2-PCLx, and all prepared membranes exhibited hydrophilic layer in the air. As shown in Figure 5a–c, the oil droplets were all dispersed on the PAN, PAN/TiO_2 and PAN/TiO_2-PCL20 surface in a spherical shape, which indicated the three samples possessed underwater oleophobic properties. As displayed in Figure 5d, the UOCA of PAN was 144.6 ± 1.5°. The UOCAs of PAN-PCLx decreased from 129 ± 1.1° to 91 ± 1.2° as the PCL electrospinning time increased. After adding TiO_2 to the PAN substrate layer, the UOCA of PAN/TiO_2 increased to 158 ± 1.5°. The UOCAs of PAN/TiO_2-PCLx decreased from 143 ± 1.3° to 105 ± 1.6° as the PCL electrospinning time increased. All the UOCAs of PAN/TiO_2-PCLx were significantly higher than corresponding PAN-PCLx, which illustrated the hydrophilic layer modified by TiO_2 enhanced the underwater oleophobic properties of the PAN/TiO_2-PCLx Janus membrane.

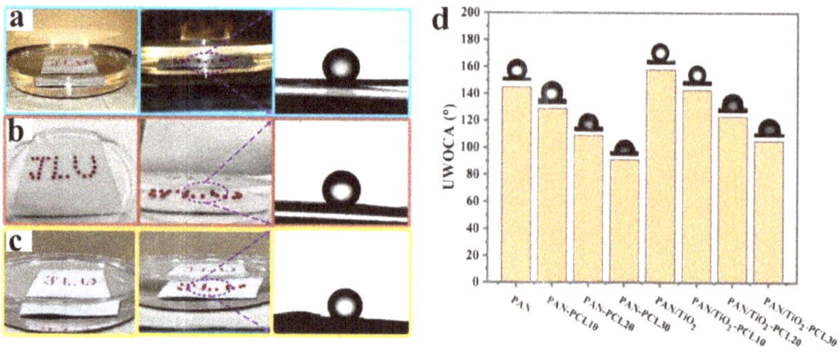

Figure 5. Oil droplets underwater onto (**a**) PAN, (**b**) PAN/TiO_2, (**c**) PAN/TiO_2-PCL20; (**d**) UOCAs onto PAN, PAN-PCLs and PAN/TiO_2-PCLs with a different electrospinning time of PCL (10 min, 20 min, 30 min).

3.6. Oil/Water Separation

Figure 6a showed the separation procedure for the immiscible n-hexane-water mixture. During the separation process, water (methyl blue dyed) permeated the membrane and rapidly flowed into the collector below, while n-hexane was retained above the membrane surface. Figure 6b shows the separation procedure for surfactant-stabilized n-hexane-in-water emulsion by PAN/TiO_2-PCL20. The emulsion showed the color of milky white before purification and became clear and transparent after filtration.

Figure 6. (**a**) Separation experiment of immiscible n-hexane-water mixture; (**b**) Separation experiment of n-hexane-in-water emulsion.

Figure 7a,b shows the separation performance of n-hexane-water mixture and n-hexane-in-water emulsion by different membranes. The permeate fluxes and separation efficiencies of PAN/TiO_2 for mixture and emulsion were all higher than PAN. For PAN/TiO_2-PCLx, the permeate fluxes and separation efficiencies for mixture and emulsion all increased rapidly and then decreased with the increase of PCL layer content. In summary, the PAN/TiO_2-PCL20 possessed the highest permeate fluxes (5340 ± 60 L m^{-2} h^{-1} for n-hexane-water mixture, 1344 ± 35 L m^{-2} h^{-1} for n-hexane-in-water emulsion) and separation efficiencies (99.95% for n-hexane-water mixture, 99.52% for n-hexane-in-water emulsion).

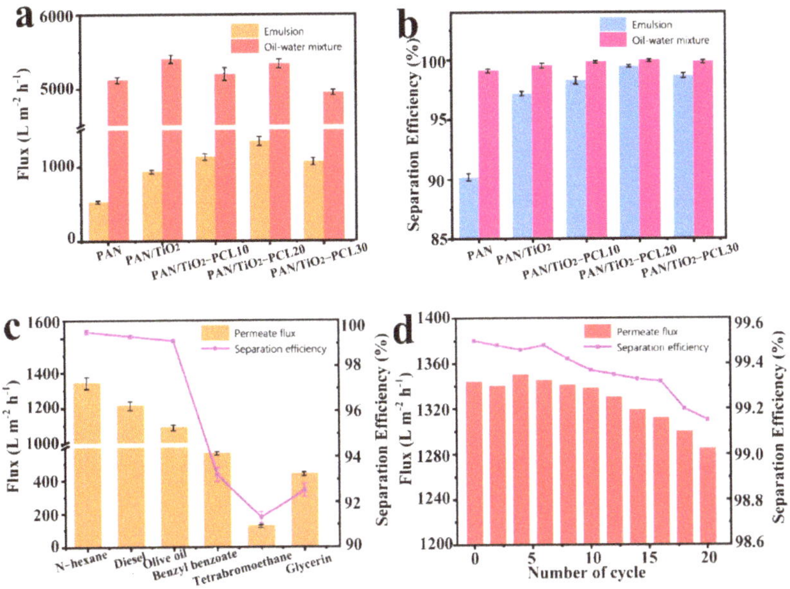

Figure 7. (**a**) Permeate fluxes and (**b**) separation efficiencies of different membranes for immiscible n-hexane-water mixture and n-hexane-in-water emulsion; (**c**) Permeate fluxes and separation efficiencies of PAN/TiO_2-PCL20 for different types of oil-in-water emulsions; (**d**) Permeate fluxes and separation efficiencies of PAN/TiO_2-PCL20 for n-hexane-in-water emulsion in 20 cycles.

Thus, PAN/TiO_2-PCL20 was selected to carry out the following separation experiments for different types of oil-in-water emulsions. N-hexane, diesel oil, olive oil, benzyl benzoate, 1,1,2,2-tetrabromoethane and glycerol were chosen to prepare the oil-in-water emulsions, and the separation efficiencies for these oil-in-water emulsions by PAN/TiO_2-PCL20 were all higher than 92.5%. Especially in the case of n-hexane-in-water emulsion, the permeate flux and efficiency reached 1344 ± 35 L m^{-2} h^{-1} and 99.52%, respectively (Figure 7c). The different separation performance for different oil-in-water emulsions might be ascribed to the viscosity and density of the oils [47]. More importantly, the durability test demonstrated that there was no significant decrease in permeate flux, and the separation efficiency of hexane-in-water emulsion for PAN/TiO_2-PCL20 still remained above 99.15%, even after 20 cycles of separation (Figure 7d). The as-prepared PAN/TiO_2-PCL20 had superior reusability of oil-in-water emulsions than other reported separation materials (see Table 2).

Table 2. Reusability of the other reported oil-in-water emulsion separation materials.

Material	Flux (L m^{-2} h^{-1})	Efficiency (%)	Reference
PAN/TiO_2-PCL20	1285	99.15	This work
Waste PET plastics	1197	95	[48]
PVDF/TiO_2	1398	99	[49]
PVP-VTES	17.45	91.47	[50]
PDMS@SiO_2@UiO66-OSiR	970	98.64	[51]
F-PPS@TiO_2	950	98.4	[52]

Figure 8 shows the optical microscope images and DLS analysis of PAN/TiO_2-PCL20m both before and after separation for n-hexane-in-water emulsion. The optical microscope images revealed that, prior to separation, the oil droplets were evenly spread in the emulsion with an average particle size of 4.03 ± 0.05 μm. After filtration, there were no apparent oil droplets in the filtrate, and few droplets were examined with an average particle size of 138.8 ± 5 nm.

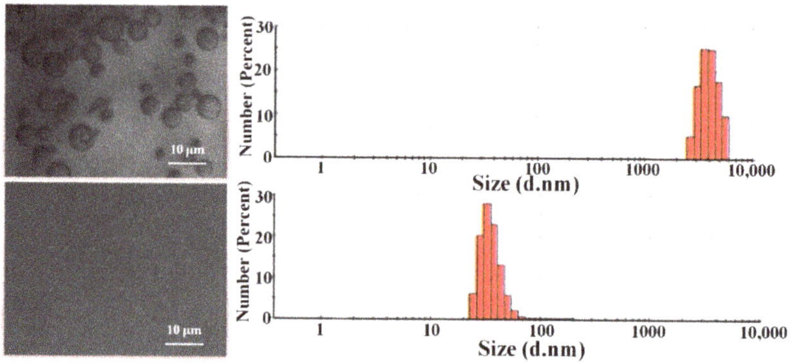

Figure 8. Optical microscope images and DLS analysis, both before and after separation of n-hexane-in-water emulsion.

3.7. Oil/Water Separation Mechanism

The mechanism of oil/water separation was summarized in accordance with the preceding discussion. Compared to PAN and PAN-PCLx membranes, the PAN/TiO_2-PCL20 exhibited excellent oil/water separation performance, which could be ascribed to its asymmetric wettability and hierarchical structure. This asymmetric wettability could be explained by the Young-Laplace Equation (5) [53]:

$$\delta P = \frac{2\gamma cos\theta}{R} \quad (5)$$

where δP was the capillary pressure, γ was the water surface tension, θ was the water contact angle and R was the pore radius.

When the PCL layer faced up to and operated in oil/water separation experiments, the hydrophobic layer provided negative capillary pressure ($F_{n\text{-}c}$) and the hydrophilic layer provided positive capillary pressure ($F_{p\text{-}c}$). When δP within the composite membrane was positive, water could penetrate successfully [54]. When water droplets were in contact with the surface of the hydrophobic PCL layer, it would be subjected to two opposing forces (the water gravity G and $F_{n\text{-}c}$). As the quantity of the water droplet increased, the value of G increased accordingly. When $G > F_{n\text{-}c}$, the water droplets could penetrate through the hydrophobic PCL layer to the PAN-based substrate layer, and the water then penetrated through the PAN/TiO_2-PCL20 by $F_{p\text{-}c}$. PAN/TiO_2 substrate had better hydrophilicity than PAN substrate due to its larger specific surface area and roughness, which could enhance asymmetric wettability for the Janus membrane and generate the larger $F_{p\text{-}c}$. Meanwhile, compared with the PAN substrate, the hydration layer formed in the PAN/TiO_2 substrate improved the oleophobicity of PAN/TiO_2-PCL20 and resisted oil penetration (Figure 9a)—all of these developments contributed to efficient emulsion separation performance.

3.8. Removal of RhB Organic Dye

RhB was chosen as the representative stain for evaluating the removal performance of the PAN/TiO_2-PCL20. As displayed in Figure 10a, PAN/TiO_2-PCL20 exhibited adsorption capacity with 5.02 mg g^{-1} for RhB within 60 min of adsorption equilibria under a dark condition (see Figure 10a). The removal rate of RhB by PAN/TiO_2-PCL20 achieved 59%t adsorption equilibria. When the removal experiment for RhB by PAN/TiO_2-PCL20 was exposed to the simulated solar irradiation, the removal rate of PAN/TiO_2-PCL20 for RhB could reach 93.2% (Figure 10b). The reason for this could be because, under simulated solar irradiation, TiO_2 generated hydroxyl radicals and superoxide radicals to degrade RhB dye molecules into carbon dioxide and water on the PAN/TiO_2-PCL20 [55,56]. This could provide evidence that PAN/TiO_2-PCL20 exhibited photocatalytic performance for RhB under the simulated solar irradiation. The inset of the Figure 10b shows the color change of the RhB solution at different time intervals, and it should be noticed that the color of the RhB solution became transparent after adsorption and photocatalytic degradation.

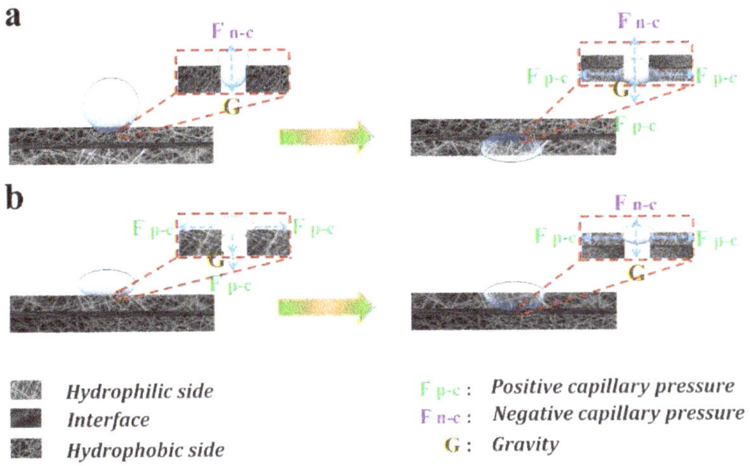

Figure 9. Schematic diagram of the oil/water separation and unidirectional permeation mechanism of PAN/TiO_2-PCL20. ((**a**): the PCL faced to water and separated emulsion, (**b**): the PAN/TiO_2 faced to water and resisted water).

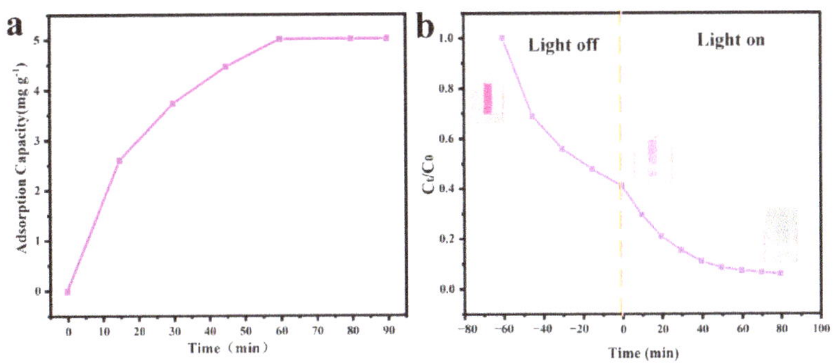

Figure 10. (**a**) Effect of adsorption time on adsorption capacity of PAN/TiO_2-PCL20; (**b**) Removal performance of PAN/TiO_2-PCL20 of RhB under simulated solar irradiation.

3.9. Unidirectional Water Transfer

The asymmetric wettability of membranes might result in unidirectional water transfer capability [53]. The unidirectional water transfer capability of PAN/TiO_2-PCL20 was further studied. When the hydrophilic PAN/TiO_2 layer faced up, water merely spread on the surface of PAN/TiO_2-PCL20 (Figure 11a). When the hydrophobic PCL layer faced up, water could spontaneously pass through PAN/TiO_2-PCL20 (Figure 11b). PAN/TiO_2-PCL20 exhibited unidirectional water transfer capability. The possible mechanism of unidirectional water transfer could also be explained, as illustrated in Figure 9: when the hydrophilic PAN/TiO_2 layer faced up, water droplets continuously diffused to the surface of the membrane and formed a thin water layer. G value per unit area of water was less than the upward force to the water droplets provided by the hydrophobic PCL layer, meaning water would be blocked on the surface of PAN/TiO_2-PCL20 (Figure 9b). When the hydrophobic PCL layer faced up, G value per unit area of water was higher than the repulsive force provided by the hydrophobic side counterpart. When G > F_{n-c}, water droplets passed through the PAN/TiO_2-PCL20 by gravity and the capillary force of the substrate (Figure 9a). The result showed that the asymmetric wettability of PAN/TiO_2-PCL20 exhibited unidirectional water transfer capability within a certain water pressure range [57,58].

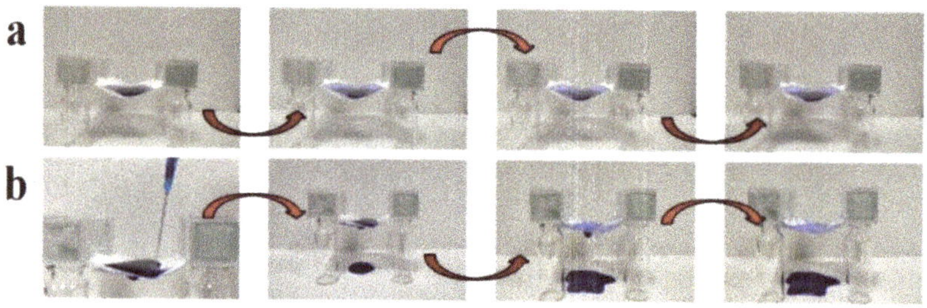

Figure 11. Unidirectional water transfer process of PAN/TiO_2-PCL20: (**a**) PAN/TiO_2 side facing water and (**b**) PCL side facing water.

3.10. Mechanical Properties

The mechanical performance of the separator was also necessary in the practical application for oil/water separation. Figure 12 shows the stress-strain curves of PAN, PAN/TiO_2 and PAN/TiO_2-PCLx. The tensile strength and train of PAN was 6.69 MPa and 28.67%, respectively. Compared with pure PAN, the tensile strength and strain of PAN/TiO_2 substrate layer were reduced to 5.68 MPa and 25.33%, respectively, which might be attributed to the existence of bead structures and porosities in the PAN/TiO_2. Compared to PAN/TiO_2, the PAN/TiO_2-PCLx possessed superior mechanical properties. With the increase of PCL content, PAN/TiO_2-PCLx exhibited a decrease in tensile strength and an increase in strain. The PAN/TiO_2-PCL30 had the most outstanding mechanical property and the maximum strain reached 53.67%. The result indicated that PAN/TiO_2-PCLx all had good mechanical performance for oil/water separation (Figure S3).

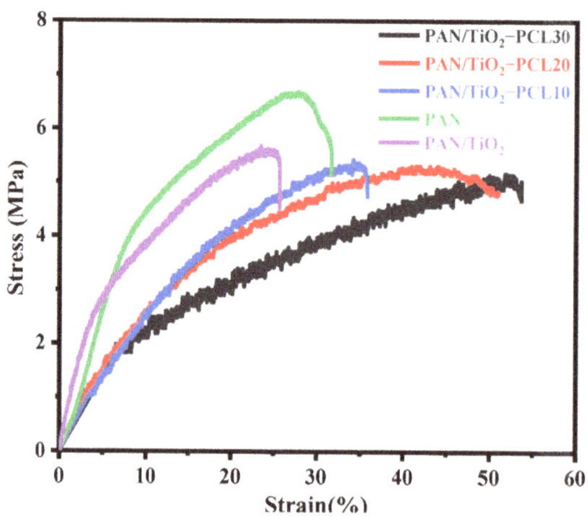

Figure 12. Stress-Strain curves of PAN, PAN/TiO_2 and PAN/TiO_2-PCLx.

4. Conclusions

In summary, an environmentally friendly PAN-based Janus composite membrane with asymmetric wettability and hierarchical structure was successfully fabricated. The addition of TiO_2 could further enhance the hydrophilicity of the hydrophilic PAN substrate by increasing the specific surface area and roughness, which further improved the asymmetric wettability of the Janus membrane. The prepared PAN/TiO_2-PCL20 exhibited excellent oil/water separation performance for diverse surfactant-stabilized oil-in-water emulsions. Even after 20 cycles, the separation efficiency for n-hexane-in-water emulsion remained above 99.15%, which exhibited the excellent reusability for an environmentally friendly separator. Meanwhile, the PAN/TiO_2-PCL20 also had superior adsorption (5.02 mg g^{-1}) and removal performance (93.2%) for RhB. In addition, the PAN/TiO_2-PCL20 had outstanding unidirectional water transfer capability and mechanical property. However, the corrosion resistance of PCL layer was expected to be further improved. To address this problem, the corrosion-resistant inorganic and organic component may be used to functionalize PCL modification by subsequent studies.

Supplementary Materials: The following supporting information can be downloaded at: https://www.mdpi.com/article/10.3390/ma17020417/s1, Figure S1: Nitrogen (N2) adsorption-desorption isotherms of (a) PAN, PAN/TiO2, (b) PAN/TiO2-PCL10, (c) PAN/TiO2-PCL20 and (d) PAN/TiO2-PCL30. Figure S2: Dynamic optical images of the wettability of a water droplet on (a) PAN, (b) PAN/TiO_2, (c) PAN/TiO_2-PCL10, (d) PAN/TiO_2-PCL20 and (e) PAN/TiO_2-PCL30. Figure S3: Stress-strain curve of PCL. References [59,60] are cited in the supplementary materials.

Author Contributions: Y.W.: Conceptualization, methodology, formal analysis, validation, data curation, roles/writing-original draft. J.H.: Methodology, software, validation, formal analysis. Y.Z.: Investigation, data curation, supervision, visualization. S.Z.: Investigation, data curation, supervision, visualization. L.L.: Validation, resources, writing-review & editing, visualization, supervision, project administration, funding acquisition. X.P.: Validation, supervision, resources, writing-review & editing, visualization. All authors have read and agreed to the published version of the manuscript.

Funding: This research was supported by the Science and Technology Development Plan of Jilin Province (20230201152GX).

Institutional Review Board Statement: Not applicable.

Informed Consent Statement: Not applicable.

Data Availability Statement: Data presented in this article are available at request from the corresponding author.

Acknowledgments: The authors would like to express their gratitude, both to the Basic Materials Laboratory of Jilin University and the Changchun Institute of Applied Chemistry at the Academia of Sciences, for providing testing instruments for this research.

Conflicts of Interest: The authors declare no conflicts of interest.

References

1. Berako Belachew, G.; Hu, C.C.; Wang, C.F.; Liao, M.Y.; Hung, W.S.; Chen, J.K.; Lai, J.Y. Preparation of citric acid-modified cellulose materials with superwetting and antibacterial characteristics for highly-efficient separation of oil/water mixtures, emulsions, and dye-containing wastewater. *Sep. Purif. Technol.* **2024**, *330*, 125389–125401. [CrossRef]
2. Sutar, R.S.; Latthe, S.S.; Jundle, A.R.; Gaikwad, P.P.; Ingole, S.S.; Nagappan, S.; Kim, Y.H.; Bhosale, A.K.; Saji, V.S.; Liu, S. A facile approach for oil-water separation using superhydrophobic polystyrene-silica coated stainless steel mesh bucket. *Mar. Pollut. Bull.* **2024**, *198*, 115790–115802. [CrossRef]
3. Sahu, P.S.; Verma, R.P.; Tewari, C.; Sahoo, N.G.; Saha, B. Facile fabrication and application of highly efficient reduced graphene oxide (rGO)-wrapped 3D foam for the removal of organic and inorganic water pollutants. *Environ. Sci. Pollut. Res.* **2023**, *30*, 93054–93069. [CrossRef] [PubMed]
4. Hasiri, M.; Kantzas, A. Efficiency of oil separation and demulsification following sonication gel degradation: Influence of Cr (III) ions, NaCl concentrations, and sodium-based retarders. *Fuel* **2024**, *357*, 129940–129953. [CrossRef]
5. Saini, H.; Otyepková, E.; Schneemann, A.; Zbořil, R.; Otyepka, M.; Fischer, R.A.; Jayaramulu, K. Hierarchical porous metal–organic framework materials for efficient oil–water separation. *J. Mater. Chem.* **2022**, *10*, 2751–2785. [CrossRef]

6. Hamedi, H.; Rezaei, N.; Zendehboudi, S. A comprehensive review on demulsification using functionalized magnetic nanoparticles. *J. Clean. Prod.* **2022**, *380*, 134868–134888. [CrossRef]
7. Iskandar, M.A.; Yahya, E.B.; Abdul Khalil, H.P.S.; Rahman, A.A.; Ismail, M.A. Recent Progress in Modification Strategies of Nanocellulose-Based Aerogels for Oil Absorption Application. *Polymers* **2022**, *14*, 849. [CrossRef]
8. Oh, S.; Bang, J.; Jin, H.J.; Kwak, H.W. Green Fabrication of Underwater Superoleophobic Biopolymeric Nanofibrous Membranes for Effective Oil–Water Separation. *Adv. Fiber Mater.* **2023**, *5*, 603–616. [CrossRef]
9. Gkinali, A.A.; Matsakidou, A.; Paraskevopoulou, A. Assessing the emulsifying properties of Tenebrio molitor larvae protein preparations: Impact of storage, thermal, and freeze-thaw treatments on o/w emulsion stability. *Int. J. Biol. Macromol.* **2023**, *250*, 126165–126175. [CrossRef]
10. Scharnberg, A.R.A.; Oliveira, H.A.; Weschenfelder, S.E.; Rubio, J.; Azevedo, A.C. Flocculation of emulsified oil-in-water with dodecylbenzene sulfonate and polyacrylamide and floc separation by dissolved air flotation. *Colloids Surf. A Physicochem. Eng. Asp.* **2023**, *669*, 131496–131505. [CrossRef]
11. Abu-Thabit, N.Y.; Kalam Azad, A.; Mezghani, K.; Akhtar, S.; Saeed Hakeem, A.; Drmosh, Q.A.; Yusuf Adesina, A. Rapid, Sustainable, and Versatile Strategy Towards Fabricating Superhydrophobic Cotton Textile Membranes for Separation of Emulsified and Stratified Oil/Water Mixtures. *Sep. Purif. Technol.* **2024**, *330*, 125352–125366. [CrossRef]
12. Ortiz-Oliveros, H.B.; Flores-Espinosa, R.M. Design of a mobile dissolved air flotation system with high rate for the treatment of liquid radioactive waste. *Process Saf. Environ. Prot.* **2020**, *144*, 23–31. [CrossRef]
13. Ihsanullah, I.; Bilal, M.; Sajid, M.; Mohammad, A.W.; Atieh, M.A.; Ghaffour, N. Emerging MXenes: Revolutionizing oily wastewater treatment-a comprehensive and critical review. *Sep. Purif. Technol.* **2024**, *329*, 125181–125203. [CrossRef]
14. Mittag, A.; Rahman, M.M.; Hafez, I.; Tajvidi, M. Development of Lignin-Containing Cellulose Nanofibrils Coated Paper-Based Filters for Effective Oil-Water Separation. *Membranes* **2022**, *13*, 1. [CrossRef] [PubMed]
15. Myeong, J.; Deshmukh, P.R.; Gyu Shin, W. Facile preparation of superhydrophilic and underwater superoleophobic stainless steel mesh for oil–water separation. *J. Ind. Eng. Chem.* **2023**, *120*, 398–409. [CrossRef]
16. Diyana Suzaimi, N.; Sean Goh, P.; Chun Wong, K.; Taniguchi, T.; Wei Lim, J.; Ahmad Nizam Nik Malek, N.; Fauzi Ismail, A. Construction of 1D akageneite-templated nanochannels in polyamide forward osmosis membrane for high flux separation and nutrient enrichment. *Sep. Purif. Technol.* **2024**, *330*, 125401–125415. [CrossRef]
17. Mallya, D.S.; Abdikheibari, S.; Dumée, L.F.; Muthukumaran, S.; Lei, W.; Baskaran, K. Removal of natural organic matter from surface water sources by nanofiltration and surface engineering membranes for fouling mitigation—A review. *Chemosphere* **2023**, *321*, 138070–138098. [CrossRef]
18. Imsong, R.; Dhar Purkayastha, D. Dual-functional superhydrophilic/underwater superoleophobic 2D $Ti_3C_2T_X$ MXene-PAN membrane for efficient oil-water separation and adsorption of organic dyes in wastewater. *Sep. Purif. Technol.* **2023**, *306*, 122636–122648. [CrossRef]
19. Han, L.; Shen, L.; Lin, H.; Huang, Z.; Xu, Y.; Li, R.; Li, B.; Chen, C.; Yu, W.; Teng, J. 3D printing titanium dioxide-acrylonitrile-butadiene-styrene (TiO_2-ABS) composite membrane for efficient oil/water separation. *Chemosphere* **2023**, *315*, 137791–137800. [CrossRef]
20. Polak, D.; Zielińska, I.; Szwast, M.; Kogut, I.; Małolepszy, A. Modification of Ceramic Membranes with Carbon Compounds for Pharmaceutical Substances Removal from Water in a Filtration-Adsorption System. *Membranes* **2021**, *11*, 481. [CrossRef]
21. Zielińska, I.; Polak, D.; Szwast, M. Analysis of the adsorption of selected pharmaceuticals on a composite material PEBAX/GO. *J. Water Process Eng.* **2021**, *44*, 102272–102279. [CrossRef]
22. Annam Renita, A.; Sathish, S.; Kumar, P.S.; Prabu, D.; Manikandan, N.; Mohamed Iqbal, A.; Rajesh, G.; Rangasamy, G. Emerging aspects of metal ions-doped zinc oxide photocatalysts in degradation of organic dyes and pharmaceutical pollutants-A review. *J. Environ. Manag.* **2023**, *344*, 118614–118636. [CrossRef]
23. Parasuraman, B.; Kandasamy, B.; Murugan, I.; Alsalhi, M.S.; Asemi, N.; Thangavelu, P.; Perumal, S. Designing the heterostructured $FeWO_4/FeS_2$ nanocomposites for an enhanced photocatalytic organic dye degradation. *Chemosphere* **2023**, *334*, 138979–138988. [CrossRef] [PubMed]
24. Zhao, Y.; Fan, X.; Zheng, H.; Liu, E.; Fan, J.; Wang, X. $Bi_2WO_6/AgInS_2$ S-scheme heterojunction: Efficient photodegradation of organic pollutant and toxicity evaluation. *J. Mater. Sci. Technol.* **2024**, *170*, 200–211. [CrossRef]
25. Rego, R.M.; Kurkuri, M.D.; Kigga, M. A comprehensive review on water remediation using UiO-66 MOFs and their derivatives. *Chemosphere* **2022**, *302*, 134845–134869. [CrossRef] [PubMed]
26. Gupta, S.; Gomaa, H.; Ray, M.B. Performance characterization of a hybrid adsorptive-photocatalytic (APC) oscillatory membrane reactor for micropollutant removal. *Sep. Purif. Technol.* **2021**, *279*, 119706–119717. [CrossRef]
27. Baig, U.; Matin, A.; Gondal, M.A.; Zubair, S.M. Facile fabrication of superhydrophobic, superoleophilic photocatalytic membrane for efficient oil-water separation and removal of hazardous organic pollutants. *J. Clean. Prod.* **2019**, *208*, 904–915. [CrossRef]
28. Wu, H.; Shi, J.; Ning, X.; Long, Y.; Zheng, J. The High Flux of Superhydrophilic-Superhydrophobic Janus Membrane of cPVA-PVDF/PMMA/GO by Layer-by-Layer Electrospinning for High Efficiency Oil-Water Separation. *Polymers* **2022**, *14*, 621. [CrossRef]
29. Afsari, M.; Shirazi, M.M.A.; Ghorbani, A.H.; Sayar, O.; Shon, H.K.; Tijing, L.D. Triple-layer nanofiber membrane with improved energy efficiency for treatment of hypersaline solution via membrane distillation. *J. Environ. Chem. Eng.* **2023**, *11*, 110638–110651. [CrossRef]

30. Selvam, A.; Raj Singh, Y.; Karna, A.; Bhatnagar, A.; Mukherjee, M.; Chakrabarti, S. Emergence of the Janus-MOF (J-MOF) Boat as a Nascent Amalgamation in the Arena of Photothermal Desalination. *ACS Appl. Mater. Interfaces* **2023**, *15*, 26918–26927. [CrossRef]
31. Li, X.; Ma, Q.; Tian, J.; Xi, X.; Li, D.; Dong, X.; Yu, W.; Wang, X.; Wang, J. Double anisotropic electrically conductive flexible Janus-typed membranes. *Nanoscale* **2017**, *47*, 18918–18930. [CrossRef]
32. Li, H.; Yang, J.; Xu, Z. Asymmetric Surface Engineering for Janus Membranes. *Adv. Mater. Interfaces* **2020**, *7*, 1902064–1902078. [CrossRef]
33. Agaba, A.; Marriam, I.; Tebyetekerwa, M.; Yuanhao, W. Janus hybrid sustainable all-cellulose nanofiber sponge for oil-water separation. *Int. J. Biol. Macromol.* **2021**, *185*, 997–1004. [CrossRef] [PubMed]
34. Nambikkattu, J.; Kaleekkal, N.J. Development of Janus membranes with asymmetric wettability for desalination of impaired seawater by direct contact membrane distillation. *J. Environ. Chem. Eng.* **2023**, *11*, 111396–111408. [CrossRef]
35. Abuhantash, F.; Abuhasheesh, Y.H.; Hegab, H.M.; Aljundi, I.H.; Al Marzooqi, F.; Hasan, S.W. Hydrophilic, oleophilic and switchable Janus mixed matrix membranes for oily wastewater treatment: A review. *J. Water Process Eng.* **2023**, *56*, 104310–104333. [CrossRef]
36. Naseem, S.; Wu, C.-M.; Motora, K.G. Novel multifunctional $Rb_xWO_3@Fe_3O_4$ immobilized Janus membranes for desalination and synergic-photocatalytic water purification. *Desalination* **2021**, *517*, 115256–115269. [CrossRef]
37. Chen, K.; Zhu, J.; Tan, Y.; Sun, F.; Gan, J.; Peng, H.; Zhan, T.; Lyu, J. Development of gradient-wetting Janus wood membrane with high-efficiency fog collection and oil-water separation. *Chem. Eng. J.* **2022**, *470*, 144356–144369. [CrossRef]
38. Pan, T.D.; Li, Z.J.; Shou, D.H.; Shou, W.; Fan, J.T.; Liu, X.; Liu, Y. Buoyancy Assisted Janus Membrane Preparation by ZnO Interfacial Deposition for Water Pollution Treatment and Self-cleaning. *Adv. Mater. Interfaces* **2019**, *6*, 1130–1161. [CrossRef]
39. Cintra, I.L.R.; Rezende, M.C.; Guerrini, L.M.; Nahra, L.R.; Lucas, R.R.; Montagna, L.S.; Botelho, E.C. Electrospinning of PAN/lignin blends aiming the production of carbon nanofibers. *MRS Commun.* **2023**, *504*, 43579–43587. [CrossRef]
40. Taghizadeh Lendeh, P.; Sarrafi, A.H.M.; Alihosseini, A.; Bahri-Laleh, N. Synthesis and characterization ZnO/TiO_2 incorporated activated carbon as photocatalyst for gas refinery effluent treatment. *Polyhedron* **2024**, *247*, 116715–116726. [CrossRef]
41. Paul, R.; Kavinarmatha, K.; Parthiban, S. Tantalum doped titanium dioxide nanoparticles for efficient photocatalytic degradation of dyes. *J. Mol. Struct.* **2023**, *1277*, 134869–134882. [CrossRef]
42. He, N.; Li, L.; Chen, J.; Zhang, J.; Liang, C. Extraordinary Superhydrophobic Polycaprolactone-Based Composite Membrane with an Alternated Micro–Nano Hierarchical Structure as an Eco-friendly Oil/Water Separator. *ACS Appl. Mater. Interfaces* **2021**, *13*, 24117–24129. [CrossRef] [PubMed]
43. Wang, X.; Li, B.; Zhou, L.; Ma, J.; Zhang, X.; Li, H.; Liang, C.; Liu, S.; Wang, H. Influence of surface structures on biocompatibility of TiO_2 /HA coatings prepared by MAO. *Mater. Chem. Phys.* **2018**, *215*, 339–345. [CrossRef]
44. Kong, L.; Li, Y.; Qiu, F.; Zhang, T.; Guo, Q.; Zhang, X.; Yang, D.; Xu, J.; Xue, M. Fabrication of hydrophobic and oleophilic polyurethane foam sponge modified with hydrophobic Al_2O_3 for oil/water separation. *J. Ind. Eng. Chem.* **2018**, *58*, 369–375. [CrossRef]
45. Liu, Z.; Liu, L.; Liu, J.; Zhang, W.; Zhou, T.; Ji, X. Enhanced photocatalytic performance of Er-doped $Bi_{24}O_{31}Br_{10}$: Facile synthesis and photocatalytic mechanism. *Mater. Res. Bull.* **2016**, *76*, 256–263. [CrossRef]
46. Hovish, M.Q.; Hilt, F.; Rolston, N.; Xiao, Q.; Dauskardt, R.H. Open Air Plasma Deposition of Superhydrophilic Titania Coatings. *Adv. Funct. Mater.* **2019**, *29*, 6421–6430. [CrossRef]
47. Parsaie, A.; Mohammadi-Khanaposhtani, M.; Riazi, M.; Tamsilian, Y. Magnesium stearate-coated superhydrophobic sponge for oil/water separation: Synthesis, properties, application. *Sep. Purif. Technol.* **2020**, *251*, 117105–117113. [CrossRef]
48. Chen, H.; Zuo, Z.; Tian, Q.; Xue, S.; Qiu, F.; Peng, X.; Zhang, T. Waste to treasure: A superwetting fiber membrane from waste PET plastic for water-in-oil emulsion separation. *J. Clean. Prod.* **2023**, *396*, 136502–136511. [CrossRef]
49. Sun, F.; Li, T.-T.; Ren, H.-T.; Shiu, B.-C.; Peng, H.-K.; Lin, J.-H.; Lou, C.-W. Dopamine-decorated lotus leaf-like PVDF/TiO_2 membrane with underwater superoleophobic for highly efficient oil-water separation. *Process Saf. Environ. Prot.* **2021**, *147*, 788–797. [CrossRef]
50. Zhang, R.; Deng, C.; Hou, X.; Li, T.; Lu, Y.; Liu, F. Preparation and Characterization of a Janus Membrane with an "Integrated" Structure and Adjustable Hydrophilic Layer Thickness. *Membranes* **2023**, *13*, 415. [CrossRef]
51. Zhai, G.; Wu, J.; Yuan, Z.; Li, H.; Sun, D. Robust Superhydrophobic PDMS@SiO_2@UiO66-OSiR Sponge for Efficient Water-in-Oil Emulsion Separation. *Inorg. Chem.* **2023**, *62*, 5447–5457. [CrossRef] [PubMed]
52. Yang, C.; Han, N.; Han, C.; Wang, M.; Zhang, W.; Wang, W.; Zhang, Z.; Li, W.; Zhang, X. Design of a Janus F-TiO_2@PPS Porous Membrane with Asymmetric Wettability for Switchable Oil/Water Separation. *ACS Appl. Mater. Interfaces* **2019**, *11*, 22408–22418. [CrossRef] [PubMed]
53. Yalishev, S.V.; Abbasi, N.A.; Iqbal, M.; Alnaser, A.S. Comparing Water Transport Properties of Janus Membranes Fabricated from Copper Mesh and Foam Using a Femtosecond Laser. *Langmuir* **2023**, *39*, 1815–1825. [CrossRef] [PubMed]
54. Balachandran, A.; Bhat, I.M.; Tabasum, R.; Mohd, G.; Wani, M.F.; Majid, K.; Wahid, M.; Lone, S. Transfer-Printed Environmental-Friendly Anisotropic Filter with Laser-Controlled Micropores for Efficient Oil/Water Separation. *ACS Appl. Polym. Mater.* **2023**, *5*, 2272–2281. [CrossRef]
55. Safajou, H.; Khojasteh, H.; Salavati-Niasari, M.; Mortazavi-Derazkola, S. Enhanced photocatalytic degradation of dyes over graphene/Pd/TiO_2 nanocomposites: TiO_2 nanowires versus TiO_2 nanoparticles. *J. Colloid Interface Sci.* **2017**, *498*, 423–432. [CrossRef]

56. Rozenman, G.G.; Peisakhov, A.; Zadok, N. Dispersion of organic exciton polaritons—A novel undergraduate experiment. *Eur. J. Phys.* **2022**, *43*, 035301. [CrossRef]
57. Chakraborty, T.; Kanth, P.C.; Trivedi, M.U.; Tripathi, B.; Pandey, M.K. Fabrication of Janus type bi-layer polymeric membranes for advance water purification. *Mater. Today Proc.* **2021**, *47*, 626–631. [CrossRef]
58. Hosseini, Z.; Kargari, A. Fabrication and characterization of hydrophobic/hydrophilic dual-layer polyphenyl sulfone Janus membrane for application in direct contact membrane distillation. *Desalination* **2024**, *571*, 117100–117127. [CrossRef]
59. Kılıç, D.; Sevim, M.; Eroğlu, Z.; Metin, Ö.; Karaca, S. Strontium oxide modified mesoporous graphitic carbon nitride/titanium dioxide nanocomposites (SrO-mpg-CN/TiO_2) as efficient heterojunction photocatalysts for the degradation of tetracycline in water. *Adv. Powder Technol.* **2021**, *32*, 2743–2757. [CrossRef]
60. Liu, G.; Wang, G.; Hu, Z.; Su, Y.; Zhao, L. Ag_2O nanoparticles decorated TiO_2 nanofibers as a p-n heterojunction for enhanced photocatalytic decomposition of RhB under visible light irradiation. *Appl. Surf. Sci.* **2019**, *465*, 902–910. [CrossRef]

materials

MDPI

Article

The Synergistic Effect of Phosphonic and Carboxyl Acid Groups for Efficient and Stable Perovskite Solar Cells

Kaihuai Du [1,†], Aili Wang [2,†], Yue Li [1], Yibo Xu [1], Lvzhou Li [2,*], Ningyi Yuan [1,*] and Jianning Ding [1,2]

1 Jiangsu Collaborative Innovation Center for Photovoltaic Science and Engineering, Jiangsu Province Cultivation Base for State Key Laboratory of Photovoltaic Science and Technology, School of Materials Science and Engineering, Changzhou University, Changzhou 213164, China; dingjn@yzu.edu.cn (J.D.)

2 School of Mechanical Engineering, Institute of Technology for Carbon Neutralization, Yangzhou University, Yangzhou 225127, China

* Correspondence: oasis@yzu.edu.cn (L.L.); nyyuan@cczu.edu.cn (N.Y.)

† These authors contributed equally to this work.

Abstract: Reducing the interfacial defects between the perovskite/electron transport layer (ETL) is the key point to improving the efficient and stable performance of perovskite solar cells (PSCs). In this study, two self-assembled molecules ((aminomethyl)phosphonic acid and glycine) with different functional groups (phosphonic acid (-H_2PO_3) and carboxylic acid (-COOH)) were mixed to form the buried bottom interface of PSCs. The synergistic effect of -H_2PO_3 with its higher anchoring ability and -COOH with its fast carrier transport improved the performance of PSCs. Additionally, the SnO_2 modified by mixed self-assembly molecules (M-SAM) showed a more appropriate energy level alignment, favoring charge transport and minimizing energy loss. In addition, the amine group (-NH_2) on the two small molecules effectively interacted with uncoordinated Pb^{2+} in perovskite and improved the quality of the perovskite films. Consequently, the $(FAPbI_3)_{0.992}(MAPbBr_3)_{0.008}$ PSCs with M-SAM reached a PCE of 24.69% (0.08 cm^2) and the perovskite modules achieved a champion efficiency of 18.57% (12.25 cm^2 aperture area). Meanwhile, it still maintained more than 91% of its initial PCE after being placed in nitrogen atmosphere at 25 °C for 1500 h, which is better than that of the single-SAM and control devices. Further reference is provided for the future commercialization of perovskite with efficient and stable characteristics.

Keywords: interface passivation; anchoring; charge transfer; self-assembled small molecule; large area

1. Introduction

Planar perovskite solar cells (PSCs) have gradually been considered to be a commercially promising device due to their small current–voltage hysteresis, high power conversion efficiency and simple preparation process [1–3]. In the past decade, the power conversion efficiency (PCE) of PSCs has increased from 3.8% to 26.1% [4–6]. The electron transport layer (ETL) in the typical n-i-p structure has electron-selective contact with the perovskite absorption layer. Concurrently, the upper surface of the ETL also affects the number of nucleation points and the grain growth of the absorption layer, and ETLs with low-defect concentrations are of decisive significance for improving the performance of PSCs devices [7,8]. The commonly reported ETLs in the n-i-p structure are titanium oxide (TiO_2) and tin oxide (SnO_2). However, compared with TiO_2, the SnO_2 substrate has the potential for the best ETL due to its high-electron mobility and easy preparation.

The common preparation methods of SnO_2 when used as electron transport layer mainly include spin-coated SnO_2 nanocrystals and chemical bath deposition (CBD) of SnO_2. Different from the spin-coated SnO_2 method, the SnO_2 film prepared via CBD is more compact and uniform. Moreover, the CBD method is not limited by the size of substrate and is more beneficial to the future commercial development of PSCs [9]. However, there are hydroxyl (-OH) and oxygen vacancy (OV) defects on the surface of SnO_2 film prepared via

CBD, which significantly increase the non-radiative recombination of the ETL/Perovskite interface, thus changing the performance of PSCs and accelerating their degradation [10]. At the same time, a large number of defects on the ETL/perovskite layer interface are not conducive to the energy level matching of the charge transfer, and lead to serious non-radiative recombination at the interface [11,12]. In order to solve these problems, researchers have made many attempts which can be divided into SnO_2 body passivation and surface passivation. For body passivation, it has been reported that the addition of phosphonate ethanolamine (PE) to the aqueous colloid of SnO_2 can significantly reduce the number of Sn dangling bonds in SnO_2, which is beneficial for improving the electrical properties of SnO_2. At the same time, the amino group (NH_2) of PE can interact with the uncoordinated Pb^{2+} in the perovskite active layer, further inhibiting interface defects and improving device performance [13]. However, some researchers have tried to passivate or modify the surface defects of SnO_2 to reduce the interface defects between SnO_2 and perovskite [12]. It has been reported that the use of ultraviolet ozone to improve the surface wettability of SnO_2 while reducing surface defects can achieve a champion efficiency of 17.21%. Some researchers have also reported that the use of KCl aqueous solution to modify the surface of SnO_2 can passivate the surface defects of SnO_2 to improve the performance of the device, and finally achieve a champion efficiency of 22.2%. At the same time, using some self-assembled molecules with anchoring groups to passivate the interface is also an effective interface passivation strategy [14]. It has been reported that an anchoring group with strong bonding strength not only shortens the assembly adsorption time and improves the uniformity of adsorption distribution, but also makes the carrier transporting layers (CTL) have a high compatibility with the deposition of the active layer of the perovskite device, thus greatly reducing interface defects [15]. Yang et al. studied a series of self-assembled molecules (SAM) with carboxylic acid groups (-COOH), such as benzoic acid, 3-aminopropionic acid, etc., and found that appropriate interface interactions can significantly reduce the density of defect states and promote charge extraction and transfer at the interface, which increases the efficiency by about 10% compared with solar cells without SAMs, and investigated different chemical interactions between SnO_2/perovskite [16]. Although small molecules containing -COOH have excellent electron transport capacities due to their special structure, their binding energy on the surface of SnO_2 is low [17–19]. The widely used anchoring group, which is similar to -COOH and has similar anchoring ability, is the phosphonic group (-H_2PO_3) [13,20,21]. Its binding energy on the surface of SnO_2 is much larger than that of the -COOH. Unfortunately, due to the geometric shape and conjugate loss of -H_2PO_3, the charge transfer rate of the -H_2PO_3 is not competitive with that of -COOH [22]. In addition, it has been reported that the amino group (-NH_2) contained in the small molecule has the effect of interacting with the uncoordinated Pb^{2+} in the perovskite, which can effectively decrease interface defects [13,23]. Sargent et al. combined ammonium fluoride with CBD-SnO_2 and found that -NH_2 could reduce the surface defect sites and match energy levels better with perovskite. This strategy showed a PCE of 23.3% and a higher open circuit voltage (V_{OC}) [20]. Meanwhile, the application of CBD SnO_2 to a large area also reflects the problem of non-uniformity. Although CBD SnO_2 has more commercial potential than spin-coated SnO_2 nanocrystals, the uniform passivation of SnO_2 defects is also a problem that must be solved first. Lin et al. developed a simple post-treatment method using periodic acid to modify the SnO_2 film and increased the proportion of tin oxide (IV) with the help of periodic acid. Consequently, the 3×3 cm^2 perovskite modules with a champion efficiency of 18.10% were shown [24]. Therefore, a strategy that can not only effectively passivate the surface defects of SnO_2 but also serve as a reference for large-scale commercialization is needed.

In this study, two self-assembled molecules ((aminomethyl)phosphonic acid and glycine) with similar molecular structures but different anchoring groups (phosphonic acid (-H_2PO_3) and carboxylic acid (-COOH)) were used as interlayers in SnO_2/perovskite. Phosphonic acid groups have a stronger anchoring ability on the surface of metal oxides than carboxylic acid groups due to their unique structure, while carboxylic acid groups have their unique characteristics of rapid charge transfer. The different anchoring groups can meet the requirements of the buried bottom interface for carrier transport (-COOH) and anchoring ability (-H_2PO_3), so as to solve the defect problem of single functional group. Photovoltaic performance may be further improved by using the complementary advantages of the two groups while improving the anchoring capacity and charge transfer. In addition, the same partial amine group (-NH_2) on the two small molecules effectively interact with uncoordinated Pb^{2+} in the perovskite layer, which can decrease interface defects. The energy level between SnO_2 and the perovskite layers was further adjusted by mixed self-assembly molecules (M-SAM). Finally, the PCE of PSC was increased from 22.72% to 24.69% (0.08 cm^2) and the V_{OC} was increased from 1.12 V to 1.16 V. Meanwhile, the stability of the device modified by M-SAM was significantly improved, which maintained more than 91% of its initial PCE after 1500 h and 100 h placed at 25 °C and 65 °C in nitrogen atmosphere. In addition, the 4 × 4 cm^2 (12.25 cm^2 aperture area) perovskite modules achieved the best efficiency of 18.57%. This study demonstrates an effective strategy for passivating the surface of SnO_2, which improves the PCE of PSCs while taking into account stability. This complementary synergistic passivation strategy provides a simple and effective method for future high-performance and durable commercial devices.

2. Experimental Section

Materials. Hydrochloric acid (HCl, 37 wt.% in water) was purchased from Sinopharm (Shanghai, China). $SnCl_2 \cdot 2H_2O$ (>99.98%) and urea (>99.0%), thioglycolic acid (TGA, ≥99.0%), chlorobenzene (CB, ≥99.9%), isopropanol (IPA, ≥99.9%) and acetonitrile (≥99.9%) were purchased from Sigma-Aldrich (Shanghai, China). Dimethyl sulfoxide (DMSO) (≥99.9%), and dimethylformamide (DMF) (≥99.9%) were purchased from Alfa Aesar (Shanghai China). FAI (≥99.5%), PbI_2 (99.99%), $MAPbBr_3$ (≥99.5%), MACl (≥99.5%), PEAI (≥99.5%), Spiro-OMeTAD (≥99.8%), 4-tert-Butylpyridine (tBP, ≥96%), Co-TFSI salt (≥99.5%) and Li-TFSI salt (≥99.5%) were purchased from Xi'an Polymer Technology Corp (Xi'an, China). (Aminomethyl)phosphonic acid (≥98%) and glycine (99%) were purchased from Adamas (Shanghai China).

CBD SnO_2. The SnO_2 layer was deposited on the surface of FTO by chemical bath deposition (CBD). The CBD solution was prepared by mixing 5.625 g of urea, 1.2375 g of $SnCl_2 \cdot 2H_2O$, 5.625 mL of HCl, 112.5 μL of TGA, and 450 mL of deionized water (DI water). The FTO substrates and the CBD solution were loaded into a glass reaction vessel and reacted at 94 °C in an oil bath for 5.5 h. Then, the SnO_2-deposited FTO were removed from the reaction vessel and cleaned via sonication with DI water and ethanol for 10 min each.

Preparation of perovskite solution. The perovskite precursor solution was prepared by mixing 1.4 M FAI, 1.53 M PbI_2, 0.5 M MACl, and 0.8 mol% $MAPbBr_3$ in 1 mL mixture solvent (volume ratio, DMF:DMSO = 8:1).

Device fabrication. The FTO/SnO_2 substrates were annealed at 170 °C for 60 min and treated with UV-Ozone for 20 min. Then the FTO/SnO_2 substrate was immersed in a small molecule aqueous solution for 20 min, then dried with nitrogen, and finally placed in a 100 °C hot stage for annealing for 5 min. The prepared FTO FTO/SnO_2/SAM were transferred to the glove box. The perovskite film was spin-coated on the substrates at 1000 rpm for 10 s, and 5000 rpm for 30 s, and chlorobenzene was dripped at 20 s after starting, then annealed at 100 °C for 60 min. For the 2D perovskite passivation, PEAI (15 mM in IPA) was deposited at 5000 rpm for 30 s. For HTL, the 25 μL Spiro-OMeTAD solution was spin-coated on the perovskite layer at 3800 rpm for 30s. The Spiro-OMeTAD solution was prepared by dissolving 101.92 mg of Spiro-OMeTAD, 24.4 μL Li-TFSI salt (520 $mg \cdot L^{-1}$ in acetonitrile solution), 49.6 μL FK209 salt (300 mg·in 1 mL acetonitrile solution), and 45.4 μL tBP solution in 1 mL chlorobenzene. Finally, 100 nm of Au was deposited via thermal evaporation.

3. Results and Discussion

The electrostatic surface potential (ESP) diagram and structural formula of P-SAM and C-SAM are shown in Figure 1a and Figure S1, respectively. The dipole moment value of P-SAM (2.11 D) was higher than that of C-SAM (1.66 D). The dipole moment can reflect the polarity of the molecule, and it is easy to obtain the relative level of the van der Waals force of the two molecules; thus P-SAM was more easily anchored on the surface of SnO_2 than C-SAM was. Figure 1b shows a flow chart of the preparation of the buried bottom interface. The glass/FTO/SnO_2 substrate was immersed in a M-SAM aqueous solution with a volume ratio of 1:1 (1 mM in deionized water) for 20 min, and we then used N_2 to blow away the residual solution on the surface of SnO_2, and annealed it at 100 °C for 5 min. Meanwhile, this immersion method using SAM has the potential to be applied to a large area. The perovskite films on SnO_2, SnO_2/P-SAM, SnO_2/C-SAM and SnO_2/M-SAM were characterized by scanning electron microscopy (SEM) and X-ray diffraction (XRD). It is evident from Figures 1c–f and S2 that the average grain size of perovskite films was similar, but the grain boundary was improved after M-SAM modification (Figures S2a–d and S3). Meanwhile, this method is also applicable to large areas; Figures S4 and S5 show M-SAM, as a buried interface, can enhance the uniformity of perovskite grains. The enhanced hydrophilicity of the SnO_2/SAM films (Figure S3) from -NH_2 in SAM molecules are helpful for the deposition of perovskite precursor solutions and the growth of better-quality perovskite films, which is consistent with the SEM results [25]. The XRD spectra in Figure 1g show that the perovskite films with an M-SAM interlayer have stronger crystallinity, which indicates that M-SAM regulates crystallinity.

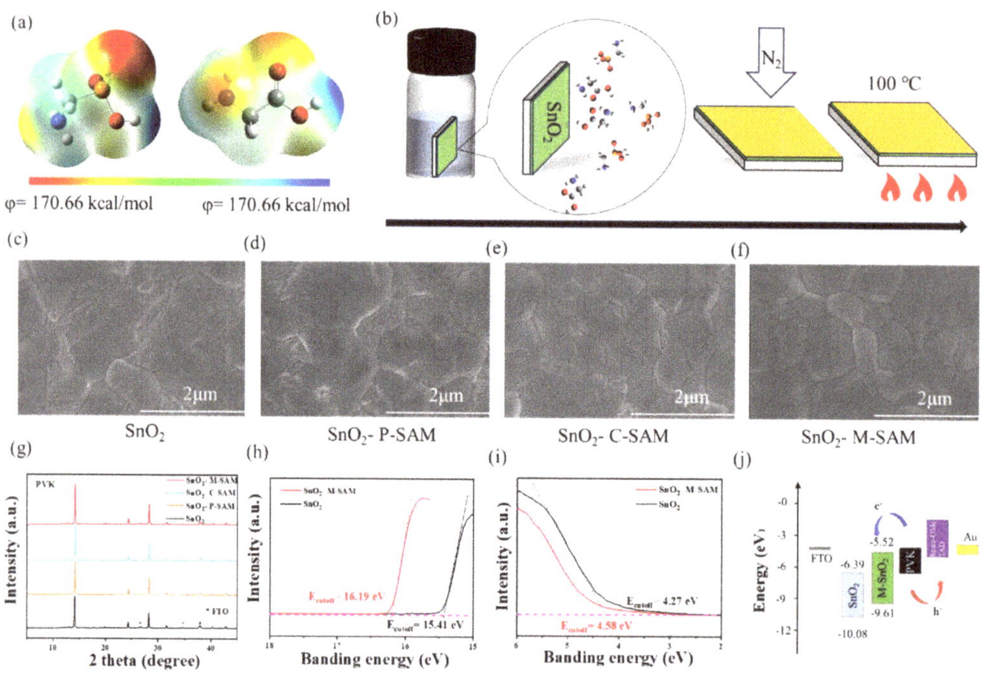

Figure 1. (**a**) ESP map of P-SAM (Left) and C-SAM (Right). (**b**) Diagram of the preparation work of M-SAM. SEM of perovskite films, (**c**) pristine SnO_2, (**d**) SnO_2/P-SAM, (**e**) SnO_2/C-SAM and (**f**) SnO_2/M-SAM as substrates. (**g**) XRD patterns of perovskite films on SnO_2, SnO_2/P-SAM, SnO_2/C-SAM and SnO_2/M-SAM. (**h**,**i**) UPS spectra of SnO_2 and SnO_2/M-SAM films. (**j**) Energy level diagram of FTO, SnO_2, SnO_2/M-SAM, and perovskite.

The effect of M-SAM modification on the band structures was further studied by ultraviolet photoelectron spectroscopy (UPS) in Figure 1h,i and ultraviolet-visible spectrophotometer (UV-Vis) in Figure S6. The band-gaps of SnO_2 and SnO_2/M-SAM were 3.94 and 3.97 eV, respectively. The Fermi level ($-E_F = E_{cutoff} - 21.22$ eV) and the conduction band (CB) were both up-shifted with the M-SAM modification and thus coordinate the interface energy level arrangement. The CB of SnO_2 and SnO_2/M-SAM were −6.39 eV and −5.52 eV, respectively (the details are showed in Table S1). In addition, the transmittance of SnO_2 films was not affected by SAMs modification. Correspondingly, a better energy band alignment was achieved between ETL/perovskite, favoring efficient electron transport (Figure 1j) [26].

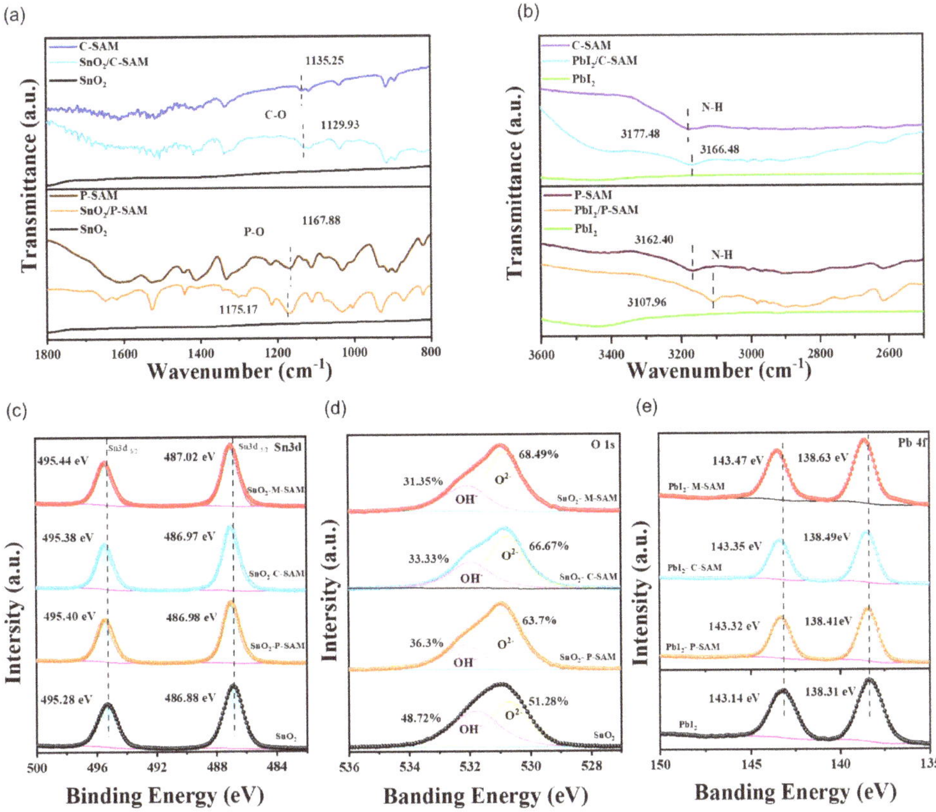

Figure 2. FTIR spectra of (**a**) SnO_2, SnO_2/P-SAM, SnO_2/C-SAM, P-SAM, and C-SAM films, (**b**) PbI_2, PbI_2/P-SAM, PbI_2/C-SAM, P-SAM, and C-SAM films. XPS spectra of (**c**) Sn 3d, and (**d**) O 1s for the pristine SnO_2, SnO_2/P-SAM, SnO_2/C-SAM and SnO_2/M-SAM films, (**e**) Pb 4f for the PbI_2, PbI_2/P-SAM, PbI_2/C-SAM and PbI_2/M-SAM films.

The interaction mechanism of P-SAM, C-SAM and M-SAM in the buried bottom interface was further explored (Figure 2). In Figure 2a, the Fourier transform infrared spectroscopy (FTIR) spectra shows the P-O vibration shifts from 1167.88 cm^{-1} in pure P-SAM to 1175.17 cm^{-1} of the SnO_2/P-SAM and the C-O vibration shifts from 1135.25 cm^{-1} in pure C-SAM to 1129.93 cm^{-1} of the SnO_2/C-SAM, confirming the interaction of P-SAM and C-SAM with SnO_2, respectively. Similarly, the interaction between the SAM and perovskites is shown in Figure 2b. It can be clearly seen from the moment of initial N-H stretching peak moves from 3162.40 (P-SAM) to 3107.96 cm^{-1} (PbI_2/P-SAM) and from 3177.48 (C-

SAM) to 3166.48 cm^{-1} (PbI_2/C-SAM), respectively, confirming that both small molecules interact with PbI_2 [23]. The X-ray photoelectron spectroscopy (XPS) measurements was further studied and the effect of SAM on surface chemical environment was discussed. The Sn 3d peaks in the pure SnO_2 at 495.28 and 486.88 eV, corresponding to Sn $3d_{3/2}$ and Sn $3d_{5/2}$, respectively, shifted to higher binding energies after SAM modifications in Figure 2c (SnO_2/P-SAM: 495.40 and 486.98 eV, SnO_2/C-SAM: 495.38 and 486.97 eV, SnO_2/M-SAM: 495.44 and 487.02 eV). The results show that both P-SAM and C-SAM had chemical interactions with SnO_2 [27]. As shown in Figure 2d, two peaks fitted at 532.5 eV and 531.1 eV corresponded to the -OH on the surface of SnO_2 and the saturated lattice oxygen in the SnO_2 film. Compared with the pure SnO_2, the calculated areas of -OH in the modified SnO_2 are all reduced. This indicates that more chemisorbed off-lattice oxygen is transferred to the lattice oxygen, and that the oxygen atom in SnO_2 exists as O^{2-}. The reduced -OH is beneficial for improving the electron transport and suppressing the nonradiative recombination [28]. Meanwhile, in Figure 2e, two peaks fitted at 143.14 and 138.31 eV of Pb 4f spectrum for pure PbI_2 film, corresponding to Pb $4f_{5/2}$ and Pb $4f_{7/2}$, respectively, shifted towards higher binding energies for all SAM-modified samples (SnO_2/P-SAM: 143.32 and 138.41 eV; SnO_2/C-SAM: 143.35 and 138.49 eV; SnO_2/M-SAM: 143.47 and 138.63 eV). The results indicate that the interaction occurs between P-SAM, C-SAM, M-SAM and perovskite.

To further investigate the effect of SAM layers on the carrier transport dynamics, steady-state photoluminescence (SSPL) and time-resolved photoluminescence (TRPL) tests were performed (Figure 3a–d). From the analysis of perovskite, the SnO_2/SAM/perovskite displays lower PL signals than that of the SnO_2/perovskite and the SnO_2/M-SAM/perovskite possesses the lowest PL intensity among the SAM-modified samples. The TRPL curves in Figure 3b exhibit how the average time constant (τ_{ave}) of perovskite film on SnO_2, SnO_2/P-SAM, SnO_2/C-SAM and SnO_2/M-SAM decreased from 1.482 μs to 1.294 μs, and 1.142 μs and 1.056 μs, respectively [29]. The detailed carrier lifetime is listed in Table S2. The decreased PL intensity and TRPL lifetime indicates that P-SAM-, C-SAM- and M-SAM-modification inhibits non-radiative recombination due to the passivated interface defects of SnO_2 and the optimized perovskite grain quality [19]. In general, the fast decay process (τ_1) is related to the charge extraction and transport, and the slow decay (τ_2) is related to the radiative recombination of the bulk perovskites. The value of τ_1 decreased from 0.020 μs (SnO_2/perovskite) to 0.013 μs (SnO_2/M-SAM/perovskite), which proves that M-SAM-modification is beneficial for the extraction of electrons at the buried interface. In addition, by comparing the SSPL and TRPL results of perovskite films' deposition on the glass substrate without ETL (Figure 3c,d), the samples show a significantly opposite trend and the corresponding fitted results are presented in Table S3. The SSPL intensity and the average lifetime of the perovskite films grown on the modified glass were significantly improved, which is attributed to the passivation of interface defects and the optimized grain quality of perovskite [29,30]. Then, the effect of SAM on perovskite films was further studied by PL intensity imaging (Figures 3e,h and S7). Compared with the control films, a weaker PL emission was observed from the SnO_2/M-SAM/perovskite, which is consistent with the SSPL spectrum shown in Figure 3a. In addition, the SnO_2/M-SAM/perovskite film exhibited more a uniform PL emission in the substrate compared to the control. These results further prove that M-SAM-modified SnO_2 can passivate the defects on the surface of SnO_2 and reduce the non-radiative recombination at the interface to improve the performance of the device. Additionally, the corresponding carrier lifetimes were summarized in Tables S4 and S5.

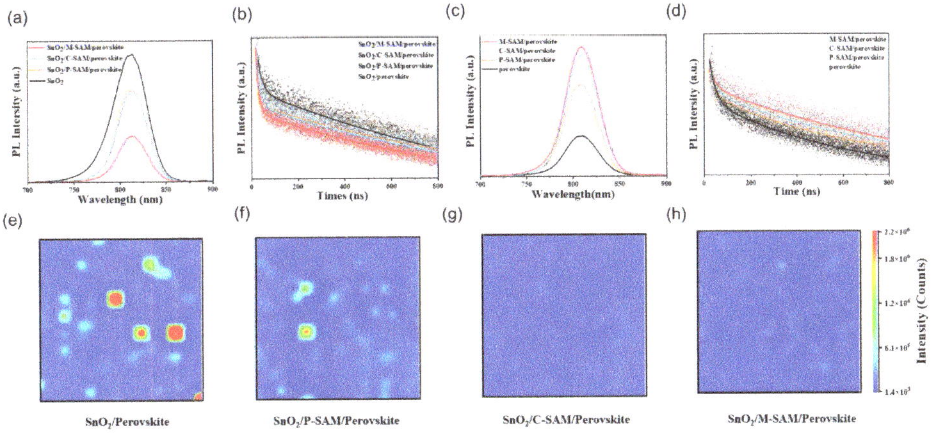

Figure 3. (**a**) SSPL and (**b**) TRPL curves of SnO_2/perovskite, SnO_2/P-SAM/perovskite, SnO_2/C-SAM/perovskite and SnO_2/M-SAM/perovskite films. (**c**) SSPL and (**d**) TRPL curves of perovskite, P-SAM/perovskite, C-SAM/perovskite and M-SAM/perovskite films. (**e**–**h**) PL intensity imaging of SnO_2/perovskite, SnO_2/P-SAM/perovskite, SnO_2/C-SAM/perovskite and SnO_2/M-SAM/perovskite films (size of 0.6×0.6 mm^2).

In short, P-SAM, C-SAM, and M-SAM interlayers in the buried bottom exhibited strong interactions with SnO_2 and perovskite, and thus suppressed the surface defects and facilitated carrier transport. Therefore, the device structure of FTO/SnO_2/SAM/perovskite/Spiro-OMeTAD/Au (Figure S8) was constructed to investigate the effect of SAM on photovoltaic performance. Figure S9 shows the statistical analysis of the PSCs performance on the devices with different molar ratios of P-SAM and C-SAM. The best PCE of the PSC was achieved at a ratio of 1:1 and the corresponding photocurrent density–voltage (*J*–*V*) curves under AM 1.5 illumination are presented in Figure 4a. The champion PCE of the device using M-SAM increased from 22.72% (control), 23.99% (with P-SAM), 23.56% (with C-SAM) to 24.69%. Compared with the control PSCs, the increase in PCE was mainly due to the obvious increase in V_{OC} (from 1.12 to 1.16 V) and J_{SC} (from 24.37 to 25.54 mA/cm^2). In addition to this, a laser-etched perovskite solar module (PSM) with an area of 4×4 cm^2 and an aperture area of 12.25 cm^2 were also made, and the related *J*–*V* curve is exhibited in Figure 4b. The PSMs with M-SAM-modifications achieved an excellent PCE of 18.57%, while the PCE of the control PSMs was only 16.07%. A stabilized photocurrent of 23.77 mA/cm^{-2} and an efficiency of 24.25% were obtained at 1.02 V (Figure 4c). The stability was further evaluated. The PCE of the unpackaged device with an M-SAM modification remained at 91% and 92.4% of the initial PCE after exposure to nitrogen atmosphere at 25 °C for 1500 h (Figure 4d) and at 65 °C for 105 h (Figure 4e), respectively. Meanwhile, the control device retained 82% and 79.2% of its original PCE. The improved stability of the M-SAM-modified device is attributed to the synergistic effect of P-SAM and C-SAM, as well as to energy level regulation and improved film quality. In addition to device stability, the perovskite films under different aging conditions were also tested. Figure 4f shows the XRD patterns of SnO_2/perovskite and SnO_2/M-SAM/perovskite films before and after aging at 60 °C and 40% relative humidity (RH). After 100 h of aging, the SnO_2/M-SAM/perovskite film showed a much weaker additional δ-$FAPbI_3$ phase diffraction peak than that of the control sample; the result implies that the presence of M-SAM could effectively suppress the phase transition induced by an exposure to moisture, which is consistent with the results in Figure S11. It is therefore further proved that the M-SAM can improve the defects of the SnO_2/Perovskite interface, and that a stable perovskite film provides an important guarantee for the stability of the later device.

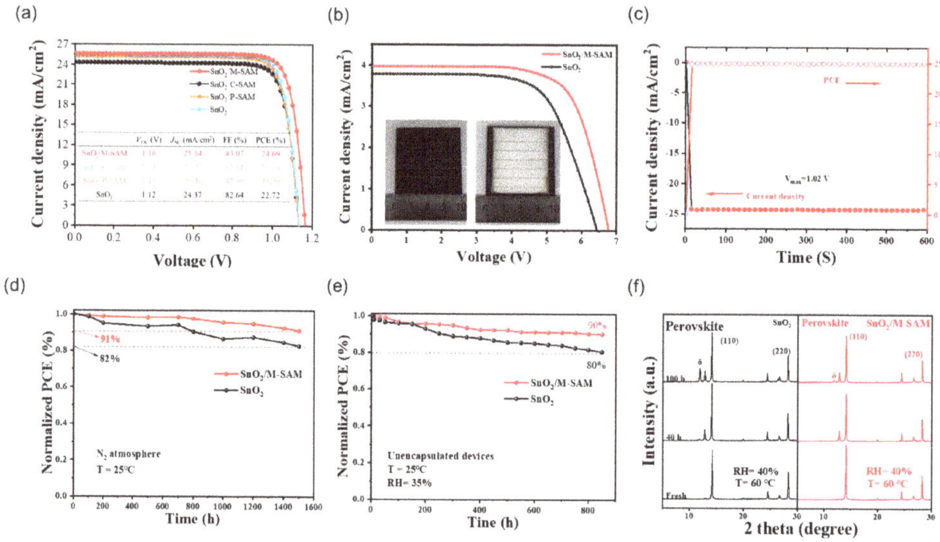

Figure 4. *J*–*V* curves of the champion PSCs for (**a**) 0.08 cm^2 and (**b**) 4 × 4 cm^2 modules (**c**) Stable power output curves under maximum power-point conditions. Long-term stability test under nitrogen atmosphere (**d**) at 25 °C and (**e**) at 65 °C. (**f**) XRD patterns for SnO_2/perovskite and SnO_2/M-SAM/perovskite films aged at ≈40%RH, ~60 °C.

Figure 5a shows the external quantum efficiency (EQE) spectra of the control and the M-SAM-modified PSC, and that the related integrated currents were 24.21 mA/cm^2 and 24.47 mA/cm^2, respectively, which are consistent with the result derived from the *J*–*V* curves. In addition, the EQE spectra of M-SAM-modified PSC was improved in the ultraviolet region, which is attributed to the passivation of Pb^{2+} via -NH_2 in M-SAM, which reduces the deep-level defects of perovskite films. The dark *J*–*V* curves (Figure 5b) show the PSCs with M-SAM modifications had a reduced leakage current, ascribing to a lower trapping density and carrier recombination at the interface of SnO_2/perovskite through M-SAM modification. A quantitative characterization of the defects was obtained by space charge limited current (SCLC) test with the electron-only device (Figure 5c). The trap filling limit voltage (V_{TFL}) of the device with the M-SAM modification was significantly reduced. The trap state densities were 2.22×10^{16} cm^{-3} and 1.94×10^{16} cm^{-3} for the control and M-SAM-modified devices, respectively, indicating that the M-SAM can passivate the defects (the parameters were listed in Table S7) [30]. The Nyquist impedance in Figure 5d shows that the recombination resistance (R_{rec}) (2117 Ω) of the device with M-SAM is higher than the control (1201 Ω), suggesting a reduced charge recombination at the ETL/PVK interface by M-SAM. Additionally, M-SAM-modified devices achieved higher extracted the built-in potentials (V_{bi}, 0.95 V) from Mott–Schottky plots (Figure 5e) compared to the control device (0.92 V). The enlarged V_{bi} contributed to the higher V_{OC}, following the same trend of V_{OC} values shown in the *J*–*V* curves [30]. The dependence of V_{OC} on light intensity (P_{light}) was fitted to obtain the ideality factor n (Figure 5f). The M-SAM-modified PSCs possessed the lower n value (1.58) compared to the control (1.93), indicating that the trap-assisted recombination of PSCs was significantly reduced after M-SAM treatment. It has been proven that using M-SAM as the buried interface has a positive effect on improving device performance.

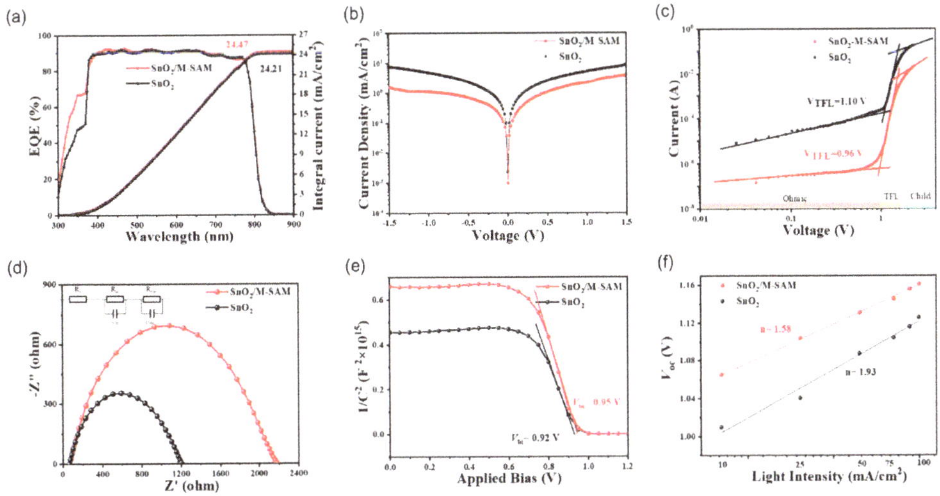

Figure 5. (**a**) The EQE spectra and integrated J_{SC} of SnO_2 and SnO_2/M-SAM devices, (**b**) dark J–V curves of SnO_2 and SnO_2/M-SAM devices, (**c**) dark J–V curves for the devices with structure of FTO/SnO_2(M-SAM)/PVK/PCBM/Ag, (**d**) Nyquist plots of SnO_2 and SnO_2/M-SAM devices, (**e**) Mott–Schottky plots of SnO_2 and SnO_2/M-SAM devices, (**f**) dependence of V_{OC} and light intensity dependence of SnO_2 and SnO_2/M-SAM devices.

4. Conclusions

Aiming at the passivation problem of surface defects commonly used in SnO_2 ETL, we developed a synergistic effect strategy by combining the strong anchoring ability of phosphonic acid (-H_2PO_3) in metal oxide surfaces and the better charge transfer ability of carboxylic acid (-COOH). At the bottom of the interface, the -H_2PO_3 in P-SAM was used to enhance the anchoring ability and the -COOH in C-SAM was used to transport carriers quickly. In addition, the molecules interacted with both SnO_2 from -H_2PO_3 and -COOH with Sn and perovskite from -NH_2 and Pb^{2+}, thus improving the energy level arrangement and quality of perovskite films. The $(FAPbI_3)_{0.992}(MAPbBr_3)_{0.008}$ PSCs with mixed SAM modification passivated the defect density and accelerated the charge extraction and transport. Consequently, the champion PCE of 24.69% and V_{OC} of 1.16 V were achieved. Meanwhile, a PCE of 18.57% was obtained for a 4 × 4 cm^2 (12.25 cm^2 aperture area) perovskite module. At the same time, more than 91% of the initial PCE was maintained after storage at 25 °C and 65 °C for 1500 h and 100 h under nitrogen atmosphere. This work improves the device efficiency by mixing two complementary self-assembled small molecules to passivate ETL/perovskite interface defects while taking into account stable improvements. This method can improve the passivation uniformity CBD SnO_2 surface, and is simple and suitable for future commercialization. The reuse of M-SAM passivation solutions need to be further researched in future works.

Supplementary Materials: The following supporting information can be downloaded at: https://www.mdpi.com/article/10.3390/ma16237306/s1, Figure S1: Structural formula of (a) P-SAM and (b) C-SAM. (-R, -CH_2); Figure S2. SEM images of perovskite surface based on (a) SnO_2, (b) SnO_2/P-SAM SnO_2, (c) SnO_2/C-SAM SnO_2 and (d) SnO_2/M-SAM. The perovskite grain size of (e) SnO_2, (f) SnO_2/P-SAM SnO_2, (g) SnO_2/C-SAM SnO_2 and (h) SnO_2/M-SAM; Figure S3. Contact angles of (a) SnO_2, (b) SnO_2/P-SAM, (c) SnO_2/C-SAM, and (d) SnO_2/M-SAM as substrates; Figure S4. SEM images for five different locations of the 4 × 4 cm^2 perovskite film deposited on SnO_2 substrate. The scale bar is 10 μm. The distribution of the grain size is plotted in histogram for comparison; Figure S5. SEM images for five different locations of the 4 × 4 cm^2 perovskite film deposited on SnO_2/M-SAM substrate. The scale bar is 10 μm. The distribution of the grain size is plotted in histogram for comparison; Figure S6. (a) Transmission and absorption spectra FTO, SnO_2 SnO_2/P-SAM, SnO_2/C-SAM and SnO_2/M-SAM films. (b) Tauc plots of pristine SnO_2 and SnO_2/M-SAM films; Figure S7. PL intensity imaging of SnO_2/perovskite and SnO_2/M-SAM/perovskite films (size of 0.02 × 0.02 mm^2); Figure S8. PSCs with the structure of FTO/SnO_2/SAM/Perovskite/Spiro-OMeTAD/Au; Figure S9. The PSCs performance (a) V_{OC}, (b) J_{SC}, (c) FF and (d) PCE of the devices with different molar ratios of P-SAM and C-SAM; Figure S10. *J–V* curves for control and M-SAM modified PSCs under reverse and forward scans (at illumination of 100 mW/cm^2); Figure S11. XRD patterns for (a) SnO_2/perovskite and (b) SnO_2/M-SAM/perovskite films aged at a RH ≈ 40%, T: ≈25 °C; Table S1: Table S1. Summarized parameters of energy band structure for SnO_2, SnO_2/M-SAM (on FTO substrates); Table S2. Fitted parameters of the TRPL curves for SnO_2/perovskite SnO_2/P-SAM/perovskite, SnO_2/C-SAM/perovskite and SnO_2/M-SAM/perovskite films (on glass substrates); Table S3. Fitted parameters of the TRPL curves for glass/perovskite, glass/P-SAM/perovskite, glass/C-SAM/perovskite and glass/M-SAM/perovskite films; Table S4. Fitted parameters of the PL intensity imaging for SnO_2/perovskite SnO_2/P-SAM/perovskite, SnO_2/C-SAM/perovskite and SnO_2/M-SAM/perovskite films (on glass substrates); Table S5. Fitted parameters of the PL intensity imaging for SnO_2/perovskite and SnO_2/M-SAM/perovskite films (on glass substrates); Table S6. Photovoltaic parameters for SnO_2 and M-SAM modified PSCs under reverse scan and forward scan (HI = ($PCE_{forward}$-$PCE_{reverse}$)/$PCE_{forward}$); Table S7. Calculated parameters and trap densities (N_{trap}) of perovskite films grown on SnO_2 and SnO_2/M-SAM substrates for electron only devices; Table S8. The modification of CBD SnO_2 for getting the improved performance of PSCs are listed below. References [31,32] are cited in the Supplementary Materials.

Author Contributions: Writing—original draft, K.D., A.W., Y.L. and Y.X.; Writing—review & editing, L.L., N.Y. and J.D.; Funding acquisition, A.W., N.Y. and J.D. All authors have read and agreed to the published version of the manuscript.

Funding: The financial support received from the Special fund for science and technology innovation of Jiangsu Province (BE2022610) and the National Natural Science Foundation of China (52303383).

Informed Consent Statement: Informed consent was obtained from all subjects involved in the study.

Data Availability Statement: The research data are contained within the article.

Conflicts of Interest: The authors declare no conflict of interest.

References

1. Kim, H.-S.; Lee, C.-R.; Im, J.-H.; Lee, K.-B.; Moehl, T.; Marchioro, A.; Moon, S.-J.; Humphry-Baker, R.; Yum, J.-H.; Moser, J.E.; et al. Lead Iodide Perovskite Sensitized All-Solid-State Submicron Thin Film Mesoscopic Solar Cell with Efficiency Exceeding 9%. *Sci. Rep.* **2012**, *2*, 591. [CrossRef]
2. Jiang, Q.; Zhang, L.; Wang, H.; Yang, X.; Meng, J.; Liu, H.; Yin, Z.; Wu, J.; Zhang, X.; You, J. Enhanced electron extraction using SnO_2 for high-efficiency planar-structure $HC(NH_2)_2PbI_3$-based perovskite solar cells. *Nat. Energy* **2017**, *2*, 16177. [CrossRef]
3. Tan, H.; Jain, A.; Voznyy, O.; Lan, X.; de Arquer, F.P.G.; Fan, J.Z.; Quintero-Bermudez, R.; Yuan, M.; Zhang, B.; Zhao, Y.; et al. Efficient and stable solution-processed planar perovskite solar cells via contact passivation. *Science* **2017**, *355*, 722–726. [CrossRef]
4. NREL Best Research-Cell Efficiencies. Available online: https://www.nrel.gov/pv/cell-efficiency.html (accessed on 20 September 2023).
5. Kojima, A.; Teshima, K.; Shirai, Y.; Miyasaka, T. Organometal Halide Perovskites as Visible-Light Sensitizers for Photovoltaic Cells. *J. Am. Chem. Soc.* **2009**, *131*, 6050. [CrossRef]
6. Min, H.; Lee, D.Y.; Kim, J.; Kim, G.; Lee, K.S.; Kim, J.; Paik, M.J.; Kim, Y.K.; Kim, K.S.; Kim, M.G.; et al. Perovskite solar cells with atomically coherent interlayers on SnO_2 electrodes. *Nature* **2021**, *598*, 444. [CrossRef]
7. Courtier, N.E.; Cave, J.M.; Foster, J.M.; Walker, A.B.; Richardson, G. How transport layer properties affect perovskite solar cell performance: Insights from a coupled charge transport/ion migration model. *Energy Environ. Sci.* **2019**, *12*, 396–409. [CrossRef]
8. Ni, Z.; Yu, Z.; Huang, J. Unveiling the ambipolar carrier transport property of SnO_{2-X} for multiple-functional interlayers in perovskite solar cells. *Appl. Phys. Lett.* **2021**, *119*, 123501. [CrossRef]
9. Yoo, J.J.; Seo, G.; Chua, M.R.; Park, T.G.; Lu, Y.; Rotermund, F.; Kim, Y.-K.; Moon, C.S.; Jeon, N.J.; Correa-Baena, J.-P.; et al. Efficient perovskite solar cells via improved carrier management. *Nature* **2021**, *590*, 587–593. [CrossRef]
10. Wu, M.; Duan, Y.; Yang, L.; You, P.; Li, Z.; Wang, J.; Zhou, H.; Yang, S.; Xu, D.; Zou, H.; et al. Multifunctional Small Molecule as Buried Interface Passivator for Efficient Planar Perovskite Solar Cells. *Adv. Funct. Mater.* **2023**, *33*, 2300128. [CrossRef]
11. Zhu, P.; Gu, S.; Luo, X.; Gao, Y.; Li, S.; Zhu, J.; Tan, H. Simultaneous Contact and Grain-Boundary Passivation in Planar Perovskite Solar Cells Using SnO_2-KCl Composite Electron Transport Layer. *Adv. Energy Mater.* **2020**, *10*, 1903083. [CrossRef]
12. Uddin, A.; Yi, H. Progress and Challenges of SnO_2 Electron Transport Layer for Perovskite Solar Cells: A Critical Review. *Solar RRL* **2022**, *6*, 2100983. [CrossRef]

13. Yi, Z.; Li, X.; Xiao, B.; Luo, Y.; Jiang, Q.; Yang, J. Large improvement of photovoltaic performance of flexible perovskite solar cells using a multifunctional phospho-ethanolamine-modified SnO_2 layer. *Sci. China Mater.* **2022**, *65*, 3392–3401. [CrossRef]
14. Park, S.Y.; Zhu, K. Advances in SnO_2 for Efficient and Stable n-i-p Perovskite Solar Cells. *Adv. Mater.* **2022**, *34*, 2110438. [CrossRef]
15. Li, E.; Liu, C.; Lin, H.; Xu, X.; Liu, S.; Zhang, S.; Yu, M.; Cao, X.M.; Wu, Y.; Zhu, W.H. Bonding Strength Regulates Anchoring-Based Self-Assembly Monolayers for Efficient and Stable Perovskite Solar Cells. *Adv. Funct. Mater.* **2021**, *31*, 2103847. [CrossRef]
16. Zuo, L.; Chen, Q.; De Marco, N.; Hsieh, Y.-T.; Chen, H.; Sun, P.; Chang, S.-Y.; Zhao, H.; Dong, S.; Yang, Y. Tailoring the Interfacial Chemical Interaction for High-Efficiency Perovskite Solar Cells. *Nano Lett.* **2017**, *17*, 269–275. [CrossRef]
17. Jiang, E.; Ai, Y.; Yan, J.; Li, N.; Lin, L.; Wang, Z.; Shou, C.; Yan, B.; Zeng, Y.; Sheng, J.; et al. Phosphate-Passivated SnO_2 Electron Transport Layer for High-Performance Perovskite Solar Cells. *Acs Appl. Mater. Interfaces* **2019**, *11*, 36727–36734. [CrossRef]
18. Nilsing, M.; Lunell, S.; Persson, P.; Ojamae, L. Phosphonic acid adsorption at the TiO_2 anatase (101) surface investigated by periodic hybrid HF-DFT computations. *Surf. Sci.* **2005**, *582*, 49–60. [CrossRef]
19. Zhu, L.; Shang, X.; Lei, K.; Wu, C.; Zheng, S.; Chen, C.; Song, H. Doping in Semiconductor Oxides-Based Electron Transport Materials for Perovskite Solar Cells Application. *Solar RRL* **2021**, *5*, 2000605. [CrossRef]
20. Jung, E.H.; Chen, B.; Bertens, K.; Vafaie, M.; Teale, S.; Proppe, A.; Hou, Y.; Zhu, T.; Zheng, C.; Sargent, E.H. Bifunctional Surface Engineering on SnO_2 Reduces Energy Loss in Perovskite Solar Cells. *Acs Energy Lett.* **2020**, *5*, 2796–2801. [CrossRef]
21. Li, M.; Zhou, J.; Tan, L.; Li, H.; Liu, Y.; Jiang, C.; Ye, Y.; Ding, L.; Tress, W.; Yi, C. Multifunctional succinate additive for flexible perovskite solar cells with more than 23% power-conversion efficiency. *Innovation* **2022**, *3*, 100310. [CrossRef]
22. Zhang, L.; Cole, J.M. Anchoring groups for dye-sensitized solar cells. *ACS Appl. Mater. Interfaces* **2015**, *7*, 3427–3455. [CrossRef]
23. Hassan, A.; Wang, Z.; Ahn, Y.H.; Azam, M.; Khan, A.A.; Farooq, U.; Zubair, M.; Cao, Y. Recent defect passivation drifts and role of additive engineering in perovskite photovoltaics. *Nano Energy* **2022**, *101*, 107579. [CrossRef]
24. Wu, Z.; Su, J.; Chai, N.; Cheng, S.; Wang, X.; Zhang, Z.; Liu, X.; Zhong, H.; Yang, J.; Wang, Z.; et al. Periodic Acid Modification of Chemical-Bath Deposited SnO2 Electron Transport Layers for Perovskite Solar Cells and Mini Modules. *Adv. Sci.* **2023**, *10*, 2300010. [CrossRef]
25. Zhou, Y.; Yang, S.; Yin, X.; Han, J.; Tai, M.; Zhao, X.; Chen, H.; Gu, Y.; Wang, N.; Lin, H. Enhancing electron transport via graphene quantum dot/SnO_2 composites for efficient and durable flexible perovskite photovoltaics. *J. Mater. Chem. A* **2019**, *7*, 1878–1888. [CrossRef]
26. Canil, L.; Cramer, T.; Fraboni, B.; Ricciarelli, D.; Meggiolaro, D.; Singh, A.; Liu, M.; Rusu, M.; Wolff, C.M.; Nga, P.; et al. Tuning halide perovskite energy levels. *Energy Environ. Sci.* **2021**, *14*, 1429–1438. [CrossRef]
27. Zhang, Y.; Xu, L.; Sun, J.; Wu, Y.; Kan, Z.; Zhang, H.; Yang, L.; Liu, B.; Dong, B.; Bai, X.; et al. 24.11% High Performance Perovskite Solar Cells by Dual Interfacial Carrier Mobility Enhancement and Charge-Carrier Transport Balance. *Adv. Energy Mater.* **2022**, *12*, 2201269. [CrossRef]
28. Cao, Z.; Li, C.; Deng, X.; Wang, S.; Yuan, Y.; Chen, Y.; Wang, Z.; Liu, Y.; Ding, L.; Hao, F. Metal oxide alternatives for efficient electron transport in perovskite solar cells: Beyond TiO_2 and SnO_2. *J. Mater. Chem. A* **2020**, *8*, 19768–19787. [CrossRef]
29. Pan, H.; Zhao, X.; Gong, X.; Li, H.; Ladi, N.H.; Zhang, X.L.; Huang, W.; Ahmad, S.; Ding, L.; Shen, Y.; et al. Advances in design engineering and merits of electron transporting layers in perovskite solar cells. *Mater. Horiz.* **2020**, *7*, 2276–2291. [CrossRef]
30. Deng, J.; Zhang, H.; Wei, K.; Xiao, Y.; Zhang, C.; Yang, L.; Zhang, X.; Wu, D.; Yang, Y.; Zhang, J. Molecular Bridge Assisted Bifacial Defect Healing Enables Low Energy Loss for Efficient and Stable Perovskite Solar Cells. *Adv. Funct. Mater.* **2022**, *32*, 2209516. [CrossRef]
31. Zhang, J.B.; Bai, C.; Dong, Y.; Shen, W.J.; Zhang, Q.; Huang, F.Z.; Cheng, Y.B.; Zhong, J. Batch chemical bath deposition of large-area SnO_2 film with mercaptosuccinic acid decoration for homogenized and efficient perovskite solar cells. *Chem. Eng. J.* **2021**, *425*, 8. [CrossRef]
32. Soe, K.T.; Thansamai, S.; Thongprong, N.; Ruengsrisang, W.; Muhammad, I.A.; Ketsombun, E.; Supruangnet, R.; Kaewprajak, A.; Kumnorkaew, P.; Saetang, V.; et al. Simultaneous Surface Modification and Defect Passivation on Tin Oxide-Perovskite Interfaces using Pseudohalide Salt of Sodium Tetrafluoroborate. *Solar RRL* **2023**, *7*, 15. [CrossRef]

materials

MDPI

Article

Flat-Band Potential Determination and Catalytical Properties of Sn_3O_4/SnO_2 Heterostructures in the Photo-Electrooxidation of Small Organic Molecules under Ultraviolet (370 nm) and Blue (450 nm) Light

Evgeny Gribov [1,*], Evgeny Koshevoy [1], Aleksey Kuznetsov [1], Maxim Mikhnenko [1], Evgeniy Losev [1,2,3] and Mikhail Lyulyukin [1,2]

[1] Boreskov Institute of Catalysis SB RAS, Pr. Lavrentieva 5, 630090 Novosibirsk, Russia; koshevoy@catalysis.ru (E.K.); kan@catalysis.ru (A.K.); m.mikhnenko@catalysis.ru (M.M.); losev.88@mail.ru (E.L.); lyulyukin@catalysis.ru (M.L.)

[2] Faculty of Natural Sciences, Chair of Solid-State Chemistry, Novosibirsk State University, Pirogova Str. 1, 630090 Novosibirsk, Russia

[3] Voevodsky Institute of Chemical Kinetics and Combustion SB RAS, Laboratory of Nanoparticles, Institutskaya Str. 3, 630090 Novosibirsk, Russia

* Correspondence: gribov@catalysis.ru; Tel.: +7-(383)-326-94-32

Abstract: Sn_3O_4 are promising semiconductor materials due to their visible light absorption ability. In this work, a series of materials, such as SnO_2, Sn_3O_4 and Sn_3O_4/SnO_2 heterostructures, with different phase ratios were prepared using hydrothermal synthesis. The materials were characterized using X-ray diffraction (XRD), Raman and diffuse reflectance spectroscopy (DRS), high resolution transmission electron microscopy (HRTEM), nitrogen adsorption (BET). Flat-band potentials (E_{FB}) of the samples were determined using the photocurrent onset potential (POP) method. It was shown that the potentials obtained with open circuit potential measurements versus illumination intensity (OCP) likely corresponded to the E_{FB} of SnO_2 nanoparticles in heterostructures due to interfacial electron transfer from the conducting band of Sn_3O_4 to that of SnO_2. The photo-electrooxidation processes of a series of organic substrates were studied in the potential range of 0.6–1.4 V vs. RHE under irradiation with ultraviolet (λ = 370 nm) and visible (λ = 450 nm) light. The Sn_3O_4 sample showed high activity in the photo-electrooxidation of acetone and formic acid in visible light. The Sn_3O_4/SnO_2 samples exhibited noticeable activity only in the oxidation of formic acid. The presence of the SnO_2 phase in the Sn_3O_4/SnO_2 samples increased the photocurrent values under ultraviolet illumination, but significantly reduced the oxidation efficiency in visible light.

Keywords: Sn_3O_4; Sn_3O_4/SnO_2; band gap; photo-electrooxidation; acetone; glycerol; methanol

1. Introduction

Tin dioxide is a promising material with high corrosion stability, high oxidation potential, low toxicity and high economic efficiency [1,2]. SnO_2 is widely used as supports for fuel cell catalysts [3], sensors [4], electrodes for supercapacitors [5], conductive transparent material [6], photocatalysts and electrocatalysts for the oxidation of organic substances [1,2,7]. However, the large band-gap energy, about 3.6 ÷ 4 eV, and the recombination of photogenerated electrons and holes [8] do not allow the material to be widely used for working in either visible or solar light and lead to low values of quantum efficiency. Approaches to broadening the wavelength range of light absorption and reducing recombination process comprise doping with various metals and nonmetals, including tin itself, mixed valency tin oxide (e.g., Sn_3O_4), heterojunction engineering, solid solution with another oxide materials and morphology control [2,9]. With these approaches, an increase in efficiency is achieved either by introducing new energy levels into the band gap or by reducing the band-gap width, which leads to a shift in the absorption edge to the visible region [2].

Sn_3O_4 and composites based on it are promising materials that arouse increased interest in various photo-stimulated and electrochemical processes, including hydrogen production [10–14], water decomposition [15], oxidation of dyes and organic compounds [16–21], sensors [22–24], lithium and sodium ion batteries [25–28], supercapacitors [29], solar cells [30], CO_2 reduction [31,32], visible light photodetectors [33], etc. The morphology of Sn_3O_4 represents a layered structure in which two layers of SnO alternate with a layer of SnO_2 [34]. The presence of Sn^{2+} ions leads to the appearance of absorption in the visible-light region [34] and also forms a morphology as nanosheets [12] or nanowires [13], which significantly increase the specific surface area and its availability for the reaction. The ability to vary the band gap and the position of the conduction band and, hence, the oxidation potential associated with the position of the valence band by adding hydrogen peroxide during synthesis has been shown theoretically and confirmed experimentally [12]. Despite this, the main disadvantage for the use of Sn_3O_4 is that the potential of the valence band is not high enough for photo-oxidation reactions. To increase the efficiency of charge separation, composites of Sn_3O_4 with electrically conductive materials based on graphene [10,35], graphene oxide [14,32], Ni foam [19], Sn [20] and heterostructures based on C_3N_4 [17], SnO [36,37], SnO_2 [22,24,38–40] and Si nanowires [41] are being studied intensively. Heterostructures based on high valence band oxidation potential of SnO_2 with an absorbance in the visible region of Sn_3O_4 can increase the efficiency of photocatalytic reactions. SnO_2-Sn_3O_4 heterostructures were used as sensors and showed high sensitivity to NO_2 due to a large number of oxygen vacancies and improved electron transfer [22]. The structures of Sn_3O_4/SnO_2 were studied in the reaction of CO_2 electroreduction to formic acid. The increased efficiency is attributed to the increased adsorption energy of intermediates at the SnO_2 and Sn_3O_4 interface [39]. SnO_2–Sn_3O_4 heterostructures were used as sensors for formaldehyde. The improved performance was attributed to the high concentration of adsorbed oxygen involved in oxidation and the characteristics of the heterophase Schottky junction [24]. SnO/Sn_3O_4 and SnO_2/Sn_3O_4 heterostructures were studied in the photooxidation of rhodamine B. The authors concluded that such heterostructures are promising in photocatalysis due to the presence of heterojunctions [36].

Despite the many works on Sn_3O_4, the processes of photo-electrooxidation of organic substances on such materials have been practically not considered. Photoelectrochemical studies are mainly devoted to the characterization of materials with photocurrent values [14,42] and the determination of the position of the flat-band potential with the Mott–Schottky method [13,14,16,32,33,41]. The photo-electrooxidation of rhodamine B on vertically aligned Sn_3O_4 nanoflakes grown on carbon paper was studied under visible light irradiation. The authors associated the increase in photocurrent with the presence of channels for electron transport and the accessibility of the surface [33]. Sn_3O_4 was studied in the hydrogen evolution reaction. It was discovered that polarization of the electrode and treatment of the material with phosphoric acid leads to an increase in the efficiency of charge separation and a decrease in photo-corrosion of the material due to a difficulty in capturing holes from Sn (II) ions [43].

In this work, a series of Sn_3O_4/SnO_2 samples with varying phase contents was synthesized. The conduction band potentials were measured using a set of photoelectrochemical methods, and the obtained data were compared. The photo-electrocatalytic properties of materials were studied in the oxidation of a number of simple organic substances (methanol, acetone, formic acid, glycerol) at different potentials under ultraviolet and visible light.

2. Materials and Methods

2.1. Materials

The following materials were used in the work: Sodium citrate pentahydrate (99.99%, special purity, LLC "Khimkraft"), $SnCl_2$*$2H_2O$ (analytical grade, JSC "Vekton"), NaOH (special purity, LLC "Component-Reaktiv"), ethanol (J. T. Baker, >99.9% reagent grade),

HCl (special purity 35–38 wt.%, Sigma-Tek LLC), acetone (EKOS-1, special purity grade), formic acid (Sigma-Aldridge, reagent grade >95%), glycerol (analytical grade, JSC Vekton), methanol (J. T. Baker, (Ultra) Gradient HPLC grade), Nafion (5 wt.% solution in ethanol, Sigma-Aldrigh).

2.2. Synthesis

Samples TO1 and TO2 were synthesized in accordance with ref. [20]. An amount of 25 mmol of sodium citrate and 10 mmol of tin (II) chloride were dissolved in 50 mL of water with stirring, and 25 mL of 0.2 M sodium hydroxide solution was slowly added. Stirring of the solution continued for 20 min (sample TO1). For sample TO2, the same procedure was followed without the addition of sodium citrate. For sample TO3, the synthesis procedure was slightly modified in accordance with ref. [10]. An amount of 0.01 mol of tin (II) chloride were dissolved in 75 mL of water with stirring, and 25 mL of 4 M sodium hydroxide solution was slowly added. The solution was stirred continuously for 20 min. Then, 6.5 mL of concentrated hydrochloric acid was slowly added to the resulting clear solution. The resulting solutions were transferred to a Teflon autoclave liner (150 mL) and placed in an oven. Hydrothermal synthesis was carried out at 180 °C for 12 h.

For comparison, SnO_2 was synthesized (sample TO4). An amount of 3.960 mL of 50 wt.% NaOH solution was diluted to 71.040 mL with water, followed by adding 1.7028 g of tin (II) chloride under vigorous stirring. The resulting clear solution was transferred to a Teflon autoclave liner (150 mL) and placed in an oven. Hydrothermal synthesis was carried out at 150 °C for 24 h.

The resulting yellow-brown (TO1 and TO3), dirty yellow (TO2) and white (TO4) precipitates were separated from the solution and washed 2 times with 0.2 M sodium hydroxide solution and 2 times with ethanol using a centrifuge. The powders were dried in an oven at 60 °C for 12 h.

2.3. Physico-Chemical Characterization

The porous structure was analyzed with low-temperature nitrogen adsorption at 77 K on an ASAP 2400 specific surface area analyzer (Micromeritics, Norcross, GA, USA). X-ray diffraction patterns were obtained using a Thermo ARL X'tra diffractometer (Thermo Fisher Scientific Inc., Ecublens, Switzerland) using a Mythen2R 1D linear detector (Dectris AG, Baden-Daettwil, Switzerland) with CuK_α radiation (λ = 1.5418 Å). The value of the average coherent scattering region (CSR) was calculated as the average value over two peaks at 24 and 37° 2θ using the Scherrer formula $D = \lambda/(\beta\cos(\theta))$, where β is the peak width at half maximum, without taking into account instrumental broadening. The lattice parameters were refined using the POLYCRYSTAL software package [44]. XPS spectra were recorded on a SPECS X-ray photoelectron spectrometer (SPECS Surface Nano Analysis GmbH, Berlin, Germany) using non-monochromatic MgK_α radiation ($h\nu$ = 1253.6 eV). UV–Vis diffuse reflectance spectra (DRS) were recorded using a Cary 300 UV–Vis spectrophotometer (Agilent Technologies Inc., Santa Clara, CA, USA). Raman spectra were recorded using a Horiba Jobin Yvon LabRAM HR spectrometer coupled with an Olympus BX41 optical microscope, an argon laser (wavelength 488 nm) and a CCD Symphony detector (Horiba Ltd., Kyoto, Japan) in backscatter geometry. The HRTEM study was carried out using ThemisZ electron microscope (Thermo Fisher Scientific, Waltham, MA, USA).

2.4. Photoelectrochemical Studies

Catalyst suspension was obtained using ultrasonication of 10 mg of sample and 0.1 mg of Nafion (as ethanol solution) in 0.6 mL of deionized water. Then, suspensions were supported on FTO with drop-casting followed by drying and thermal treatment at 100 °C for 30 min to obtain a uniform layer with mass ratio of ~1.5 mg cm^{-2} and good stability in Na_2SO_4 electrolyte. Before the measurements, the sample was mounted to the cell and irradiated in air for 30 min with UV light to completely oxidize the remaining organic impurities. Before each experiment, the potential of mercury sulfate electrode was measured relative to reversible hydrogen electrode (RHE), and the potentials presented in the work were recalculated relative to RHE.

Experiments were performed in homemade three-electrode cell using potentiostat Autolab PG302N with a frequency response analyzer. Mercury sulfate electrode, Pt foil and fluorinated tin oxide (FTO) glass coated with photocatalyst were applied as reference, counter and working electrodes, respectively. An amount of 1 M Na_2SO_4 and 1 M Na_2SO_4 + 1 M EtOH were used as electrolytes. Ethanol as hole trap was added in order to reduce recombination effect. High power (100 W) light emitting diodes (LEDs) with a maximum irradiation at 370 nm and 450 nm were used as light sources.

Photo-electrocatalytic oxidation of a series of organic substrates (formic acid, methanol, glycerol and acetone) with concentration of 0.1 M was carried out at constant photon flux of 2×10^{16} cm^{-2} s^{-1} in the potential range of 0.6–1.4 V vs. RHE by alternating light phase (30 s)–dark phase (30 s). Impedance measurements were recorded at different potentials using frequency range of 1–50,000 Hz with 10 mV of voltage amplitude. Capacity was calculated according to Equation (1):

$$C = \frac{1}{2\pi f Z''}, \quad (1)$$

where C was capacity (F/g), f was frequency (Hz), Z'' was imaginary part of impedance (Ohm).

Flat-band potential values (E_{FB}) were obtained with three independent Mott–Schottky (MS) plot, open-circuit-potential (OCP) and photocurrent-onset-potential (POP) methods. The values obtained will be referred to as E_{MS}, E_{OCP} and E_{POP}, respectively. Since these values obtained are quite different, as will be shown later, the correct value of flat-band potential proposed in the work will be denoted as E_{FB}.

E_{MS} was determined using Equation (2):

$$C_{sc}^{-2} = \frac{2}{N_d e_0 \varepsilon_0 \varepsilon_s S^2}\left(E - E_{MS} - \frac{kT}{e_0}\right), \quad (2)$$

where C_{sc} (F/g) was the depletion layer capacity, N_d (m^{-3}) was the charge carriers density, e_0 was the electron charge (1.6×10^{-19} C), ε_0 was the vacuum permittivity (8.85×10^{-12} F m^{-1}), ε_s was the dielectric constant of material, S (m^2) was the electrode surface area, E (V) was the potential of electrode, k ($1.381 \cdot \times 10^{-23}$ J K^{-1}) was the Boltzmann constant, T (K) was the temperature.

E_{POP} was determined from chopped illumination (3 s) of the catalysts at 370 nm when recording cyclic voltammograms with scan rate of 1 mV/s. E_{POP} was obtained at potential when positive and negative spikes of photocurrents become equal. E_{OCP} was determined from the dependence of OCP of catalyst irradiated by UV light (λ = 370 nm) on light intensity, which ranged from 3 to 190 mW/cm^2. E_{OCP} was estimated by extrapolation of measured potentials data to infinite light power density.

3. Results

3.1. Physico-Chemical Properties

The XRD analysis (Figure 1a) showed that TO1 represents the triclinic phase Sn_3O_4 (P_{-1}, PDF #16-0737), while TO4 only showed the SnO_2 phase (P42/mnm, PDF #41-1445). Samples TO2 and TO3 contain SnO_2 and Sn_3O_4 phases in different proportions. Table S1 shows the calculated cell parameters. For sample TO1, all observed peaks are slightly shifted to smaller angles relative to those of the Sn_3O_4 phase, which indicates an increase in unit cell parameters due to the possible presence of water molecules. For samples TO2 and TO3, the observed positions coincide quite well with the literature's data. The coherent scattering region (CSR) values and phase composition estimates are presented in Table 2 and discussed below.

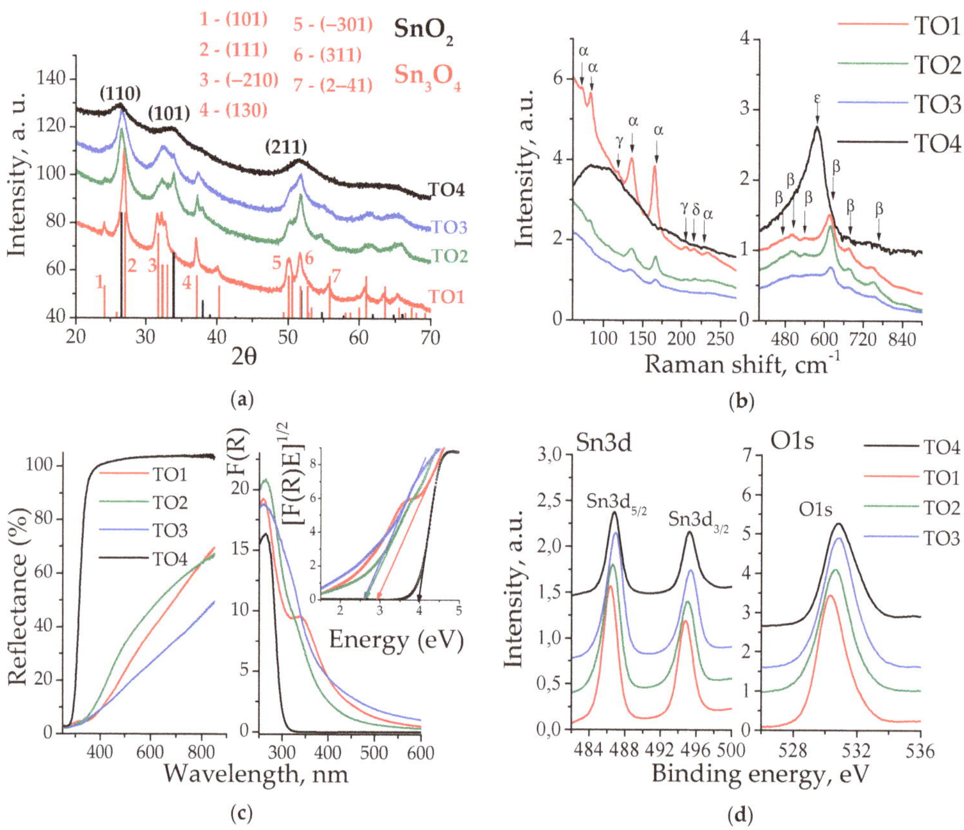

Figure 1. (**a**) XRD patterns. Bars show reflections and relative intensities of SnO_2 (black) and Sn_3O_4 (red) phases. Indexes of main reflections are also presented; (**b**) Raman shifts in the wavenumber range of 60–270 cm^{-1} (**left**) and 400–900 cm^{-1} (**right**). Greek letters indicate phases found: α—Sn_3O_4, β—SnO_2, γ—SnO, δ—$Sn_3^{II}O_2(OH)_2$ [45], ε—SnO_2; (**c**) Spectra of DRS (**left**), Kubelka–Munk transformed (**right**) and Tauc plots (inset); (**d**) XPS spectra in Sn3d (**left**) and O1s (**right**) regions.

The Raman spectra (Figure 1b) of a SnO_2 phase are characterized by the presence of main peaks at 631 cm^{-1} (A_{1g}), 773 cm^{-1} (B_{2g}) and 474 cm^{-1} (E_g). [46–48]. For sample TO4, a large band is observed at 576 cm^{-1} (ε), characteristic of an amorphous phase with small particle sizes and surface defects [49–51]. The presence of 3 nm nanoparticles in TO4 sample is confirmed with XRD data (Table 2). A large number of β peaks with low intensity corresponds to surface phonon vibrations [47]. TO2 and TO3 samples exhibit α peaks at 135 cm^{-1} and 167 cm^{-1}. For the pure Sn_3O_4 phase (sample TO1), the intensity of these peaks increases, and new ones appear at 72 cm^{-1}, 83 cm^{-1}. All these peaks are assigned to the Sn_3O_4 phase [16,19,20,23]. Of the three low intensity peaks observed at 205 cm^{-1}, 215 cm^{-1} and 233 cm^{-1}, the last one belongs to the Sn_3O_4 (α) phase. A similar peak at 241 cm^{-1} was observed for Sn_3O_4 in refs. [23,25]. The peak at 215 cm^{-1} (δ) may arise due to the presence of hydroxyl groups. A similar band was reported at 224 cm^{-1} for $Sn_3^{II}O_2(OH)_2$ in ref. [45]. Impurities of the SnO phase (γ) were noted at 118 cm^{-1} and 205 cm^{-1} [48]. Since the SnO phases for all samples and the SnO_2 phase for TO1 were not observed in the XRD spectrum, it can be assumed that these species may be present in small quantities in the form of amorphous impurities.

In the DRS spectra (Figure 1c), the presence of the Sn_3O_4 phase results in the visible-light absorption appearance. In Kubelka–Munk coordinates, the spectra are characterized by a wide absorption tail in the visible region extended to more than 600 nm (less than 2 eV). The complex structure of the spectra in Tauc coordinates is most likely caused by the complex structure of energy levels in the band gap due to the interaction of the SnO_2 and SnO layers. From the Tauc equation, the band-gap values for indirect transitions were determined in accordance with the literature's data [15,27,52]. The results are present in Table 2. The XPS spectra are shown in Figure 1d, while the decomposition into spectral components are shown in Figure S1. The TO4 sample in the Sn3d region is characterized by the presence of a doublet at 486.8 eV and at 495.2 eV, which correspond to $Sn3d_{5/2}$ and $Sn3d_{3/2}$ bands, respectively. Decomposition of the band at 486.8 eV shows only the Sn^{4+} form of ions, of which the energy, according to the literature's data, lies at 486.9 eV [53]. In the oxygen region, decomposition produces bands at 530.7, 532.2 and 535.5 eV associated with lattice oxygen [53–55], surface OH groups and/or chemisorbed oxygen [56,57] and with adsorbed water, respectively. The $Sn3d_{5/2}$ bands of other samples consist of two components at 486.6–486.9 eV and 485.9–486.1 eV, corresponding to Sn^{4+} and Sn^{2+} ions, respectively [25,26]. The O1s band decomposes into two components at 530.2 eV (O1) and at 531.5 eV (O2). The assignment of bands of oxygen groups is quite difficult. In some studies, the O1 and O2 bands are assigned to oxygen groups associated with Sn^{2+} and Sn^{4+}, respectively [21,28]. Since in this work the intensity of the O1 and O2 bands does not correlate with the content of these ions, we attribute O1 to lattice oxygen groups and O2 to chemisorbed oxygen in accordance with ref. [24,25]. The calculated surface composition of the samples is presented in Table 1.

Table 1. Surface chemical composition of the samples according to decomposition of XPS spectra.

Sample	Sn^{2+}, % [1]	Sn^{4+}, % [1]	O1, % [2]	O2, % [2]	Surface Composition
TO1	54	46	81	19	$SnO_{1.25}$
TO2	38	62	94	6	$SnO_{1.3}$
TO3	28	72	91	9	$SnO_{1.43}$
TO4	0	100	81 [3]	19 [4]	$SnO_{1.88}$

[1] In relation to the Sn content, [2] In relation to the O content, [3] Lattice oxygen, [4] Defective oxygen.

According to Table 1, the surface chemical composition of the TO1 sample is $Sn_3O_{3.75}$, which is close to the ideal Sn_3O_4. A higher Sn^{4+} content (46%) than expected (33%) indicates the presence of small amounts of amorphous SnO_2 on the surface, which is confirmed by Raman spectra. For samples TO2 and TO3, an increase in the amount of Sn^{4+} is observed, which correlates with the XRD data, where the content of the SnO_2 phase increases. Sample TO4 has the formula $SnO_{1.88}$ and corresponds to the SnO_2 phase. The small size of TO4 particles (CSR 3 nm, Table 2) results in a high content of defective surface oxygen groups. Figure 2 shows the HRTEM data for the TO3 sample. Differently oriented Sn_3O_4 nanosheets with dimensions of the order of several tens of nanometers and a thickness of about 5 nm are well observed.

Table 2. Results obtained with XRD, DRS and low-temperature nitrogen adsorption methods. The phase composition was estimated from the ratio of the integral intensities of the diffraction maxima (111) for Sn_3O_4 and (110) for SnO_2.

Sample	CSR [1], nm		Phase Composition, %		Band-Gap Width, eV	S_{BET} [2], m^2/g	V_{pore} [3], cm^3/g
	Sn_3O_4	SnO_2	Sn_3O_4	SnO_2			
TO1	27	-	100	0	2.94	38.5	0.077
TO2	24	9	46	54	2.64	63	0.131
TO3	12	5.5	38	62	2.86	95.9	0.098
TO4	-	3	0	100	3.98	2.05	0.008

[1] Coherent scattering region; [2] specific surface area determined by Brunauer–Emmet–Teller method; [3] specific pore volume.

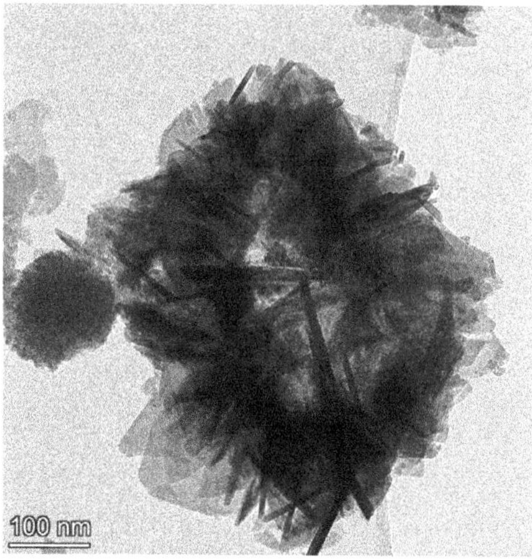

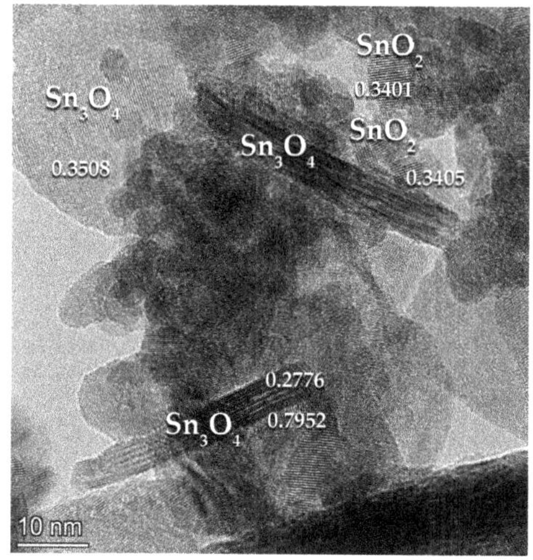

Figure 2. High-resolution TEM microphotographs of TO3 sample in low (**left**) and high (**right**) resolution.

The SnO_2 phase consists of nanoparticles of 5–10 nm in size, which are in close contact with the surface of Sn_3O_4 nanosheets. The found nanoparticle sizes correlate with the CSR sizes obtained for this sample with XRD (Table 2). Table 2 summarizes the data obtained with the methods discussed above.

TO1 and TO4 samples represent Sn_3O_4 and SnO_2 phases with CSR sizes of 27 and 3 nm, respectively. The band-gap values obtained are consistent with the literature for these phases: 2.94 eV for Sn_3O_4 [10] and 3.98 eV for SnO_2 [58]. The specific surface area of the TO4 sample is very low (2.05 m^2/g), which could be associated with the rigid structure of small-sized SnO_2 particles (3 nm). It is noteworthy that the incorporation of SnO_2 enhances the specific surface area of Sn_3O_4/SnO_2 heterostructures, as reflected by an increase in the specific surface area from 38.5 m^2/g to 95.9 m^2/g with an increase in the SnO_2 phase content from 0% to 62%.

3.2. Flat-Band Potential Determination

The Mott–Schottky (MS), photocurrent onset (POP) and open-circuit potential (OCP) methods were used to determine flat-band potentials in pure electrolyte and with ethanol additives. The obtained data are presented in Figures S2 and S3 and in Figure 3 for TO1 sample. The potentials values obtained are compared in Table 3.

Table 3. Values of potentials (V vs. RHE) measured by MS (E_{MS}), POP (E_{POP}) and OCP (E_{OCP}) methods as well as E_{FB} values corresponding to E_{POP} obtained in ethanol-containing electrolyte E_{POP}(Et).

Sample	E_{MS}, V	E_{POP}, V	E_{OCP}, V	E_{FB}, V
TO1	−0.06 ± 0.02	−0.38	0.127	−0.42
TO1 + EtOH	−0.04 ± 0.02	−0.42	−0.133	
TO2	−0.04 ± 0.04	−0.39	0.163	−0.44
TO2 + EtOH	−0.01 ± 0.03	−0.44	−0.155	
TO3	0.04 ± 0.03	−0.14	0.191	−0.44
TO3 + EtOH	0.01 ± 0.03	−0.44	−0.131	
TO4	−0.11 ± 0.03	−0.05	0.295	−0.15
TO4 + EtOH	−0.26 ± 0.04	−0.15	0.121	

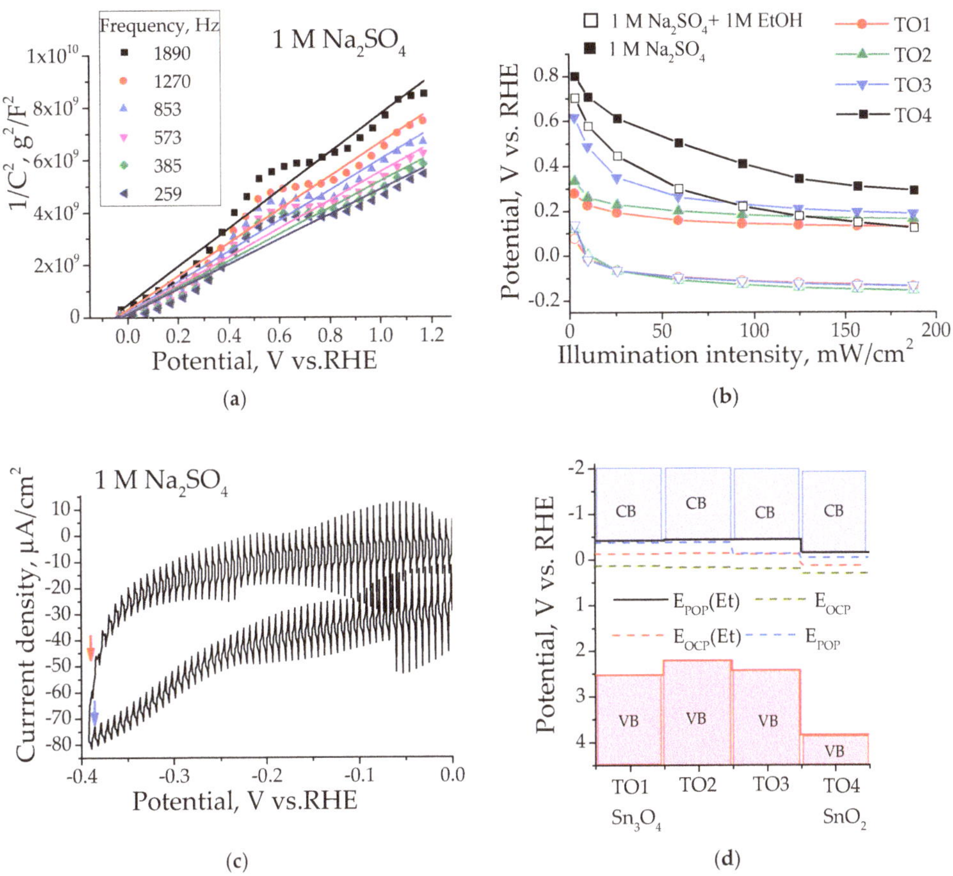

Figure 3. (**a**) MS plot obtained for TO1 sample; (**b**) Dependence of the open-circuit potential on the light intensity (OCP method) in pure electrolyte (full symbols) and with the addition of ethanol (empty symbols). The errors in determining potentials are ± 15 mV for all samples; (**c**) POP method for TO1 sample in pure electrolyte; (**d**) Potential values of the conduction band (CB), calculated valence band (VB) and potential levels, obtained by POP and OCP methods in pure electrolyte and with the addition of ethanol (Et).

For the MS method, several frequencies were used, and the resulting E_{MS} values were averaged. The observed differences in potential values for different methods may be due to various reasons. E_{MS} values were found to differ significantly from those of other methods. Apparently, the MS method is not applicable in our case for various reasons, which may include the influence of particle size, the influence of the FTO substrate, the presence of two phases with different characteristics and morphology and others [59–61]. In the OCP method, as was shown earlier [60], the determined potential can be limited by defect levels below the conduction band, at which rapid recombination occurs. We previously suggested that, for TiO_2-N samples, methanol additions reduce the degree of recombination due to an interaction with holes, and the detected E_{OCP} becomes close to the E_{POP} corresponding to E_{FB} [60]. Previously, we also showed that the POP method for monophase samples can be recommended as the principal one for measuring E_{FB} [60]. Based on these considerations, we assume that the E_{FB} values for the studied samples correspond to the E_{POP} values measured in the presence of the sacrificial agent ethanol. The obtained E_{FB} values for samples containing the Sn_3O_4 phase showed similar values of about −0.43 V and did not depend on the SnO_2 content. For pure SnO_2 (TO4), the E_{FB} value was found to be −0.15 V.

In the literature, E_{FB} values were estimated either from data on the position of the Sn_3O_4 valence band obtained with XPS or using the MS method, and they were not studied with photoelectrochemical methods. In the first case, such an assessment leads to clearly overestimated values. Thus, the theoretically calculated value was about −2 V vs. $E_{H+/H2}$ [15], −1.14 eV vs. NHE [17], −1.2 V vs. RHE [62]. Data obtained with the MS method show higher values and a large spread: −0.97 V vs. NHE (=−0.56 V vs. RHE) [13], −0.94 V vs. NHE (=−0.53 V vs. RHE) [32], −0.6 V vs. NHE (=−0.19 V vs. RHE) [33], −1.1 V vs. NHE (=−0.69 V vs. RHE) [16], −1.1 V vs. RHE [41], −0.23 V vs. NHE at pH 0 (=−0.23 V vs. RHE) [14]. Samples synthesized using citrate ions showed values close to ours: −0.3 V vs. SHE at pH 0 (=−0.3 V vs. RHE) [11].

E_{OCP}(Et) values obtained with the addition of ethanol were found to be similar for all samples containing the Sn_3O_4 phase, which prompted us to plot these values in an energy diagram calculated from the data on the band gap and E_{FB} potentials (Figure 3d). Close E_{OCP} (dark yellow lines) values found for all samples (0.13–0.19 V) and a higher one (0.29 V) found for the TO4 sample having the SnO_2 phase suggest that these potentials correspond to defect levels at which rapid recombination of the electron–hole pair occurs [60]. The presence of oxygen vacancies was reported for Sn_3O_4 at a potential of −0.3 V vs. NHE at pH 7 (=0.09 V vs. RHE) [33], which is close to the value found in our work. The E_{OCP}(Et) (red lines) in the presence of ethanol do not reach the E_{FB}, in contrast to the results obtained for this method in our previous work [60]. However, the values obtained for all samples containing the Sn_3O_4 phase are similar and coincide with the E_{FB} of the SnO_2 (TO4) sample. We assume that interfacial electron transfer may occur from the E_{FB} levels from Sn_3O_4 in TO2 and TO3 samples to the E_{FB} level of SnO_2. Unlike the OCP method, in the POP method, the electrode is polarized, and due to the excess of electrons, the influence of the recombination process is significantly reduced. For TO1, TO2 and TO4 samples, the effect of ethanol is not observed, and the E_{POP} potentials correspond to E_{FB}. For the TO3 sample with a high content of the SnO_2 phase, E_{POP} in a pure electrolyte is fixed at the level of E_{FB} SnO_2 due to interfacial transfer and high recombination, and with the addition of ethanol, it coincides the E_{FB} level of Sn_3O_4 due to a decrease in recombination.

3.3. *Photoelectrocatalytic Oxidation of Organic Substrates*

The samples were studied in the photoelectrochemical oxidation of small organic molecules (methanol, acetone, glycerol and formic acid). Experiments were carried out at potentials of 0.6, 1 and 1.4 V and at wavelengths of 370 and 450 nm. Dependencies of the current density on time at potential of 1 V vs. RHE are presented in Figure S4. The photocurrent values showed an insignificant increase with the potential (Figure S5). The exception is the TO2 sample, where the current in the oxidation reaction of formic acid under visible light irradiation sharply decreases with increasing potential, which may be due to the deactivation of the catalyst that resulted from the oxidation of Sn^{2+} ions. Under the influence of UV radiation, the catalysts operate stably and a slight increase in the photocurrent from the potential is observed. Photocurrents were compared at a potential of 1 V vs. RHE. The data are presented in Figure 4.

In general, the photocurrents of samples under UV illumination is several times higher than that found under visible light, which is associated with higher light absorption by the sample. For most materials, it increases in the series water = acetone < methanol < formic acid < glycerol. A comparison of the materials with each other shows that SnO_2 (TO4) exhibits the greatest activity in all reactions, except for the oxidation of formic acid, which is explained by the highest potential of the TO4 valence band (Figure 3d). Under visible light irradiation, for all substrates (except formic acid), the activity correlates with the level of the valence band: TO1 (2.52 eV) > TO3 (2.42 eV) > TO2 (2.20 eV). The TO4 sample, representing SnO_2, does not absorb visible light, so its photocurrent values are at the detection levels. The photocurrents of TO2 and TO3 samples in the oxidation under visible light of all substrates do not depend on the addition of an organic substrate and correspond to the photocurrents of water oxidation. TO1 is the most active one, and its photocurrent falls in the order acetone > HCOOH ≫ glycerol > methanol = water. The oxidation of formic acid probably proceeds with a different mechanism since all samples are active under both UV and visible light, and the series of photocurrent values differs depending on the type of illumination: TO2 > TO4 > TO3 > TO1 (UV irradiation) and TO1 > TO2 > TO3 (visible light).

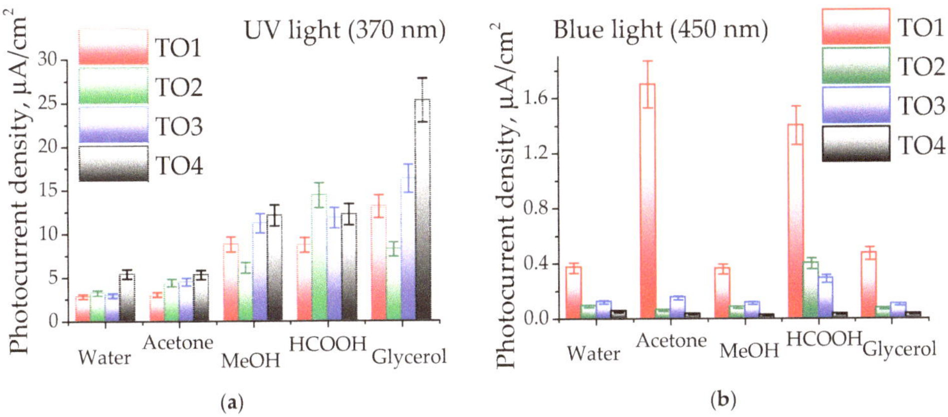

Figure 4. Photocurrents density in the electrooxidation of various organic substrates at a potential of 1 V RHE under (**a**) UV illumination (λ = 370 nm) and (**b**) visible light illumination (λ = 450 nm). Electrolyte—1 M Na_2SO_4 + 0.1 M organic substrate.

4. Discussion

Additions of organic compounds as a sacrificial agent—electron donors—are widely used to increase the yield of hydrogen in photoreforming reactions [63]. The mechanism is the efficient trapping of photoinduced holes, which leads to a sharp decrease in recombination and an increase in the number of free electrons that participate in the formation of hydrogen from water [64]. The amount of hydrogen released during ultraviolet irradiation in an alcohol solution is determined by the number of hydrogen atoms located near the OH groups and decreases in the order glycerol > ethylene glycol > methanol > ethanol [63,65], which is consistent with the results obtained in our work for alcohol compounds (methanol and glycerol) under UV illumination. Oxidation of organic molecules on a surface occurs through a series of successive stages of electron transfer from the adsorbed molecules to the semiconductor. The ref. [66] presents the oxidation pathways of various molecules. Methanol is adsorbed to form methoxy groups, which are successively oxidized to formaldehyde and formate groups. Acetone is successively converted into acetate groups, then into formate groups. Formic acid is adsorbed as formate and oxidized to CO_2 and H_2O [66].

The low rate of oxidation of acetone under UV irradiation is due to the lower affinity for the surface adsorption of acetone than of alcohols [67,68]. The desorption of acetone from the surface of TiO_2 as an intermediate product of the photooxidation of isopropanol was observed in work [68]. The high efficiency of HCOOH oxidation is due to the low isoelectric point of Sn_3O_4, which lies in the acidic region at pH = 2.5–3 [18] and the acidic nature of formic acid, which leads to strong adsorption of formate ions on the surface of the material, which are more reactive than molecular particles [69]. The resulting series of activity in the oxidation of HCOOH is most likely associated with the influence of the surface pH of the samples. Using formic acid as a sacrificial agent for hydrogen evolution under UV light on strontium titanates resulted in more hydrogen evolution than using methanol [70], which is consistent with our results.

To clarify the influence of the valence band potential on the measured photocurrents, the oxidation potential of organic compounds was assessed, and the dependence was plotted in coordinates photocurrent-exp(−(E^0_{ox}−E_{VB})) (Figure 5), as proposed for alcohols in work [65]. A detailed description of potential calculations is provided in the Supplementary File and Table S2.

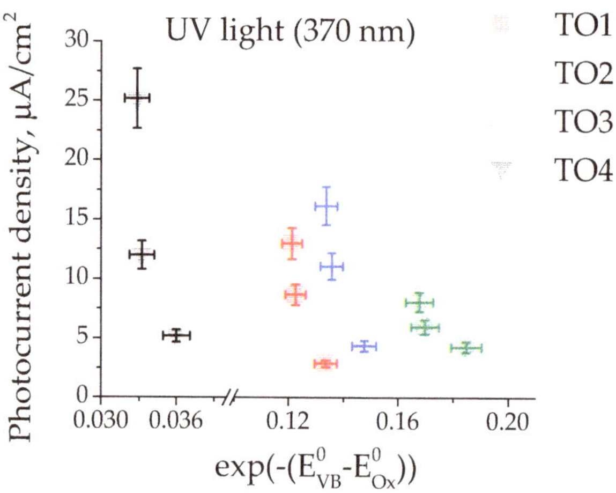

Figure 5. Dependence of photocurrents density in ultraviolet light regions on the difference in the oxidation potentials of organic substrates (E^0_{ox}) and the valence band of the samples (E_{VB}). Symbols and colors indicate the following samples: red squares—TO1, olive circles—TO2, blue triangles—TO3, black inverted triangles—TO4. Photocurrents of formic acid are not shown here.

With the increase in difference between the valence band of the samples and the oxidation potential of organic compounds, an increase in photocurrents is observed, which is consistent with the hydrogen evolution increase reported in ref. [65]. For each substrate and each sample, the photocurrent under UV illumination increases with increasing valence band potential or decreasing substrate oxidation potential, respectively. The behaviour of acetone and formic acid are different. In the first case, low photocurrent values are associated with a low degree of acetone adsorption. In the second case, the surface acidic properties of the samples play a more decisive role in oxidation. Under visible light, a similar trend is observed.

5. Conclusions

In this work, a series of Sn_3O_4, SnO_2 and Sn_3O_4/SnO_2 samples with different phase ratios were prepared using hydrothermal synthesis. The structure of the samples was confirmed with XRD and Raman spectroscopy. The HRTEM method showed that SnO_2 nanoparticles are in close contact with Sn_3O_4 nanosheets. DRS data showed that all samples containing the Sn_3O_4 phase absorb light in visible region. Flat-band potential measurements showed that the intermittent illumination (photocurrent onset potential) method with the addition of a sacrificial agent (ethanol) makes it possible to determine the E_{FB} for all samples, while the potential determined with the OCP method most likely corresponds to either the E_{FB} of SnO_2 nanoparticles in the Sn_3O_4/SnO_2 samples due to the possibility of interfacial electron transfer to tin dioxide or the potential of defective centers, where charge recombination occurs. The flat-band potentials for samples containing Sn_3O_4 were determined to be −0.43 V vs. RHE, while that for SnO_2 was −0.15 V vs. RHE. The samples were studies in photo-electrocatalytic oxidation of a series of organic substrates (acetone, methanol, glycerol, formic acid) at a potential of 1 V vs. RHE under irradiation with ultraviolet (λ = 370 nm) and visible (λ = 450 nm) light. The presence of the SnO_2 phase was shown to increase oxidation photocurrents in the ultraviolet region and to decrease those in the visible region. It was found that the photocurrent magnitudes can be related to the potentials of both the valence band of materials and oxidation of the substrate. For acetone, low photocurrent values were supposed to be associated with a low adsorption value. In the case of formic acid, the acidic properties of the surface are of decisive importance. Under visible light, the Sn_3O_4 sample showed high activities in the acetone and formic acid photo-electrooxidation, while the Sn_3O_4/SnO_2 samples exhibited noticeable activity only in the oxidation of formic acid.

Supplementary Materials: The following supporting information can be downloaded at: https://www.mdpi.com/article/10.3390/ma16237300/s1, Figure S1: Decomposition of the obtained XPS bands in the Sn3d and O1s energy regions; Figure S2: Dependence of capacitance on potential in Mott–Schottky coordinates (MS method) in pure electrolyte and with ethanol additives; Figure S3. Intermittent irradiation (photocurrent onset potential, POP method) of samples in pure electrolyte and with the addition of ethanol; Figure S4. Dependence of current densities on the time at a potential of 1 V vs. RHE in the electrooxidation of water, acetone, methanol, formic acid and glycerol under irradiation with ultraviolet at 370 nm and visible at 450 nm light; Figure S5. Dependence of photocurrents on the potential in the electrooxidation of water, acetone, methanol, formic acid and glycerol under irradiation with ultraviolet at 370 nm and visible at 450 nm light. Table S1: Comparison of the literature's data and experimental results of lattice parameters of the Sn_3O_4 phase in sample TO1 and the SnO_2 phase in other samples; Table S2: Standard oxidation potentials (versus NHE) calculated for the different organic substrates; Table S3: The E_{FB}, band-gap and E_{CB} values obtained in this work for TO1–TO4 samples.

Author Contributions: Conceptualization, E.G. and E.K.; methodology, E.G. and E.K.; software, M.L.; investigation, E.G., E.K., A.K., M.M. and E.L.; resources, E.G.; writing—review and editing, E.G., E.K., M.L. and A.K.; visualization, E.G. and E.K.; funding acquisition, E.G. All authors have read and agreed to the published version of the manuscript.

Funding: This research was funded by RUSSIAN SCIENCE FOUNDATION, grant number 22-23-20208 and NOVOSIBIRSK REGION, agreement No. p-21 dated 6 April 2022.

Institutional Review Board Statement: Not applicable.

Informed Consent Statement: Not applicable.

Data Availability Statement: Data are contained within the article and supporting information.

Acknowledgments: Authors express gratitude to Evgenii Gerasimov (BIC SB RAS) for the HRTEM studies and to Artem Ayupov (BIC SB RAS) for low-temperature nitrogen adsorption studies. The studies were performed using facilities of the Shared Research Center "National center of investigation of catalysts" at Boreskov Institute of Catalysis". Raman spectroscopy was carried out using equipment of the chair of solid-state chemistry and laboratory «MDEST» of the REC «INChT» in Novosibirsk State University.

Conflicts of Interest: The authors declare no conflict of interest.

References

1. Wang, H.; Li, X.; Zhao, X.; Li, C.; Song, X.; Zhang, P.; Huo, P.; Li, X. A review on heterogeneous photocatalysis for environmental remediation: From semiconductors to modification strategies. *Chin. J. Catal.* **2022**, *43*, 178–214. [CrossRef]
2. Sun, C.; Yang, J.; Xu, M.; Cui, Y.; Ren, W.; Zhang, J.; Zhao, H.; Liang, B. Recent intensification strategies of SnO_2-based photocatalysts: A review. *Chem. Eng. J.* **2022**, *427*, 131564. [CrossRef]
3. Chikunova, I.O.; Semeykina, V.S.; Kuznetsov, A.N.; Kalinkin, P.N.; Gribov, E.N.; Parkhomchuk, E.V. Template-assisted synthesis and electrochemical properties of SnO_2 as a cathode catalyst support for PEMFC. *Ionics* **2020**, *26*, 1861–1873. [CrossRef]
4. Sharma, A.; Ahmed, A.; Singh, A.; Oruganti, S.K.; Khosla, A.; Arya, S. Review—Recent Advances in Tin Oxide Nanomaterials as Electrochemical/Chemiresistive Sensors. *J. Electrochem. Soc.* **2021**, *168*, 027505. [CrossRef]
5. Reddy, M.V.; Linh, T.T.; Hien, D.T.; Chowdari, B.V.R. SnO_2 Based Materials and Their Energy Storage Studies. *ACS Sustain. Chem. Eng.* **2016**, *4*, 6268–6276. [CrossRef]
6. Dalapati, G.K.; Sharma, H.; Guchhait, A.; Chakrabarty, N.; Bamola, P.; Liu, Q.; Saianand, G.; Sai Krishna, A.M.; Mukhopadhyay, S.; Dey, A.; et al. Tin oxide for optoelectronic, photovoltaic and energy storage devices: A review. *J. Mater. Chem. A* **2021**, *9*, 16621–16684. [CrossRef]

7. Mishra, S.R.; Ahmaruzzaman, M. Tin oxide based nanostructured materials: Synthesis and potential applications. *Nanoscale* **2022**, *14*, 1566–1605. [CrossRef]
8. Dhanalakshmi, M.; Saravanakumar, K.; Lakshmi Prabavathi, S.; Abinaya, M.; Muthuraj, V. Fabrication of novel surface plasmon resonance induced visible light driven iridium decorated SnO_2 nanorods for degradation of organic contaminants. *J. Alloys Compd.* **2018**, *763*, 512–524. [CrossRef]
9. Xiong, Y.; Lin, Y.; Wang, X.; Zhao, Y.; Tian, J. Defect engineering on SnO_2 nanomaterials for enhanced gas sensing performances. *Adv. Powder Mater.* **2022**, *1*, 100033. [CrossRef]
10. Sethi, Y.A.; Kulkarni, A.K.; Ambalkar, A.A.; Panmand, R.P.; Kulkarni, M.V.; Gosavi, S.W.; Kale, B.B. Efficient solar light-driven hydrogen generation using an Sn_3O_4 nanoflake/graphene nanoheterostructure. *RSC Adv.* **2021**, *11*, 29877–29886. [CrossRef]
11. Tanabe, T.; Hashimoto, M.; Mibu, K.; Tanikawa, T.; Gunji, T.; Kaneko, S.; Abe, H.; Miyauchi, M.; Matsumoto, F. Synthesis of single phase Sn_3O_4: Native visible-light-sensitive photocatalyst with high photocatalytic performance for hydrogen evolution. *J. Nanosci. Nanotechnol.* **2017**, *17*, 3454–3459. [CrossRef]
12. Li, L.; Zhang, K.; Jin, W.; Xia, W.; He, J.; Zeng, X. An excellent H_2 production photoelectrode based on mixed valence Sn_3O_4 nanoflake arrays treated by H_2O_2 hydrothermal reaction. *Int. J. Hydrogen Energy* **2022**, *47*, 10594–10602. [CrossRef]
13. Mone, P.; Mardikar, S.; Balgude, S. Morphology-controlled synthesis of Sn_3O_4 nanowires for enhanced solar-light driven photocatalytic H_2 production. *Nano-Struct. Nano-Objects* **2020**, *24*, 100615. [CrossRef]
14. Yu, X.; Zhao, Z.; Ren, N.; Liu, J.; Sun, D.; Ding, L.; Liu, H. Top or Bottom, Assembling Modules Determine the Photocatalytic Property of the Sheetlike Nanostructured Hybrid Photocatalyst Composed with Sn_3O_4 and rGO (GQD). *ACS Sustain. Chem. Eng.* **2018**, *6*, 11775–11782. [CrossRef]
15. Manikandan, M.; Tanabe, T.; Li, P.; Ueda, S.; Ramesh, G.V.; Kodiyath, R.; Wang, J.; Hara, T.; Dakshanamoorthy, A.; Ishihara, S.; et al. Photocatalytic water splitting under visible light by mixed-valence Sn_3O_4. *ACS Appl. Mater. Interfaces* **2014**, *6*, 3790–3793. [CrossRef]
16. He, Y.; Li, D.; Chen, J.; Shao, Y.; Xian, J.; Zheng, X.; Wang, P. Sn_3O_4: A novel heterovalent-tin photocatalyst with hierarchical 3D nanostructures under visible light. *RSC Adv.* **2014**, *4*, 1266–1269. [CrossRef]
17. Li, C.; Yu, S.; Dong, H.; Liu, C.; Wu, H.; Che, H.; Chen, G. Z-scheme mesoporous photocatalyst constructed by modification of Sn_3O_4 nanoclusters on g-C_3N_4 nanosheets with improved photocatalytic performance and mechanism insight. *Appl. Catal. B Environ.* **2018**, *238*, 284–293. [CrossRef]
18. Huda, A.; Suman, P.H.; Torquato, L.D.M.; Silva, B.F.; Handoko, C.T.; Gulo, F.; Zanoni, M.V.B.; Orlandi, M.O. Visible light-driven photoelectrocatalytic degradation of acid yellow 17 using Sn_3O_4 flower-like thin films supported on Ti substrate ($Sn_3O_4/TiO_2/Ti$). *J. Photochem. Photobiol. A Chem.* **2019**, *376*, 196–205. [CrossRef]
19. Yang, R.; Ji, Y.; Zhang, J.; Zhang, R.; Liu, F.; Chen, Y.; Liang, L.; Han, S.; Yu, X.; Liu, H. Efficiently degradation of polyacrylamide pollution using a full spectrum Sn_3O_4 nanosheet/Ni foam heterostructure photoelectrocatalyst. *Catal. Today* **2019**, *335*, 520–526. [CrossRef]
20. Yang, R.Q.; Liang, N.; Chen, X.Y.; Wang, L.W.; Song, G.X.; Ji, Y.C.; Ren, N.; Lü, Y.W.; Zhang, J.; Yu, X. Sn/Sn_3O_{4-x} heterostructure rich in oxygen vacancies with enhanced visible light photocatalytic oxidation performance. *Int. J. Miner. Metall. Mater.* **2021**, *28*, 150–159. [CrossRef]
21. Zhang, L.; Liu, X.; Zhang, X.; Zhang, W.; Ma, J.; Wang, Q.; Su, S. Sulfur-doped Sn_3O_4 nanosheets for improved photocatalytic performance. *J. Alloys Compd.* **2023**, *961*, 170904. [CrossRef]
22. Chen, G.; Lu, S.; Shang, Y.; Zeng, W.; Zhan, H.; Mei, J.; He, Z. Sn_3O_4-SnO_2 heterostructure for highly sensitive and selective NO_2 sensing at low operating temperature. *Nano Express.* **2021**, *2*, 020021. [CrossRef]
23. Masteghin, M.G.; Silva, R.A.; Cox, D.C.; Godoi, D.R.M.; Silva, S.R.P.; Orlandi, M.O. The role of surface stoichiometry in NO_2 gas sensing using single and multiple nanobelts of tin oxide. *Phys. Chem. Chem. Phys.* **2021**, *23*, 9733–9742. [CrossRef]
24. Yang, X.-Y.; Shi, Y.-T.; Xie, K.-F.; Wang, J.-Y.; Wang, Y.-F.; Zheng, Y.; Fang, S.-M.; Zhang, Y.-H. Engineering of in-plane SnO_2-Sn_3O_4 hierarchical nanoflower heterojunctions for enhanced formaldehyde sensing. *Appl. Surf. Sci.* **2023**, *614*, 156110. [CrossRef]
25. Chen, X.; Huang, Y.; Zhang, K.; Feng, X.; Wei, C. Novel hierarchical flowers-like Sn_3O_4 firstly used as anode materials for lithium ion batteries. *J. Alloys Compd.* **2017**, *690*, 765–770. [CrossRef]
26. Chothe, U.P.; Ambalkar, A.A.; Ugale, C.K.; Kulkarni, M.V.; Kale, B.B. Synergy of a heteroatom (P-F) in nanostructured Sn_3O_4 as an anode for sodium-ion batteries. *Sustain. Energy Fuels* **2021**, *5*, 2678–2687. [CrossRef]
27. Furukawa, M.; Iwamoto, D.; Inamori, K.; Tateishi, I.; Katsumata, H.; Kaneco, S. Synthesis of Tungsten-Modified Sn_3O_4 through the Cetyltrimethylammonium Bromide-Assisted Solvothermal Method for Dye Decolorization under Visible Light Irradiation. *Catalysts* **2023**, *13*, 1179. [CrossRef]
28. Jia, H.; Cao, X.; Pan, A.; Huang, L.; Yin, B.; Chen, J.; Tan, X.; Tang, Y.; Han, M.; Liang, S. Vertically oriented Sn_3O_4 nanoflakes directly grown on carbon fiber cloth for high-performance lithium storage. *Inorg. Chem. Front.* **2019**, *6*, 1468–1474. [CrossRef]
29. Reddy Pallavolu, M.; Tanaya Das, H.; Anil Kumar, Y.; Naushad, M.; Sambasivam, S.; Hak Jung, J.; Joo, S.W. Marigold flower-like Sn_3O_4 nanostructures as efficient battery-type electrode material for high-performing asymmetric supercapacitors. *J. Electroanal Chem.* **2022**, *920*, 116641. [CrossRef]
30. Fadaam, S.A.; Ali, H.M.; Salih, A.A.; Obaid, M.A.; Sabeeh Ali, A.; Habubi, N.F. Synthesis and characterization of metastable phases of SnO and Sn_3O_4 thin films for solar cells applications. *J. Phys. Conf. Ser.* **2021**, *1963*, 012003. [CrossRef]

31. Liu, L.X.; Zhou, Y.; Chang, Y.C.; Zhang, J.R.; Jiang, L.P.; Zhu, W.; Lin, Y. Tuning Sn_3O_4 for CO_2 reduction to formate with ultra-high current density. *Nano Energy* **2020**, *77*, 105296. [CrossRef]
32. Romeiro, F.C.; Silva, B.C.; Martins, A.S.; Zanoni, M.V.B.; Orlandi, M.O. Superior performance of rGO-tin oxide nanocomposite for selective reduction of CO_2 to methanol. *J. CO2 Util.* **2021**, *46*, 101460. [CrossRef]
33. Xia, W.; Qian, H.; Zeng, X.; Dong, J.; Wang, J.; Xu, Q. Visible-Light Self-Powered Photodetector and Recoverable Photocatalyst Fabricated from Vertically Aligned Sn_3O_4 Nanoflakes on Carbon Paper. *J. Phys. Chem. C* **2017**, *121*, 19036–19043. [CrossRef]
34. Berengue, O.M.; Simon, R.A.; Chiquito, A.J.; Dalmaschio, C.J.; Leite, E.R.; Guerreiro, H.A.; Guimarães, F.E.G. Semiconducting Sn_3O_4 nanobelts: Growth and electronic structure. *J. Appl. Phys.* **2010**, *107*, 033717. [CrossRef]
35. Zhang, W.; Zheng, M.; Li, F.; You, Y.; Jiang, D.; Yuan, H.; Ma, L.; Shen, W. SnO_x/graphene anode material with multiple oxidation states for high-performance Li-ion batteries. *Nanotechnology* **2021**, *32*, 195407. [CrossRef]
36. Xia, W.; Wang, H.; Zeng, X.; Han, J.; Zhu, J.; Zhou, M.; Wu, S. High-efficiency photocatalytic activity of type II SnO/ Sn_3O_4 heterostructures via interfacial charge transfer. *Crystengcomm* **2014**, *16*, 6841–6847. [CrossRef]
37. Zeng, W.; Liu, Y.; Chen, G.; Zhan, H.; Mei, J.; Luo, N.; He, Z.; Tang, C. SnO- Sn_3O_4 heterostructural gas sensor with high response and selectivity to parts-per-billion-level NO_2 at low operating temperature. *RSC Adv.* **2020**, *10*, 29843–29854. [CrossRef]
38. Zeng, W.; Liu, Y.; Mei, J.; Tang, C.; Luo, K.; Li, S.; Zhan, H.; He, Z. Hierarchical SnO_2– Sn_3O_4 heterostructural gas sensor with high sensitivity and selectivity to NO_2. *Sens. Actuators B Chem.* **2019**, *301*, 127010. [CrossRef]
39. Wu, J.; Xie, Y.; Du, S.; Ren, Z.; Yu, P.; Wang, X.; Wang, G.; Fu, H. Heterophase engineering of SnO_2/ Sn_3O_4 drives enhanced carbon dioxide electrocatalytic reduction to formic acid. *Sci. China Mater.* **2020**, *63*, 2314–2324. [CrossRef]
40. Huda, A.; Mahendra, I.P.; Ichwani, R.; Handoko, C.T.; Ngoc, H.M.; Yudono, B.; Bustan, M.D.; Gulo, F. High efficient visible-light activated photo catalytic semiconductor SnO_2/ Sn_3O_4 heterostructure in direct blue 71 (DB71) degradation. *Rasayan J. Chem.* **2019**, *12*, 308–318. [CrossRef]
41. Yang, R.; Ji, Y.; Li, Q.; Zhao, Z.; Zhang, R.; Liang, L.; Liu, F.; Chen, Y.; Han, S.; Yu, X.; et al. Ultrafine Si nanowires/ Sn_3O_4 nanosheets 3D hierarchical heterostructured array as a photoanode with high-efficient photoelectrocatalytic performance. *Appl. Catal. B Environ.* **2019**, *256*, 117798. [CrossRef]
42. Jana, S.; Konar, S.; Mitra, B.C.; Mondal, A.; Mukhopadhyay, S. Fabrication of a new heterostructure Au/Pt/SnO_2: An excellent catalyst for fast reduction of para-nitrophenol and visible light assisted photodegradation of dyes. *Mater. Res. Bull.* **2021**, *141*, 111351. [CrossRef]
43. Chen, L.; Yue, S.; Wang, J.; Chen, W.; Zhang, Y.; Xie, M.; Han, W. Overall water splitting on surface-polarized Sn_3O_4 through weakening the trap of Sn(II) to holes. *Appl. Catal. B Environ.* **2021**, *299*, 120689. [CrossRef]
44. Tsybulya, S.V.; Cherepanova, S.V.; Soloviyova, L.P. Polycrystal software package for IBM/PC. *J. Struct. Chem.* **1996**, *37*, 332–334. [CrossRef]
45. Eckold, P.; Rolff, M.; Niewa, R.; Hügel, W. Synthesis, characterization and in situ Raman detection of $Sn_3O_2(OH)_{2-x}Cl_x$ phases as intermediates in tin corrosion. *Corros. Sci.* **2015**, *98*, 399–405. [CrossRef]
46. Gaur, L.K.; Chandra Mathpal, M.; Kumar, P.; Gairola, S.P.; Agrahari, V.; Martinez, M.A.R.; Aragon, F.F.H.; Soler, M.A.G.; Swart, H.C.; Agarwal, A. Observations of phonon anharmonicity and microstructure changes by the laser power dependent Raman spectra in Co doped SnO_2 nanoparticles. *J. Alloys Compd.* **2020**, *831*, 154836. [CrossRef]
47. Yu, K.N.; Xiong, Y.; Liu, Y.; Xiong, C. Microstructural change of nano-SnO_2 grain assemblages with the annealing temperature. *Phys. Rev. B* **1997**, *55*, 2666. [CrossRef]
48. Zaraska, L.; Gawlak, K.; Wiercigroch, E.; Malek, K.; Kozieł, M.; Andrzejczuk, M.; Marzec, M.M.; Jarosz, M.; Brzózka, A.; Sulka, G.D. The effect of anodizing potential and annealing conditions on the morphology, composition and photoelectrochemical activity of porous anodic tin oxide films. *Electrochim. Acta* **2019**, *319*, 18–30. [CrossRef]
49. Ilka, M.; Bera, S.; Kwon, S.H. Influence of surface defects and size on photochemical properties of SnO_2 nanoparticles. *Materials* **2018**, *11*, 904. [CrossRef]
50. Kar, A.; Olszówka, J.; Sain, S.; Sloman, S.R.I.; Montes, O.; Fernández, A.; Pradhan, S.K.; Wheatley, A.E.H. Morphological effects on the photocatalytic properties of SnO_2 nanostructures. *J. Alloys Compd.* **2019**, *810*, 151718. [CrossRef]
51. Liu, L.Z.; Wu, X.L.; Gao, F.; Shen, J.C.; Li, T.H.; Chu, P.K. Determination of surface oxygen vacancy position in SnO_2 nanocrystals by Raman spectroscopy. *Solid State Commun.* **2011**, *151*, 811–814. [CrossRef]
52. Wang, D.; Lin, Z.; Miao, C.; Jiang, W.; Li, H.; Liu, C.; Che, G. An S-scheme photocatalyst constructed by modifying Ni-doped Sn_3O_4 micro-flowers on g-C_3N_4 nanosheets for enhanced visible-light-driven hydrogen evolution. *J. Ind. Eng. Chem.* **2022**, *113*, 380–388. [CrossRef]
53. Babu, B.; Neelakanta Reddy, I.; Yoo, K.; Kim, D.; Shim, J. Bandgap tuning and XPS study of SnO_2 quantum dots. *Mater. Lett.* **2018**, *221*, 211–215. [CrossRef]
54. Sun, M.; Su, Y.; Du, C.; Zhao, Q.; Liu, Z. Self-doping for visible light photocatalytic purposes: Construction of SiO_2/SnO_2/SnO_2:Sn^{2+} nanostructures with tunable optical and photocatalytic performance. *RSC Adv.* **2014**, *4*, 30820–30827. [CrossRef]
55. Davar, F.; Salavati-Niasari, M.; Fereshteh, Z. Synthesis and characterization of SnO_2 nanoparticles by thermal decomposition of new inorganic precursor. *J. Alloys Compd.* **2010**, *496*, 638–643. [CrossRef]
56. Chuvenkova, O.A.; Domashevskaya, E.P.; Ryabtsev, S.V.; Yurakov, Y.A.; Popov, A.E.; Koyuda, D.A.; Nesterov, D.N.; Spirin, D.E.; Ovsyannikov, R.Y.; Turishchev, S.Y. XANES and XPS investigations of surface defects in wire-like SnO_2 crystals. *Phys. Solid State* **2015**, *57*, 153–161. [CrossRef]

57. Nagasawa, Y.; Choso, T.; Karasuda, T.; Shimomura, S.; Ouyang, F.; Tabata, K.; Yamaguchi, Y. Photoemission study of the interaction of a reduced thin film SnO_2 with oxygen. *Surf. Sci.* **1999**, *433*, 226–229. [CrossRef]
58. Gribov, E.N.; Koshevoy, E.I.; Chikunova, I.O.; Parmon, V.N. Template-Assisted SnO_2: Synthesis, Composition, and Photoelectrocatalytical Properties. *Catalysts* **2023**, *13*, 168. [CrossRef]
59. Hankin, A.; Bedoya-Lora, F.E.; Alexander, J.C.; Regoutz, A.; Kelsall, G.H. Flat band potential determination: Avoiding the pitfalls. *J. Mater. Chem. A* **2019**, *7*, 26162–26176. [CrossRef]
60. Koshevoy, E.; Gribov, E.; Polskikh, D.; Lyulyukin, M.; Solovyeva, M.; Cherepanova, S.; Kozlov, D.; Selishchev, D. Photoelectrochemical Methods for Determination of Flat-Band Potential in Semiconducting Photocatalysts: A Comparison Study. *Langmuir* **2023**, *39*, 13466–13480. [CrossRef]
61. Patel, M.Y.; Mortelliti, M.J.; Dempsey, J.L. A compendium and meta-analysis of flatband potentials for TiO_2, ZnO, and SnO_2 semiconductors in aqueous media. *Chem. Phys. Rev.* **2022**, *3*, 011303. [CrossRef]
62. Liu, Y.S.; Yamaguchi, A.; Yang, Y.; Aisnada, A.N.E.; Uchida, S.; Abe, H.; Ueda, S.; Yamaguchi, K.; Tanabe, T.; Miyauchi, M. Synthesis and Characterization of the Orthorhombic Sn_3O_4 Polymorph. *Angew. Chem. Int. Ed. Engl.* **2023**, *62*, e202300640. [CrossRef] [PubMed]
63. Fajrina, N.; Tahir, M. A critical review in strategies to improve photocatalytic water splitting towards hydrogen production. *Int. J. Hydrogen Energy* **2019**, *44*, 540–577. [CrossRef]
64. Corredor, J.; Rivero, M.J.; Rangel, C.M.; Gloaguen, F.; Ortiz, I. Comprehensive review and future perspectives on the photocatalytic hydrogen production. *J. Chem. Technol. Biotechnol.* **2019**, *94*, 3049–3063. [CrossRef]
65. Chen, W.-T.; Chan, A.; Al-Azri, Z.H.N.; Dosado, A.G.; Nadeem, M.A.; Sun-Waterhouse, D.; Idriss, H.; Waterhouse, G.I.N. Effect of TiO_2 polymorph and alcohol sacrificial agent on the activity of Au/TiO_2 photocatalysts for H_2 production in alcohol–water mixtures. *J. Catal.* **2015**, *329*, 499–513. [CrossRef]
66. Kohl, D. Surface processes in the detection of reducing gases with SnO_2-based devices. *Sens. Actuators* **1989**, *18*, 71–113. [CrossRef]
67. Xu, W.; Raftery, D.; Francisco, J.S. Effect of Irradiation Sources and Oxygen Concentration on the Photocatalytic Oxidation of 2-Propanol and Acetone Studied by in Situ FTIR. *J. Phys. Chem. B* **2003**, *107*, 4537–4544. [CrossRef]
68. Gu, Q.; Fu, X.; Wang, X.; Chen, S.; Leung, D.Y.C.; Xie, X. Photocatalytic reforming of C_3-polyols for H_2 production. *Appl. Catal. B Environ.* **2011**, *106*, 689–696. [CrossRef]
69. Li, L.; Guo, W.; Zhu, Y.; Wu, Y. Hydrogen production by photoelectrochemically splitting solutions of formic acid. *ChemSusChem* **2011**, *4*, 1475–1480. [CrossRef]
70. Zielinska, B.; Borowiakpalen, E.; Kalenczuk, R. Photocatalytic hydrogen generation over alkaline-earth titanates in the presence of electron donors. *Int. J. Hydrogen Energy* **2008**, *33*, 1797–1802. [CrossRef]